Two-Dimensional Calculus

Two-Dimensional Calculus

ROBERT OSSERMAN
STANFORD UNIVERSITY

ROBERT E. KRIEGER PUBLISHING COMPANY, INC.
MALABAR, FLORIDA

To My Parents

Original Edition 1968
Reprint 1977, 1986

Printed and Published by
ROBERT E. KRIEGER PUBLISHING COMPANY, INC.
KRIEGER DRIVE
MALABAR, FLORIDA 32950

Copyright © 1968 by
HARCOURT, BRACE & WORLD, INC.
Copyright transferred to Robert Osserman Oct. 20, 1975
Reprinted by Arrangement

Printed in the United States of America

Library of Congress Cataloging in Publication Data

Osserman, Robert.
  Two-dimensional calculus.

  Reprint of the edition published by Harcourt, Brace &
World, New York, issued in Harbrace college mathematics
series.
  Includes index.
  1. Calculus.   I. Title.
[QA303.086   1977]   515   76-50613
ISBN 0-88275-473-4

10 9 8 7 6

Foreword

The Harbrace College Mathematics Series has been undertaken in response to the growing demands for flexibility in college mathematics curricula. This series of concise, single-topic textbooks is designed to serve two primary purposes: First, to provide basic undergraduate text materials in compact, coordinated units. Second, to make available a variety of supplementary textbooks covering single topics.

To carry out these aims, the series editors and the publisher have selected as the foundation of the series a sequence of six textbooks covering functions, calculus, linear algebra, multivariate calculus, theory of functions, and theory of functions of several variables. Complementing this sequence are a number of other planned volumes on such topics as probability, statistics, differential equations, topology, differential geometry, and complex functions.

By permitting more flexibility in the construction of courses and course sequences, this series should encourage diversity and individuality in curricular patterns. Furthermore, if an instructor wishes to devise his own topical sequence for a course, the Harbrace College Mathematics Series provides him with a set of books built around a flexible pattern from which he may choose the elements of his new arrangement. Or, if an instructor wishes to supplement a full-sized textbook, this series provides him with a group of compact treatments of individual topics.

An additional and novel feature of the Harbrace College Mathematics Series is its continuing adaptability. As new topics gain emphasis in the curricula or as promising new treatments appear, books will be added to the series or existing volumes will be revised. In this way, we will meet the changing demands of the instruction of mathematics with both speed and flexibility.

SALOMON BOCHNER
W. G. LISTER

Preface to Reprint Edition

The preface to the first edition of this book in 1968 describes the advantages of a thorough discussion of the two-dimensional case of multi-variable calculus over the more conventional treatments. Among those advantages are the ease of incorporating linear algebra, and the possibility of geometric visualization. The principal change in the standard undergraduate curriculum since that time is the increasingly widespread and early introduction of linear algebra. As a result, it is much more common now for students to approach advanced calculus with the algebraic background needed to insure a deeper understanding of the material. On the other hand, what is still generally missing is the intuitive and geometric insight into both the algebra and the calculus. A principal goal of this book is to assist its readers in achieving that insight.

Some instructors may feel that by starting with the two-dimensional case, a student will have a much easier time approaching the subject in three and higher dimensions. Others may prefer to start with the general case, and refer back periodically to the more accessible description in two dimensions. In either case, two-dimensional calculus remains the most basic component of several-variable calculus, without which no student can acquire a mastery of the subject.

ROBERT OSSERMAN

vii

Preface

In 1958 there appeared a treatment of advanced calculus by Nickerson, Spencer, and Steenrod. In their preface they observed that "the standard treatises on this subject, at any rate those available in English, tend to be omnibus collections of seemingly unrelated topics. The presentation of vector analysis often degenerates into a list of formulas and manipulative exercises, and the student is not brought to grips with the underlying mathematical ideas."

This view of the situation was widely shared by those who were familiar with the available textbooks. The most important parts of the subject were often obscured in a jumble of disconnected topics. In particular, the theory of functions of several variables, which is a clearly delineated unit, was frequently scattered piecemeal among other topics in elementary and advanced calculus.

There were several reasons for this disorder. To some degree it can be ascribed to historical accident in the development of the undergraduate mathematics curriculum. But more fundamentally, it stems from certain inherent difficulties. The main problem is that without a knowledge of linear algebra it is possible to attain a purely manipulative mastery of the subject, but no real understanding. On a more elementary level, it is possible to discuss functions of several variables without using vector methods, but many important insights will be missed.

The solution proposed by Nickerson, Spencer, and Steenrod, and since adopted by a number of others, was to present first a full-scale treatment of linear algebra, and then to use it freely in their development of higher-dimensional calculus. This approach certainly solves some of the basic problems, but it also creates others.

This text presents an alternative approach. The decision to restrict the entire discussion to two variables was based on several considerations, the most important of which are:

1. The basic problem concerning linear algebra is easily solved. The subjects of quadratic forms and linear transformations in two dimensions are

viii

presented in a relatively few pages. They appear as a natural part of the general theory of differentiable functions and transformations and are then used to give further insight into that theory.

2. Geometric intuition operates as a powerful aid to understanding the fundamental concepts, and it also casts light on the reasons for introducing those concepts. Geometric visualization is clearest in two dimensions, still possible in three dimensions, and absent in higher dimensions. Indeed, the later development of geometric insight in n dimensions depends on a thorough familiarity with the lower-dimensional models.

3. The conceptual understanding acquired in the two-dimensional case allows the student to move rapidly through the higher dimensional cases, distinguishing the underlying ideas from the technical complications which arise in the transition from two to higher dimensions.

4. A thorough familiarity with two-dimensional calculus is an ideal preparation for the theory of functions of a complex variable. Indeed, complex functions are interpreted geometrically as plane transformations. Chapter Three of this book concentrates on the ability to visualize these transformations and to work with them effectively.

5. The unity achieved by detaching functions of several variables from the rest of advanced calculus is further consolidated by the elimination of dimension hopping. Unless the student is ready for a general treatment of n-dimensional calculus, he is probably more confused than edified by the usual proliferation of cases—one function of three variables, two functions of four variables, etc.

It should be made clear that the idea of a unified presentation of two-dimensional calculus is certainly not original with the author. Just such a course has been given for a number of years at Stanford. Experience has shown that a student is then prepared to go very rapidly through the classical treatment of three-dimensional calculus, or else to go directly to the general n-dimensional case if his over-all mathematical level is sufficiently advanced. There are a number of good treatments available for each of these alternatives. (See the list of references on page 430.)

The two-dimensional aspect of this book should be viewed in perspective as subsidiary to the principal goal—a presentation of the fundamental ideas that distinguish several-variable from one-variable calculus. It is the clearest possible exposition of these underlying ideas, and wherever relevant their geometric or physical content, that has motivated the selection of topics, their order, and the accompanying examples.

Another, and related goal, is a treatment that is sufficiently elementary to be accessible to a student with a minimum of preparation, but sufficiently rich in insights to be stimulating for a more advanced student who may have seen some of the topics presented, perhaps in a more superficial way, as part of a basic course in calculus.

The prerequisite for reading this book is a knowledge of elementary one-variable calculus. Karel de Leeuw's *Calculus,* which precedes this book in the Harbrace College Mathematics Series, is an excellent, concise treatment of all that is needed in the way of preparation.

Since some students will have had more extensive training in elementary calculus, this book is designed so that it may be used at various levels.

For a brief introduction to functions of several variables, it would suffice to cover the material in the first two chapters, together with selected sections in Chapters 3 and 4. The first few exercises at the end of each section are generally of a more routine nature, and may be used to provide a check on elementary comprehension of the text.

For students who already have some familiarity with the background material in Chapter 1, the main body of the text could be completed in a standard course, lasting roughly one semester.

Finally, those students who may have been previously exposed to some version of several-variable calculus, and who wish to deepen their understanding, will find in addition to the topics discussed in the main text, a large number of applications and elaborations in the exercise sections.

The exercises form an important component of the book. They include various topics which could have been presented in the body of the text but are relegated to the exercises in order not to disrupt the main flow of ideas. One example is the treatment of simple connectivity, which uses some of the material in the exercises to Section 25 and appears as an addendum to that section. Another is the discussion of some of the more subtle properties of mappings, connected with the notion of the degree of a map; these may be handled using special methods available in two dimensions. Thus, winding numbers, the argument principle, and Rouché's theorem are seen in a more general setting which puts into perspective their usual treatment involving functions of a complex variable.

The importance of the exercises makes it advisable to read entirely through each exercise section, at least summarily. The exercises often contain facts worth having seen and may occasionally provide just the insight needed to illuminate a part of the text. Those exercises which appear to be specially relevant may be explored in greater detail. A few exercises are starred to indicate that they may for one reason or another pose greater difficulties than the rest.

It is a pleasure to acknowledge my gratitude to Blaine Lawson, Charles Micchelli, and Tom Savits for their assistance in the preparation of the exercises; to Rosemarie Stampfel and Gail Lemmond for their excellent typing services; to Karel de Leeuw, who read through two entire versions of the manuscript and made innumerable helpful suggestions; to my son, Paul, for his diligent aid with the index; and to my wife, Maria, who good humoredly bore the brunt of months of writing, revising, and preoccupation.

ROBERT OSSERMAN

Notation

General:

[3]	number 3 in list of references on pages 430–32
Ex. 12.4	fourth exercise at the end of Section 12
Eq. (3.5), Th. 7.2	similar references to numbered equations and theorems
◆	indicates the end of a proof
$\Rightarrow$	implies
$\Leftrightarrow$	if and only if
$x \in A$	x is an element of the set A
$A \subset B$	the set A is included in the set B
$\log x$	natural logarithm of x (base e)
$\exp x$	e^x
$f(x) \mid_a^b$ or $[f(x)]_a^b$	$f(b) - f(a)$
$f(x, y) \mid_{(x_0,\, y_0)}$	$f(x, y)$ evaluated at (x_0, y_0)

Specific:		*page introduced*
$\langle a, b \rangle$	vector with components a, b	7
$\mathbf{v}$	boldface indicates a vector quantity	12
$|\mathbf{v}|$	magnitude of the vector $\mathbf{v}$	13

Contents

Two-Dimensional
Calculus

Introduction

Examples of functions of several variables are encountered at every turn in mathematics and its applications. By way of illustration, consider the following.

Algebra

1. The expressions for the roots of the general quadratic equation $ax^2 + bx + c = 0$

$$\frac{-b + \sqrt{b^2 - 4ac}}{2a} \quad \text{and} \quad \frac{-b - \sqrt{b^2 - 4ac}}{2a}$$

represent functions of the three variables a, b, c.

2. The simultaneous linear equations

$$ax + by = e$$
$$cx + dy = f$$

have a determinant

$$\begin{vmatrix} a & b \\ c & d \end{vmatrix} = ad - bc,$$

which is a function of four variables.

For three equations in three unknowns, the corresponding determinant is a function of nine variables.

Trigonometry

1. The "addition formula" for the sine of a sum of two angles is given by the expression

$$\sin (x + y) = \sin x \cos y + \cos x \sin y,$$

which is a function of two variables.

1

2. The cosine law for a triangle

$$c^2 = a^2 + b^2 - 2ab \cos C$$

gives c as a function of the three variables a, b, C.

Geometry

1. The distance from the point (X, Y) to a fixed line $ax + by + c = 0$ is

$$\frac{|aX + bY + c|}{\sqrt{a^2 + b^2}},$$

which is a function of the two variables X, Y.

2. The volume of a cylinder of height h and elliptical base is

$$V = \pi abh,$$

which is a function of the three variables a, b, h.

Calculus of One Variable

1. For a given differentiable function $f(x)$ the equation of the tangent line to the curve $y = f(x)$ at $x = x_0$ is

$$Y = f'(x_0)(X - x_0) + f(x_0),$$

a function of the two variables x_0 and X.

2. For a fixed continuous function $f(x)$, the definite integral

$$\int_a^b f(x)\, dx$$

is a function of the two variables a and b.

It is quite true that in most of these examples we tend to regard the expressions as "formulas" involving certain constants, rather than "functions" involving certain variables. Thus, in the last illustration, the definite integral is defined by considering a and b as having fixed values. However, the whole theory of integration is based on the fundamental theorem of calculus, which is derived precisely by allowing the limits of integration to vary, and observing the change in the value of the integral, considered as a function of the variable endpoints.

Again, the formula for the distance from a point to a line can be much more easily derived, as we shall show, by allowing the point to vary, and considering the distance to be a function of the variable coordinates, rather than by the usual static approach using a fixed point and a fixed line.

If we turn to the functions that arise in physical applications, we find that they are almost invariably functions of more than one variable. For example, temperature and pressure in a gas (such as the earth's atmosphere) are, in general, functions of the four variables x, y, z, t, where x, y, z are the coordinates of position in space, and t represents time.

In view of the fact that "almost all" the functions we encounter seem to involve more than one variable, we may wonder how the calculus of functions of a single variable is at all applicable. One answer is that a function of several variables can always be artificially reduced to be dependent upon a single variable by fixing all the other variables. Thus we may consider the variation of temperature at a given instant of time along a vertical line above a point on the earth's surface. The temperature then becomes a function of the single variable height. Or we may consider the variation of temperature with time at a fixed point in space.

The question is how satisfactory a description of the entire function do we obtain by this process of studying it separately with respect to each variable? A brief answer is that in some cases we get an adequate description, while in others this approach fails completely. From the mathematical point of view, there are a number of new and important concepts that arise when we allow the variables to vary simultaneously. These will be discussed in detail in Chapter 2. From the physical point of view, the artificiality of holding all but one variable fixed can be realized from the example of measuring the temperature along a vertical line. As we move along the line and take readings, the time is in fact varying too. Although this may seem to be a purely practical difficulty rather than a theoretical one, it was pointed out by Einstein in his first paper on relativity that basic problems arise in trying to define what is meant by "the same time" in different places.[1]

Without going as far afield as relativity, let us consider a very concrete physical problem. If a sound is emitted underwater, what path does it take? It is known, just as in the case of light, that the path is not a straight line but is refracted if the velocity is not constant. The velocity of sound underwater (in meters per second) is given approximately by the following formula[2]:

$$v = 1448 + 4.637T - 0.0538T^2 + (1.307 - 0.009T)(S - 35) + 0.018D,$$

where T is temperature in degrees centigrade, D is depth in meters, and S is salinity in parts per thousand. Note that each of the variables T, D, S is

---

[1] See, for example, reference [20]. Still better, see Einstein's original paper on relativity ([15], 37–65), and in particular, Sections 1 and 2, dealing with the definition of simultaneity and the relativity of lengths and times. We may note that there is a widespread misconception that advanced mathematical training is required in order to approach the theory of relativity. As a matter of fact, high school algebra is all that is needed to understand the two sections referred to above, and Einstein's entire paper uses no mathematical concepts beyond those discussed in the first three chapters of the present book.

[2] For more precise versions of this formula, see references [4] and [26].

itself a function of the four variables x, y, z, t. As we move along the path of a "sound wave," the variables x, y, z, t are all changing. They, in turn, determine the values of T, D, S, which determine the velocity. A typical question that we may encounter is the following: if T, D, and S are not known explicitly as functions of x, y, z, t, but if we have determined their values at some point and their rate of change in various directions, can we determine the rate of change of the velocity v in an arbitrary direction?

The purpose of this book is to provide the basic tools for treating problems of this general nature. We may note that an important feature of the above problem is that we have to consider a whole set of functions of several variables. We shall see later (Chapter 3) that although some information may be derived by considering each function separately, there are important advantages to be gained from dealing with the set of functions simultaneously.

In its most general form, the subject introduced in this book may be described as the study of systems of functions of several variables:

$$y_1 = f_1(x_1, x_2, \ldots, x_n)$$
$$y_2 = f_2(x_1, x_2, \ldots, x_n)$$
$$\vdots$$
$$y_m = f_m(x_1, x_2, \ldots, x_n)$$

Here we have n variables, denoted as $x_1, \ldots, x_n$, and m functions, $f_1, \ldots, f_m$.

Our approach will be to gradually increase the number of variables and functions to be considered. We shall also find a variety of ways in which we may reduce a problem to the consideration of a function of a single variable, to which we may then apply the results of elementary calculus.

A final word concerning theorems and their proofs. The statements of theorems comprise the core of basic information about the subject matter. You should make every effort to understand the meaning of each theorem. Many examples and comments are given to assist you in acquiring this understanding. You should also attempt to follow the reasoning in the proofs of the theorems. In some cases the proofs will help shed light on the content of the theorem. There are some cases, however, where a proof will be too difficult to understand fully on a first reading. Only by reading subsequent sections and then returning to the proof will you be able to grasp completely the material. We are dealing here with a fundamental and almost paradoxical difficulty. Stated briefly, it is that learning is sequential but knowledge is not. A branch of mathematics (or any other body of knowledge) consists of an intricate network of interrelated facts, each of which contributes to the understanding of those around it. When confronted with this network for the first time, we are forced to follow a particular path, which involves a somewhat arbitrary ordering of the facts. It is the large degree of choice in the path to be followed that accounts for the many different presentations possible for a given subject. One can be a very efficient tourist and race from one landmark to the next, or one can stop to investigate interesting side paths. One can choose between direct routes and scenic routes. One can enjoy the happy glint of recognition when an earlier point on the path is approached from a new direction.

This book has been arranged in a way intended to provide a path that affords a commanding view and at the same time a firm foothold for the student first discovering his way through the intricacies of higher-dimensional calculus. You are urged to make repeated return trips over parts of the path that have already been traveled; you will surely be happily surprised at how much easier it is to traverse the same stretch the second time, and at how often what at first seemed only a confused tangle of underbrush later assumes a clearly defined form and pattern. This transformation from an amorphous to a crystalline structure is a process that must take place inside each individual. Besides being an esthetically satisfying experience, it is the only process that can be truly called learning.

CHAPTER ONE

Background

1 Vectors in the plane

Our purpose in the present section is to recall the elementary properties of vectors in two dimensions, and to establish the notation that we shall use in the sequel.[1]

Although other aspects of vectors may be more important in other contexts, the basic property of vectors in connection with the calculus is that they may be characterized by a magnitude and a direction. Examples of vector quantities arise in the most diverse parts of mathematics and physics, as we shall see shortly, but the archetype of a vector is the representation of a "displacement." The classic example of "5 miles to the Northeast" describes a vector whose magnitude is 5, and whose direction is Northeast. It provides a clear illustration of the fact that a vector is not, in general, associated with any fixed point. The vector "5 miles Northeast" represents the notion of starting at an arbitrary point and moving a distance of 5 miles in the Northeast direction from there (Fig. 1.1).

In general, a displacement is a motion through a given distance in a given direction. Any pair of points p, q in the plane can be used to define a dis-placement—the one described by moving from p to q. Of course, many pairs of points may be used to describe the same displacement (Fig. 1.2).

For computation with vectors, it is often convenient to describe them in terms of their *components*. For displacement vectors, this description is particularly simple. We may describe the position of a point q relative to another point p as being "3 miles West and 4 miles South." This information is clearly equivalent to giving the distance and direction from p to q, and it may be used to describe the same displacement from any other point p' to the corresponding point q'.

---

[1] For further discussion of vectors from various points of view, see [5], [9], and [10].

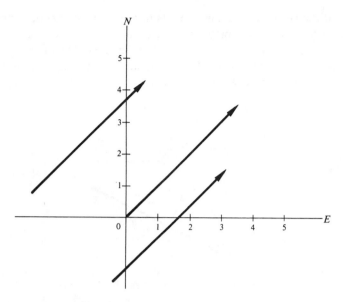

FIGURE 1.1   The vector "5 miles to the Northeast"

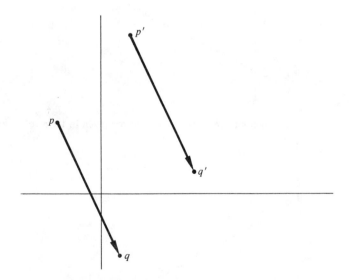

FIGURE 1.2   Two displacements corresponding to the same vector

To represent vectors analytically, we set up a rectangular coordinate system in the plane. We use the notation

$$\langle a, b \rangle$$

to represent the vector whose components in the x and y directions are a and b, respectively. This vector may be pictured as the displacement from any point (x_1, y_1) to the point $(x_1 + a, y_1 + b)$ (Fig. 1.3). Thus if p and q have coordinates (x_1, y_1) and (x_2, y_2) (Fig. 1.4) and if the displacement vector from p to q is $\langle a, b \rangle$, then we must have $x_1 + a = x_2$, $y_1 + b = y_2$, or

$$\langle a, b \rangle = \langle x_2 - x_1, y_2 - y_1 \rangle.$$

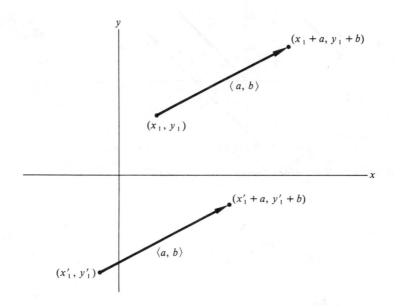

FIGURE 1.3 Displacements corresponding to a vector given in terms of components

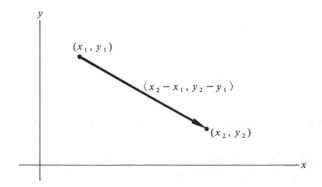

FIGURE 1.4 The components of a displacement vector

Note that the vector $\langle a, b \rangle$ describes the *relative* position of q to p. The first component describes how far q is to the right of p (or left, if a is negative), and the second component describes how far q is above p (or below, if b is negative).

Example 1.1

The vector $\langle 2, -3 \rangle$ describes a displacement of 2 units to the right and 3 units down (Fig. 1.5).

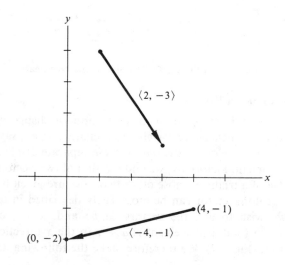

FIGURE 1.5 Examples of vectors

Example 1.2

The displacement vector from the point $(4, -1)$ to $(0, -2)$ is given by $\langle 0 - 4, -2 - (-1) \rangle = \langle -4, -1 \rangle$. Thus the second point is 4 units to the left and 1 unit below the first (Fig. 1.5).

Example 1.3

If the x axis points to the East and the y axis to the North, then a displacement of 5 units in the Northeast direction describes the hypothenuse of a right isosceles triangle whose sides have length $5/\sqrt{2}$. Thus, in terms of components, this vector is (Fig. 1.6)

$$\langle \tfrac{5}{2}\sqrt{2}, \tfrac{5}{2}\sqrt{2} \rangle.$$

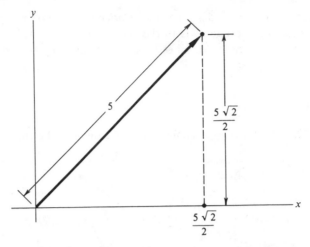

FIGURE 1.6 Components of the vector "5 miles to the Northeast"

Algebraic Operations on Vectors

Addition of vectors is modeled on a succession of displacements. If we perform two displacements successively, each characterized, say, by a length and a direction, then we define the *sum* of these separate displacements to be the total displacement. Pictorially, we find the sum of two vectors represented as the third side of a triangle whose other two sides are given by the original vectors. The resulting vector can be most easily described in terms of components. If we wish to add the vectors $\langle a, b \rangle$ and $\langle c, d \rangle$, considered as displacements, the total displacement is $a + c$ in the x direction and $b + d$ in the y direction (Fig. 1.7). We therefore make the following definition.

Definition 1.1 Addition of Vectors The *sum* of two vectors $\langle a, b \rangle$, $\langle c, d \rangle$ is defined by the rule

$$\langle a, b \rangle + \langle c, d \rangle = \langle a + c, b + d \rangle.$$

All the basic properties of vector addition may be derived from this formula. For example, the fact that the value of the sum is independent of the order in which we add. This property is not completely obvious from our original description in terms of successive displacements (Fig. 1.8), but it becomes clear if we use elementary geometry and note that in both cases the sum is the diagonal of a parallelogram whose pairs of opposite sides are the given vectors (Fig. 1.9). This is known as the *parallelogram law* for addition of vectors.

Besides adding vectors, we may consider any multiple of a given vector. "Twice a vector" would consist of going twice as far in the same direction (Fig. 1.10). In general, a multiple of a vector has components that are the same multiple of the original components. We use this as our definition.

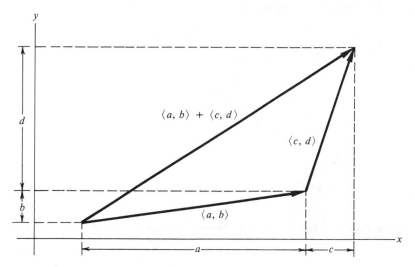

FIGURE 1.7 Addition of vectors

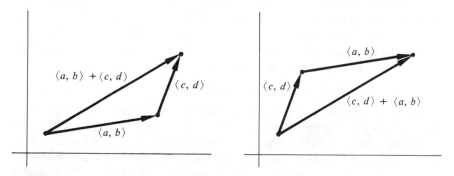

FIGURE 1.8 Adding vectors in different orders

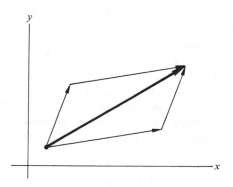

FIGURE 1.9 Parallelogram law for addition of vectors

11

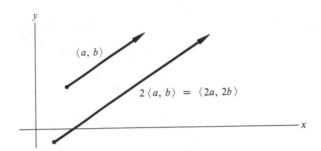

FIGURE 1.10 Multiplication of a vector by 2

Definition 1.2 Multiplication of Vectors by Scalars If $\langle a, b \rangle$ is any vector and λ is any number, we define their product to be

$$\lambda\langle a, b \rangle = \langle \lambda a, \lambda b \rangle.$$

The term *scalar* used in this connection has the following significance. Vectors, like numbers, are mathematical objects that may be combined according to certain rules. In elementary algebra we learn the basic rules for operations with numbers. In order to state these, we use letters to represent arbitrary numbers, and we can then state general rules, such as

$$a(b + c) = ab + ac.$$

In vector algebra, we proceed analogously, but there is a new feature. When dealing with a vector we are also interested in certain quantities such as its magnitude and its components, which are not vectors but numbers. Thus, we are dealing simultaneously with two kinds of mathematical objects, vectors and numbers. The word *scalar* is simply a synonym used for *number* in this context, when we wish to distinguish between "vector quantities" and purely numerical or "scalar" quantities. Thus, Def. 1.2 of multiplication of a vector by a scalar describes an algebraic operation that combines a scalar and vector to produce a new vector.

When we wish to state basic rules for algebraic operations that involve both vectors and scalars, we use certain letters for vectors and others for scalars. We could precede each formula by a statement explaining which letters represent which type of object, but it is much more convenient to adopt a uniform notation so that one may distinguish at a glance. We shall adopt the convention of using letters in **boldface** type to represent vectors, and those in ordinary type to represent numbers. Thus the equation

$$\mathbf{v} = \langle a, b \rangle$$

means that the letter $\mathbf{v}$ stands for the vector whose components are the numbers a, b.

We list some of the basic rules of vector algebra as follows:

$$\mathbf{v} + \mathbf{w} = \mathbf{w} + \mathbf{v}$$
$$(\mathbf{v} + \mathbf{w}) + \mathbf{z} = \mathbf{v} + (\mathbf{w} + \mathbf{z})$$
$$(\lambda + \mu)\mathbf{v} = \lambda\mathbf{v} + \mu\mathbf{v}$$
$$\lambda(\mathbf{v} + \mathbf{w}) = \lambda\mathbf{v} + \lambda\mathbf{w}.$$

In these equations $\mathbf{v}$, $\mathbf{w}$, and $\mathbf{z}$ represent arbitrary vectors, and λ and μ arbitrary scalars. Each equation asserts an equality between vectors and is verified by simply comparing the components of the vectors on each side of the equation. In general, a vector equation is equivalent to two ordinary scalar equations.

For the magnitude of the vector $\mathbf{v}$ we use the notation

$$|\mathbf{v}| = \text{magnitude of } \mathbf{v}.$$

By the Pythagorean theorem

$$|\mathbf{v}| = \sqrt{a^2 + b^2}$$

if $\mathbf{v} = \langle a, b \rangle$ (Fig. 1.11). To describe the direction of a vector, it is convenient to use the angle from the positive x direction to the vector, measured in the counterclockwise direction (Fig. 1.12). If we denote this angle by α, then the components a, b of the vector are given by

$$a = |\mathbf{v}| \cos \alpha \qquad \text{and} \qquad b = |\mathbf{v}| \sin \alpha.$$

Thus we may express any vector $\mathbf{v}$ in the form

$$\mathbf{v} = \langle |\mathbf{v}| \cos \alpha, |\mathbf{v}| \sin \alpha \rangle = |\mathbf{v}|\langle \cos \alpha, \sin \alpha \rangle,$$

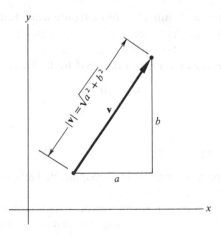

FIGURE 1.11   Magnitude of a vector

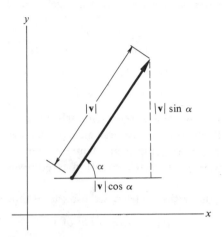

FIGURE 1.12 Direction of a vector

that is, in terms of its magnitude $|\mathbf{v}|$ and its direction α. Note that the vector $\langle\cos\alpha, \sin\alpha\rangle$ has magnitude 1, and, conversely, if $|\mathbf{v}| = 1$, then $\mathbf{v} = \langle\cos\alpha, \sin\alpha\rangle$. We use the notation $\mathbf{T}_\alpha$ for this vector. Thus,

$$\mathbf{T}_\alpha = \langle\cos\alpha, \sin\alpha\rangle.$$

A vector of magnitude 1 is called a *unit vector*.

Note that if $\mathbf{v} = \langle a, b\rangle$, then α is uniquely defined by the equations

$$\cos\alpha = \frac{a}{\sqrt{a^2 + b^2}} \qquad \sin\alpha = \frac{b}{\sqrt{a^2 + b^2}}, \qquad 0 \le \alpha < 2\pi,$$

provided $(a^2 + b^2)^{1/2} \ne 0$. But $a^2 + b^2 = 0$ only when both $a = 0$ and $b = 0$.

When describing vectors in terms of components, it is natural to consider any pair of numbers a, b and, in particular, the pair 0, 0. The corresponding vector is called the *zero vector* and denoted by $\mathbf{0}$. Thus

$$\mathbf{0} = \langle 0, 0\rangle.$$

Note that

1. $|\mathbf{0}| = 0$
2. If $|\mathbf{v}| = 0$, then $\mathbf{v} = \mathbf{0}$.

Thus the zero vector has a well-defined magnitude (zero) and is characterized by this property. It is, however, the only vector that does not have a well-defined direction. As a displacement vector, it corresponds to staying at the same point, which consists in going no distance and in no particular direction.

Corresponding to each vector $\mathbf{v} = \langle a, b \rangle$, we have the *opposite vector* $\langle -a, -b \rangle$, which we denote by $-\mathbf{v}$.

Note that

1. $-\mathbf{v} = -1\mathbf{v}$
2. $|-\mathbf{v}| = |\mathbf{v}|$
3.   if $\mathbf{v} = \lambda \langle \cos \alpha, \sin \alpha \rangle$,
     then $-\mathbf{v} = \lambda \langle \cos (\alpha \pm \pi), \sin (\alpha \pm \pi) \rangle$ (Fig. 1.13).

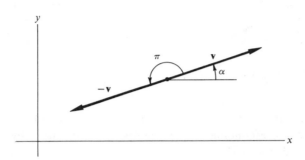

FIGURE 1.13   Opposite vectors

For each vector $\mathbf{v}$, the vector $-\mathbf{v}$ may be characterized algebraically by the property

$$\mathbf{v} + (-\mathbf{v}) = \mathbf{0}.$$

This equation may be described by saying that "$-\mathbf{v}$ is the additive inverse of $\mathbf{v}$."

Similarly, we may introduce the operation of subtraction as the inverse operation of addition. For any vectors $\mathbf{v}$, $\mathbf{w}$, we define the vector $\mathbf{v} - \mathbf{w}$ by

$$\mathbf{v} - \mathbf{w} = \mathbf{v} + (-\mathbf{w}).$$

Then $\mathbf{v} - \mathbf{w}$ can be characterized as the unique vector which, when added to $\mathbf{w}$, gives $\mathbf{v}$, that is,

$$\mathbf{v} = \mathbf{w} + (\mathbf{v} - \mathbf{w}).$$

Geometrically, the above two equations are depicted in Fig. 1.14, which shows equivalent representations of the vector $\mathbf{v} - \mathbf{w}$. In terms of components, we have simply

$$\langle a, b \rangle - \langle c, d \rangle = \langle a - c, b - d \rangle.$$

The most important single formula in elementary vector algebra allows us to compare the directions of two vectors $\mathbf{v}$, $\mathbf{w}$, or equivalently, to find the

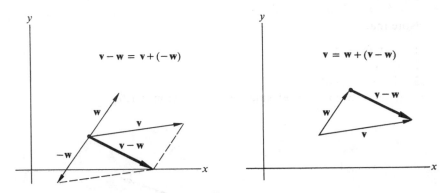

FIGURE 1.14 Subtraction of vectors

angle between them. If we let $\mathbf{v} = \langle a, b \rangle$, $\mathbf{w} = \langle c, d \rangle$, and if θ is the angle between the corresponding displacements, then by applying the law of cosines we find (Fig. 1.15)

$$|\mathbf{v} - \mathbf{w}|^2 = |\mathbf{v}|^2 + |\mathbf{w}|^2 - 2|\mathbf{v}|\,|\mathbf{w}| \cos \theta,$$

or

$$(a - c)^2 + (b - d)^2 = (a^2 + b^2) + (c^2 + d^2) - 2|\mathbf{v}|\,|\mathbf{w}| \cos \theta$$

$$a^2 - 2ac + c^2 + b^2 - 2bd + d^2 = a^2 + b^2 + c^2 + d^2 - 2|\mathbf{v}|\,|\mathbf{w}| \cos \theta$$

$$(1.1) \qquad\qquad ac + bd = |\mathbf{v}|\,|\mathbf{w}| \cos \theta.$$

The expression on the left-hand side of Eq. (1.1) arises so often when dealing with pairs of vectors, that a special designation is used for it (in fact, at least three different designations are currently used).

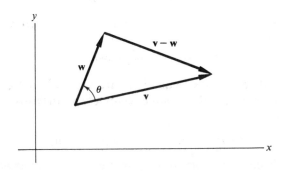

FIGURE 1.15 Angle between vectors

Definition 1.3 Given two vectors $\mathbf{v} = \langle a, b \rangle$ and $\mathbf{w} = \langle c, d \rangle$, their *dot product* is denoted by $\mathbf{v} \cdot \mathbf{w}$ and defined by

$$\mathbf{v} \cdot \mathbf{w} = ac + bd.$$

The terms *scalar product* and *inner product* are also frequently used for this expression.

Example 1.4

$$\langle 2, 3 \rangle \cdot \langle 4, -1 \rangle = 2 \cdot 4 + 3 \cdot (-1) = 8 - 3 = 5.$$

Note that the dot product of any two vectors is a scalar (whence the term *scalar product*).

We may now rewrite Eq. (1.1) as follows. If $\mathbf{v}$ and $\mathbf{w}$ are any two nonzero vectors, then the angle θ between them is given by

$$\cos \theta = \frac{\mathbf{v} \cdot \mathbf{w}}{|\mathbf{v}| \, |\mathbf{w}|}, \qquad 0 \leq \theta \leq \pi.$$

Example 1.5

$$\mathbf{v} = \langle 1, 2 \rangle, \qquad \mathbf{w} = \langle 2, 1 \rangle.$$

Then $\mathbf{v} \cdot \mathbf{w} = 4$, $|\mathbf{v}| = \sqrt{5}$, $|\mathbf{w}| = \sqrt{5}$, and $\cos \theta = \frac{4}{5}$.

There are two special cases of importance. The first is the dot product of a vector with itself. We have the elementary, but basic, formula

$$\mathbf{v} \cdot \mathbf{v} = |\mathbf{v}|^2.$$

The second is the case in which we form the dot product of a vector $\mathbf{v}$ with a unit vector $\mathbf{T}$. We have $|\mathbf{T}| = 1$, and hence

$$\mathbf{v} \cdot \mathbf{T} = |\mathbf{v}| \, |\mathbf{T}| \cos \theta = |\mathbf{v}| \cos \theta.$$

Geometrically, $\mathbf{v} \cdot \mathbf{T}$ represents the *projection of the vector* $\mathbf{v}$ *in the direction of the unit vector* $\mathbf{T}$. Note that this projection is considered positive if $\theta < \frac{1}{2}\pi$ and negative if $\theta > \frac{1}{2}\pi$ (Fig. 1.16). If $\theta = \frac{1}{2}\pi$, the projection is of course zero.

Definition 1.4 Two vectors $\mathbf{v}$, $\mathbf{w}$ are called *orthogonal* if $\mathbf{v} \cdot \mathbf{w} = 0$. We write $\mathbf{v} \perp \mathbf{w}$.

From Eq. (1.1) we see that there are precisely three cases in which $\mathbf{v} \perp \mathbf{w}$. Either $\mathbf{v} = \mathbf{0}$, or $\mathbf{w} = \mathbf{0}$, or else neither is zero and the angle θ between them satisfies $\cos \theta = 0$, that is, $\theta = \frac{1}{2}\pi$. When we speak of orthogonal vectors,

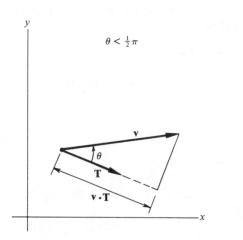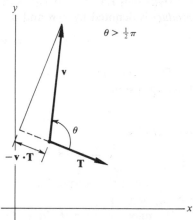

FIGURE 1.16 Projection of vectors

we usually have in mind the third case, but it is convenient to include the case of zero vectors. From our definition the zero vector is then orthogonal to every vector.

We conclude this section by proving two elementary lemmas concerning orthogonal vectors. These lemmas are often considered to be "self-evident," but they are frequently used and it is worthwhile to give their proofs.

Lemma 1.1 *Let* $\mathbf{v} = \langle a, b \rangle$ *be an arbitrary nonzero vector. Then a vector* $\mathbf{w}$ *is orthogonal to* $\mathbf{v}$ *if and only if* $\mathbf{w}$ *is of the form* $\lambda \langle -b, a \rangle$.

PROOF. If $\mathbf{w} = \lambda \langle -b, a \rangle$, then $\mathbf{v} \cdot \mathbf{w} = \lambda(-ba + ab) = 0$. Conversely, if $\mathbf{w} = \langle c, d \rangle$ and $\mathbf{v} \cdot \mathbf{w} = 0$, we have $ac + bd = 0$ or $ac = -bd$. Hence,

$$a\mathbf{w} = \langle ac, ad \rangle = \langle -bd, ad \rangle = d \langle -b, a \rangle$$

and

$$b\mathbf{w} = \langle bc, bd \rangle = \langle bc, -ac \rangle = -c \langle -b, a \rangle.$$

Since $\mathbf{v} \neq \mathbf{0}$, a and b cannot both be zero, and we may therefore write $\mathbf{w}$ in one of the two forms

$$\mathbf{w} = \frac{d}{a} \langle -b, a \rangle \qquad \text{or} \qquad \mathbf{w} = -\frac{c}{b} \langle -b, a \rangle. \qquad \blacklozenge$$

Lemma 1.2 *Let* $\mathbf{v} = \langle a, b \rangle$ *and* $\mathbf{w} = \langle c, d \rangle$ *be orthogonal nonzero vectors. Then every vector* $\mathbf{u} = \langle e, f \rangle$ *can be written in the form*

(1.2) $$\mathbf{u} = \lambda \mathbf{v} + \mu \mathbf{w}.$$

PROOF. Equation (1.2) represents the two scalar equations

$$e = \lambda a + \mu c$$
$$f = \lambda b + \mu d,$$

which have the (unique) solution $\lambda = (ed - fc)/(ad - bc)$, $\mu = (af - be)/(ad - bc)$, provided $ad - bc \neq 0$. But by Lemma 1.1 $\mathbf{w} = \langle c, d \rangle = \nu \langle -b, a \rangle$, for some scalar ν. In other words, $c = -\nu b$, $d = \nu a$, and hence $ad - bc = \nu(a^2 + b^2) \neq 0$, since $\mathbf{w} = \nu \langle -b, a \rangle \neq 0$. ◆

It is most convenient to apply Lemma 1.2 in the case where the vectors $\mathbf{v}$ and $\mathbf{w}$ have unit length. We have then

$$|\mathbf{v}| = 1, \qquad |\mathbf{w}| = 1, \qquad \mathbf{v} \perp \mathbf{w},$$

or equivalently,

$$\mathbf{v} \cdot \mathbf{v} = 1, \qquad \mathbf{w} \cdot \mathbf{w} = 1, \qquad \mathbf{v} \cdot \mathbf{w} = 0.$$

If we use Eq. (1.2) for an arbitrary vector $\mathbf{u}$, and take the dot product of both sides with $\mathbf{v}$ and $\mathbf{w}$, respectively, we find

$$\mathbf{u} \cdot \mathbf{v} = \lambda \mathbf{v} \cdot \mathbf{v} + \mu \mathbf{w} \cdot \mathbf{v} = \lambda \cdot 1 + \mu \cdot 0 = \lambda$$
$$\mathbf{u} \cdot \mathbf{w} = \lambda \mathbf{v} \cdot \mathbf{w} + \mu \mathbf{w} \cdot \mathbf{w} = \lambda \cdot 0 + \mu \cdot 1 = \mu.$$

Thus the coefficients λ, μ represent the projections of the vector $\mathbf{u}$ in the directions of $\mathbf{v}$ and $\mathbf{w}$, respectively (Fig. 1.17).

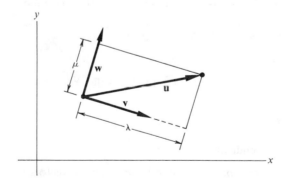

FIGURE 1.17 Components of a vector with respect to a pair of orthogonal unit vectors

It is sometimes convenient to introduce the special unit vectors $\mathbf{i} = \langle 1, 0 \rangle$, $\mathbf{j} = \langle 0, 1 \rangle$, which are clearly orthogonal. Then an arbitrary vector $\mathbf{v} = \langle a, b \rangle$ may be written as

$$\mathbf{v} = a\mathbf{i} + b\mathbf{j}.$$

We see that the components of a vector are precisely the projections of the vector in the direction of the x and y axes, respectively, and that the representation of a vector by its components is merely a special case of Eq. (1.2).

Exercises

1.1 For each of the following pairs of points, write down (in terms of components) the displacement vector from the first point to the second, and illustrate with a sketch.

a. (2, 3), (4, 4)   *d.* (3, −1), (−1, 3)
b. (4, 4), (2, 3)   *e.* (−2, −1), (0, 0)
c. (2, 3), (3, 2)   *f.* (3, 2), (−1, 2)

1.2 For each of the following vectors $\mathbf{v}$, find $|\mathbf{v}|$ (the magnitude of $\mathbf{v}$) and α (the direction of $\mathbf{v}$), and illustrate with a sketch.

a. $\langle 3, 0 \rangle$ *d.* $\langle -2, -2 \rangle$
b. $\langle -1, 0 \rangle$ *e.* $\langle 0, -\sqrt{2} \rangle$
c. $\langle -1, \sqrt{3} \rangle$ *f.* $\langle 4, -4 \rangle$

1.3 Find the components of a vector $\mathbf{v}$ whose magnitude and direction are given by

a. $|\mathbf{v}| = 5, \alpha = \frac{3}{4}\pi$ *d.* $|\mathbf{v}| = 1, \alpha = \frac{3}{2}\pi$
b. $|\mathbf{v}| = 2, \alpha = \pi$ *e.* $|\mathbf{v}| = 5, \alpha = \tan^{-1}\frac{4}{3}, \quad 0 < \alpha < \pi$
c. $|\mathbf{v}| = 2, \alpha = \frac{7}{6}\pi$ *f.* $|\mathbf{v}| = 1, \alpha = \tan^{-1}\left(-\frac{2}{3}\right), \ \pi < \alpha < 2\pi$

1.4 Let $\mathbf{v}_1 = \langle 2, 1 \rangle$, $\mathbf{v}_2 = \langle -2, 4 \rangle$, $\mathbf{v}_3 = \langle 0, -5 \rangle$. Find

a. $\mathbf{v}_1 + \mathbf{v}_2 + \mathbf{v}_3$ *c.* $\mathbf{v}_1 \cdot \mathbf{v}_2$
b. $|\mathbf{v}_1|^2 + |\mathbf{v}_2|^2 - |\mathbf{v}_3|^2$ *d.* $3\mathbf{v}_1 + \frac{1}{2}\mathbf{v}_2 + \mathbf{v}_3$

Interpret your answers with a sketch.

1.5 Let $\mathbf{v} = \langle 2, 5 \rangle$, $\mathbf{w} = \langle a, 4 \rangle$. Find a value of a such that the vectors $\mathbf{v}$ and $\mathbf{w}$

a. have the same length
b. have the same direction
c. are perpendicular

1.6 Find the angle between the following pairs of vectors.

a. $\langle 1, 3 \rangle$ and $\langle 6, -2 \rangle$ *c.* $\langle \sqrt{3}, 1 \rangle$ and $\langle -3, \sqrt{3} \rangle$
b. $\langle 2, 3 \rangle$ and $\langle 10, 2 \rangle$ *d.* $\langle -\sqrt{2}, \sqrt{3} \rangle$ and $\langle 2, -\sqrt{6} \rangle$

1.7 Write out the following vector statements in terms of components and verify

a. $\mathbf{v} \cdot \mathbf{w} = \mathbf{w} \cdot \mathbf{v}$ *d.* $|\mathbf{v} \cdot \mathbf{w}| \leq |\mathbf{v}| |\mathbf{w}|$
b. $(\lambda \mathbf{v}) \cdot \mathbf{w} = \lambda(\mathbf{v} \cdot \mathbf{w})$ *e.* $|\lambda \mathbf{v}| = |\lambda| |\mathbf{v}|$
c. $\mathbf{u} \cdot (\mathbf{v} + \mathbf{w}) = \mathbf{u} \cdot \mathbf{v} + \mathbf{u} \cdot \mathbf{w}$

1.8 Deduce each of the following from the relations given in Ex. 1.7a, b, c, d.

a. $\mathbf{u} \cdot (\lambda \mathbf{v} + \mu \mathbf{w}) = \lambda(\mathbf{u} \cdot \mathbf{v}) + \mu(\mathbf{u} \cdot \mathbf{w})$

b. $|\mathbf{v} + \mathbf{w}|^2 = |\mathbf{v}|^2 + 2\mathbf{v} \cdot \mathbf{w} + |\mathbf{w}|^2$ (*Hint:* $|\mathbf{u}|^2 = \mathbf{u} \cdot \mathbf{u}$).

c. $|\mathbf{v} + \mathbf{w}| \le |\mathbf{v}| + |\mathbf{w}|$ (*Hint:* square both sides)

d. $|\mathbf{v} - \mathbf{w}| \le |\mathbf{v}| + |\mathbf{w}|$

e. $|\mathbf{v}|^2 - |\mathbf{w}|^2 = (\mathbf{v} - \mathbf{w}) \cdot (\mathbf{v} + \mathbf{w})$

*1.9 *a.* Show that if $\langle a, b \rangle$ and $\langle c, d \rangle$ are any two vectors, then $ad - bc = 0$ if and only if one of the vectors is a scalar multiple of the other. (*Hint:* apply Lemma 1.1 to the vectors $\langle a, b \rangle$ and $\langle d, -c \rangle$.) See also Ex. 1.29 for another proof of this fact.

b. Show that equality holds in Ex. 1.7d if and only if one of the vectors is a scalar multiple of the other.

c. Under what conditions does equality hold in Ex. 1.8c?

d. Each of the inequalities in Ex. 1.8c, d is referred to as "the triangle inequality." Show the reason for this terminology by appropriate sketches illustrating each of the inequalities. Show by sketches how equality may occur in each case.

1.10 Let $\mathbf{v} = \langle a, b \rangle$, $\mathbf{w} = \langle b, a \rangle$. Show that $(\mathbf{v} - \mathbf{w}) \perp \langle 1, 1 \rangle$. Interpret geometrically in terms of the triangle with vertices $(0, 0)$, (a, b), and (b, a).

1.11 Let $\mathbf{V}$ and $\mathbf{W}$ be any two orthogonal unit vectors, and let $\mathbf{v} = a\mathbf{V} + b\mathbf{W}$, $\mathbf{w} = c\mathbf{V} + d\mathbf{W}$. Show that $\mathbf{v} \cdot \mathbf{w} = ac + bd$. (*Hint:* use Ex. 1.7.)

Exercises 1.12–1.18 illustrate various ways in which vectors may be used to derive results in plane geometry.

1.12 Let A, B, C, be the vertices of a triangle T, and let $\mathbf{v}$, $\mathbf{w}$ be the displacement vectors from A to B and from A to C, respectively. Write down conditions, in terms of the vectors $\mathbf{v}$, $\mathbf{w}$, and $\mathbf{u} = \mathbf{v} - \mathbf{w}$, such that

a. T is isosceles
b. T is equilateral
c. T is a right triangle
d. the angle at A is obtuse

1.13 Use the notation of the preceding exercise.

a. Prove that the angle at B is a right angle if and only if $\mathbf{v} \cdot \mathbf{w} = |\mathbf{v}|^2$.

b. Find a vector representing the median dropped from A to side BC.

c. Draw a picture indicating that the median dropped from B to side AC is represented by the vector $\frac{1}{2}\mathbf{w} - \mathbf{v}$.

d. Verify that $\mathbf{v} + \frac{2}{3}(\frac{1}{2}\mathbf{w} - \mathbf{v}) = \frac{2}{3}[\frac{1}{2}(\mathbf{v} + \mathbf{w})]$. Interpret geometrically, and deduce that the three medians of a triangle intersect at a point.

1.14 Let Q be a quadrilateral with vertices A, B, C, D. Let $\mathbf{u}$, $\mathbf{v}$, $\mathbf{w}$ be the displacement vectors from A to B, C, and D, respectively. Express the following displacement vectors in terms of $\mathbf{u}$, $\mathbf{v}$, and $\mathbf{w}$.

a. The sides of Q, from B to C and from D to C
b. The diagonal from B to D

c. From the midpoint of AB to the midpoint of BC

d. From the midpoint of AD to the midpoint of DC

e. From A to the midpoint of AC

f. From A to the midpoint of BD

1.15 Use the notation of the preceding exercise.

a. Show that if one pair of opposite sides of Q are parallel and equal in length, then the same is true of the other pair, and Q is a parallelogram.

b. Show that the midpoints of the sides of Q are the vertices of a parallelogram.

c. Show that Q is a parallelogram if and only if the diagonals bisect each other.

d. Show that if Q is a parallelogram, then the sums of the squares of the diagonals equals the sum of the squares of the sides.

1.16 Show that the equation of a straight line, $ax + by = c$, can be written in vector form as

$$\langle a, b \rangle \cdot \langle x - x_0, y - y_0 \rangle = 0,$$

where (x_0, y_0) is any point on the line. Interpret geometrically. What geometric meaning can you ascribe to the pair of numbers a, b, which appear as coefficients in the equation of the line?

1.17 Let $ax + by + c = 0$ be the equation of a line L.

a. Show that $\mathbf{N} = \langle a, b \rangle / (a^2 + b^2)^{1/2}$ is a unit vector orthogonal to L. (See Ex. 1.16.)

b. Show that if (x_1, y_1) is an arbitrary point, and if (x_0, y_0) is a point on L, then $|\langle x_1 - x_0, y_1 - y_0 \rangle \cdot \mathbf{N}|$ represents the perpendicular distance from (x_1, y_1) to L.

c. Using part b, show that the distance d from (x_1, y_1) to L is given by

$$d = \frac{|ax_1 + by_1 + c|}{\sqrt{a^2 + b^2}}.$$

1.18 Draw a sketch showing the vectors $\langle \cos \alpha, \sin \alpha \rangle$ and $\langle -\sin \alpha, \cos \alpha \rangle$ as displacement vectors starting at the origin.

a. How are these vectors placed with respect to the coordinate axes and with respect to each other?

b. Find the projections of the vector $\langle x, y \rangle$ in the direction of each of these vectors.

c. Given a point whose original coordinates are (x, y), derive the formulas

$$X = x \cos \alpha + y \sin \alpha$$

$$Y = -x \sin \alpha + y \cos \alpha$$

for its coordinates (X, Y) with respect to a pair of axes making an angle α with the x and y axes.

If a velocity is represented as a vector by its magnitude and direction, then algebraic operations on these vectors have physical significance. (This is not surprising, since velocities are merely displacements per unit time.) For example, if one motion is superimposed on another, the resultant velocity is represented as a vector sum. Also, the apparent velocity relative to a moving observer is given by the difference of the actual velocity and the velocity of the observer. Both cases are illustrated in Exs. 1.19–1.24.

1.19 A train is traveling along a straight track at 40 miles per hour. A boy in the train throws a ball at 30 miles per hour in a direction perpendicular to the motion of the train and parallel to the ground. Find the speed and direction of the ball relative to the ground, and illustrate with a vector diagram.

1.20 Answer Ex. 1.19 under the assumption that the ball is thrown at the same speed but at an angle of 60° with the forward direction of the train.

1.21 A plane flying toward the Northeast at an airspeed of 120 miles per hour is subjected to a 50 mile an hour wind from the Southeast. What is the speed of the plane relative to the ground?

1.22 A river flows due South at 2 miles per hour. A swimmer whose speed is 4 miles an hour wishes to cross to a point directly opposite. In which direction should he head?

1.23 A passenger on a boat traveling due East notices that a flag on the boat is pointing directly to the South. When the speed of the boat is doubled, the flag points toward the Southwest. What is the direction of the wind?

1.24 An important factor in astronomical measurements is the *aberration of light*. This phenomenon consists in a displacement in position due to the fact that the apparent velocity vector of light reaching the earth from a star is equal to the difference between the actual velocity vector of the light and the velocity vector of the earth's motion around the sun. Show by a vector diagram how the apparent direction of a star in the plane of the earth's orbit varies according to the position of the earth in its orbit. (Assume that the star is sufficiently far away so that light rays coming from it may be considered parallel, independent of the earth's position in its orbit. It is a fact that the "parallax," or displacement in direction caused by the light rays not being exactly parallel, is considerably smaller than the effect of the aberration of light.)

A *force* in physics is a quantity having a given magnitude and direction. If a number of forces act on a point, and if each is represented by a vector having the given magnitude and direction, then the total effect is the same as that of a single force called the *resultant*, corresponding to the vector sum of the given forces. Exs. 1.25–1.28 are illustrations.

1.25 Four forces of magnitude 2, 3, 4, and 5, respectively, act on a point at the origin. These forces are directed inward along the diagonals through the first, second, third, and fourth quadrants, respectively. Find the magnitude and direction of the resultant force.

1.26 A force **F**, of magnitude 8, is directed downward and to the right at an angle of 30° with the vertical. Find a pair of forces, one horizontal and one vertical, whose resultant is **F**.

1.27 A system of forces is in *equilibrium* if their resultant is zero.

   a. Under what conditions are two forces in equilibrium?

   b. Show that three forces are in equilibrium if and only if the corresponding vectors, suitably translated, represent the three sides of a triangle described in succession.

1.28 *Newton's law of gravity* states that the gravitational effect of a body of mass m_1 on a body of mass m_2 at a distance d is a force whose direction is along the line joining the two bodies, and whose magnitude is Gm_1m_2/d^2, where G is a fixed constant.
Suppose that two bodies of masses m_1 and m_2 are a fixed distance d apart, and that both of them act on a third body of mass m.

   a. Find the position of the third body such that the two forces are in equilibrium, and show that this position depends on the distance d and the ratio $\lambda = m_2/m_1$, but not on the mass m or the constant G.

   b. Show that in the case of the earth and the moon, where λ is approximately 0.012, the equilibrium position is approximately $\frac{9}{10}$ of the way from the earth to the moon.

The basic facts concerning a pair of simultaneous linear equations are of great importance and are used throughout this book. They are the following. The equations

$$ax + by = e$$

$$cx + dy = f$$

have one and only one solution for every choice of e and f provided the determinant

$$\begin{vmatrix} a & b \\ c & d \end{vmatrix} = ad - bc$$

is not zero. This unique solution may be written as

$$x = \frac{\begin{vmatrix} e & b \\ f & d \end{vmatrix}}{\begin{vmatrix} a & b \\ c & d \end{vmatrix}}, \qquad y = \frac{\begin{vmatrix} a & e \\ c & f \end{vmatrix}}{\begin{vmatrix} a & b \\ c & d \end{vmatrix}}.$$

In case the determinant is zero, either there is no solution or there are infinitely many solutions, depending on the values of e and f. These facts are related to certain properties of vectors as is shown in Exs. 1.29 and 1.30.

1.29   Let $v = \langle a, b \rangle$, $w = \langle c, d \rangle$ be nonzero vectors, and let θ be the angle between them.

a. Show that $|v|^2|w|^2 - (v \cdot w)^2 = (|v||w| \sin \theta)^2$.

b. Writing out the left-hand side of the above equation in terms of components, show that

$$|ad - bc| = |v||w||\sin \theta|.$$

c. Let T be a triangle with vertices A, B, C and let v, w be the vectors from A to B and from A to C, respectively. Show that the altitude from the vertex C to the side AB is equal to $|w||\sin \theta|$.

d. Show that $\frac{1}{2}|ad - bc|$ = area of triangle T.
(*Note:* this is an important geometric interpretation of the determinant.)

e. Use part d to show geometrically that $ad - bc = 0$ if and only if the vectors v and w have the same or opposite directions. (Compare with Ex. 1.9a.)

1.30   Let $v = \langle a, b \rangle$ and $w = \langle c, d \rangle$ be nonzero vectors.

a. Describe the totality of straight lines $ax + by = e$, for different values of the constant e, in terms of the vector v. (*Hint:* see Ex. 1.16.)

b. Using part a and Ex. 1.29e, show that the condition $ad - bc \neq 0$ implies that the two straight lines whose equations are $ax + by = e$ and $cx + dy = f$ are not parallel, and hence, no matter how e and f are chosen, there is a unique point of intersection.

c. Show that if $ad - bc = 0$, the two lines in part b are either parallel and distinct or else identical, so that the two equations either have no common solutions or else the simultaneous solutions consist of all points lying on the line $ax + by = e$.

2   Plane curves

Let $f(t)$ be a function of the single variable t. Then $f(t)$ is said to be *differentiable* in an interval $a \le t \le b$ if the derivative $f'(t)$ exists at each point in the interval. It is *continuously differentiable* if the derivative $f'(t)$ is a continuous function of t.

Definition 2.1   A *regular (plane) curve* C is defined by a pair of continuously differentiable functions $x(t)$, $y(t)$ in an interval $a \le t \le b$, satisfying

(2.1) $$x'(t)^2 + y'(t)^2 \neq 0.$$

The significance of condition (2.1) will be explained later. Example 2.4 below is an illustration of what may happen if it does not hold. (See also Example 2.5 and Exs. 2.14 and 2.15.)

For the present we have a number of remarks to make concerning Def. 2.1.[2] The chief observation to be made is that a curve, by our definition, is not simply a set of points in the plane, but is always described in terms of an additional variable t, called the *parameter*. The reason for defining curves in this way is that it is important to distinguish, for example, between a circle described once in the clockwise direction and the same circle described once in the counterclockwise direction or twice in the clockwise direction. The easiest way to make this distinction is to consider these as three different curves, even though they correspond to the same set of points in the plane. Intuitively we should think of a curve in the plane as a set of points together with a parameter, which in essence describes how the curve can be traced with a pencil.

The set of points defined by the equation

$$(2.2) \qquad\qquad y = f(x), \qquad\qquad a \le x \le b,$$

where $f(x)$ is continuously differentiable, can be described as above by simply using x as the parameter. If we set

$$(2.3) \qquad\qquad x = t, \qquad y = f(t), \qquad\qquad a \le t \le b,$$

then

$$x'(t)^2 + y'(t)^2 = 1 + f'(t)^2 > 0$$

and condition (2.1) is satisfied. Thus Eq. (2.3) defines a regular curve. Equation (2.2) is often referred to as the *explicit form* or *nonparametric form* of this curve, in contrast to Eq. (2.3), which describes the curve in *parametric form*.

An equation of the form

$$(2.4) \qquad\qquad F(x, y) = 0$$

is frequently said to define a curve in *implicit form*. The fact is that "in

---

[2] It may be argued that our definition of a curve is somewhat ambiguous, since it essentially describes how a curve is represented without stating clearly what a curve *is*. A reader who is bothered by this lack of precision may find it clearer to say that a curve *is* the pair of functions $x(t)$, $y(t)$. The remainder of the text may then be read without change. Although this way of gaining precision is frequently used, we have preferred a less formal formulation, which embodies the intuitive notion of what a curve consists of. This intuitive notion has in fact been made rigorous in a variety of ways, each one abstracting some particular aspect of curves. There are a number of current definitions of curves in different branches of modern mathematics. Thus, in differential geometry a curve is often defined by starting with pairs of functions $x(t)$, $y(t)$ and considering two pairs to be equivalent if one is obtained from the other by a suitable change of parameter. In algebraic geometry a curve is generally regarded as a set of points, corresponding to what we have called the "implicit form" of a curve. For more sophisticated discussions of curves from these two points of view see [17], pp. 1–6, and [35], Chapter 3.

general" the set of points that satisfy such an equation can be traced out in the manner indicated. We shall return to this subject in Sect. 8, where we give a precise meaning to the phrase "in general."

Example 2.1

Let

$$(2.5) \qquad x = a(1 - t) + ct, \qquad y = b(1 - t) + dt, \qquad 0 \le t \le 1.$$

Here a, b, c, d are arbitrary numbers. As t goes from 0 to 1, the point (x, y) defined by Eq. (2.5) moves along a straight line from (a, b) to (c, d) (Fig. 2.1). The geometric significance of the parameter t becomes clearer if we rewrite Eq. (2.5) in the form

$$(2.5a) \qquad x - a = t(c - a), \qquad y - b = t(d - b), \qquad 0 \le t \le 1.$$

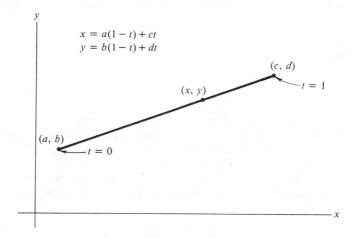

$$x = a(1 - t) + ct$$
$$y = b(1 - t) + dt$$

FIGURE 2.1   Straight-line segment from (a, b) to (c, d)

These equations assert the equality of the ratios $(x - a)/(c - a)$ and $(y - b)/(d - b)$, with t as their common value (Fig. 2.2). When $c - a = 0$ or $d - b = 0$ we do not have this interpretation. However, when $c - a = 0$, the first of Eqs. (2.5a) yields $x - a = 0$ for all t, so that we obtain a segment of the vertical line $x = a$. If, on the other hand, $c - a \ne 0$, then substituting $t = (x - a)/(c - a)$ from the first of Eqs. (2.5a) into the second, we may rewrite Eqs. (2.5a) in the form

$$(2.5b) \qquad y = \frac{d - b}{c - a}(x - a) + b,$$

where x lies between a and c. This represents the same line segment in the explicit form of Eq. (2.2).

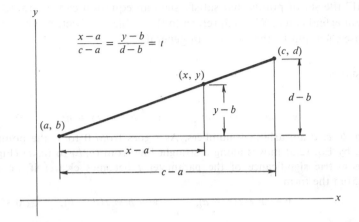

FIGURE 2.2 Geometric interpretation of parametric equations for a straight-line segment

Example 2.2

Let

(2.6) $$x = c + r \cos t, \qquad y = d + r \sin t, \qquad 0 \le t \le 2\pi.$$

Here c, d, and r are constants. Equation (2.6) defines the circle of radius r about the point (c, d), described once in counterclockwise direction and starting and ending at $(c + r, d)$ (Fig. 2.3). The parameter t in this case is the angle, indicated in Fig. 2.3, measured in radians. This time we cannot expect to represent the curve in the explicit form of Eq. (2.2), since all values of x between $c - r$ and $c + r$ correspond to two distinct values of y. However, from Eq. (2.6) we find that

(2.6a) $$(x - c)^2 + (y - d)^2 = r^2,$$

which is in the form of Eq. (2.4). Furthermore, given any point (x, y) satisfying Eq. (2.6a), we can find a value of the parameter t so that (x, y) satisfies Eq. (2.6). Thus Eq. (2.6a) can be regarded as the implicit form of a curve (in our sense); namely, the circle defined by Eq. (2.6).

Example 2.3

Let

(2.7) $$x = (v_0 \cos \alpha)t, \qquad y = (v_0 \sin \alpha)t - \tfrac{1}{2}gt^2, \qquad 0 \le t \le t_1,$$

where v_0, α, and g are constants. These equations represent the position after time t of a projectile having initial speed v_0 and initial inclination α, where g is the constant of gravitational acceleration (Fig. 2.4). This example is typical of many problems arising in physics, where the parameter t represents time (rather than

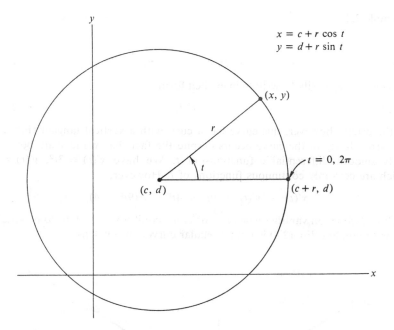

$$x = c + r \cos t$$
$$y = d + r \sin t$$

FIGURE 2.3 Circle of radius r

any geometrical quantity), and the functions $x(t)$ and $y(t)$ describe the position of a moving point at any time t.

If the quantity $v_0 \cos \alpha$ is positive, then we may solve the first of Eqs. (2.7) as $t = x/(v_0 \cos \alpha)$, and, substituting into the second, we find

$$(2.7a) \qquad y = (\tan \alpha)x - \left(\frac{1}{2}\frac{g}{v_0^2} \sec^2 \alpha\right)x^2, \qquad 0 \le x \le x_1,$$

which is, in explicit form, the equation of a parabolic arc.

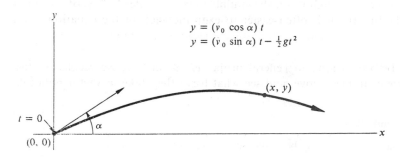

$$y = (v_0 \cos \alpha)\, t$$
$$y = (v_0 \sin \alpha)\, t - \tfrac{1}{2}gt^2$$

FIGURE 2.4 Arc of parabola

Example 2.4

Let

(2.8) $x = t^3,$ $y = t^2,$ $-1 \le t \le 1.$

This curve may easily be written in explicit form:

(2.8a) $y = x^{2/3},$ $-1 \le x \le 1.$

At the origin, however, this curve has a cusp with a vertical tangent (Fig. 2.5). This irregularity in the curve occurs despite the fact that x and y are both perfectly smooth differentiable functions of t. We have $x'(t) = 3t^2,$ $y'(t) = 2t,$ which are certainly continuous functions of t. However,

$$x'(t)^2 + y'(t)^2 = 9t^4 + 4t^2 = t^2(9t^2 + 4),$$

and this expression vanishes when $t = 0$. Thus, condition (2.1) fails to be satisfied at the origin, and Eq. (2.8) is not a "regular curve" in our sense.

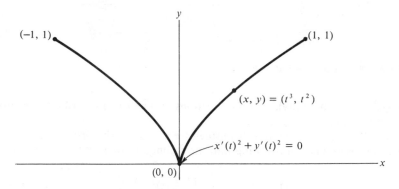

FIGURE 2.5   Example of a nonregular curve

It is worth noting that condition (2.1) is equivalent to the statement that $x'(t)$ and $y'(t)$ do not vanish simultaneously. Rather than verifying condition (2.1) directly, it is often easier to examine each of the equations $x'(t) = 0$ and $y'(t) = 0$ separately, and then check for common zeros.

Before turning to general properties of curves, we reconsider the first three examples above, and see what form they take in vector notation.

Example 2.1a

Equations (2.5a) may be written as a single vector equation

(2.9) $\langle x - a, y - b \rangle = t\mathbf{v}_0,$ $0 \le t \le 1,$

where v_0 is a fixed vector:

$$v_0 = \langle c - a, d - b \rangle.$$

Equation (2.9) expresses the fact that the displacement vector from (a, b) to (x, y) is a multiple of the fixed vector v_0 (Fig. 2.6).

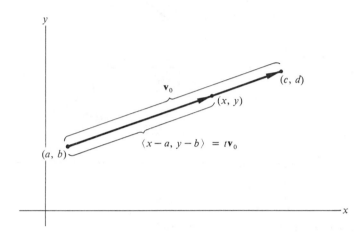

FIGURE 2.6   Straight-line segment in vector form

Example 2.2a

Equations (2.6) may be written in the form

$$(2.10) \qquad \langle x - c, y - d \rangle = r\langle \cos t, \sin t \rangle, \qquad 0 \le t \le 2\pi.$$

This states that the displacement vector from (c, d) to (x, y) is a constant r times the unit vector $\langle \cos t, \sin t \rangle$ and hence has constant magnitude r (Fig. 2.7).

Example 2.3a

Equations (2.7) may be written as

$$(2.11) \qquad \langle x, y \rangle = t v_0 + \tfrac{1}{2} g t^2 w,$$

where v_0 and w are the fixed vectors

$$v_0 = \langle v_0 \cos \alpha, v_0 \sin \alpha \rangle, \qquad w = \langle 0, -1 \rangle.$$

The vector v_0 is known as the *initial velocity vector*. Equation (2.11) represents the displacement vector from the origin to the point (x, y) as the sum of two vectors. The first of these has the direction of the initial velocity vector and a magnitude that increases uniformly with t; the second is vertically downward in direction, and its magnitude grows at an increasing rate (Fig. 2.8).

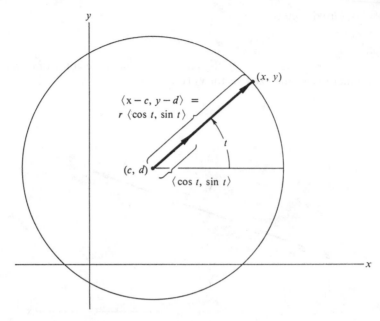

FIGURE 2.7 Circle of radius *r* in vector form

Remark It is worth noting that the device used in Eq. (2.11) to write the curve (2.7) in vector form may be applied to an arbitrary curve. The *position vector* or *radius vector* of a point (x, y) is the vector $\langle x, y \rangle$, which may be thought of as the displacement vector from the origin to the point (x, y). To each curve $x(t)$, $y(t)$ corresponds the *vector function* $\langle x(t), y(t) \rangle$, which assigns to each value of t the position vector of the point $(x(t), y(t))$.

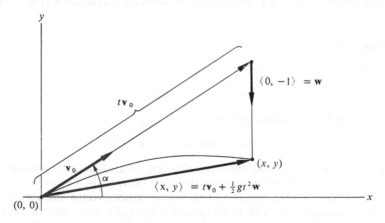

FIGURE 2.8 Arc of parabola in vector form

There are some formal advantages to this notation, but there is also a certain amount of confusion, which may result from identifying the point (x, y) with the vector $\langle x, y \rangle$. We prefer to reserve vector notation for quantities that are fundamentally vectorial.

With the above examples in mind, we turn to a more general discussion of plane curves.

Definition 2.2   The *length* of the regular curve

(2.12) $$C: x(t), y(t), \qquad\qquad a \leq t \leq b,$$

is the number

(2.13) $$L = \int_a^b \sqrt{x'(t)^2 + y'(t)^2}\, dt.$$

It can be shown that this is the same value we obtain by a limit process if we approximate the curve by inscribed polygons.[3] However, we simply use the integral in Eq. (2.13) as our definition of arc length.

If we consider only the part of the curve C corresponding to values of t between a and a fixed value t_0, then the arc length of that part is equal to

(2.14) $$s(t_0) = \int_a^{t_0} \sqrt{x'(t)^2 + y'(t)^2}\, dt.$$

Thus the arc length s is a function of the parameter value t_0 (Fig. 2.9). By the fundamental theorem of calculus, the derivative of s is

(2.15) $$s'(t_0) = \sqrt{x'(t_0)^2 + y'(t_0)^2}.$$

Equation (2.15) expresses the rate of change of arc length with respect to the parameter t. When t represents time, the quantity $s'(t_0)$ is the *speed* at which the point is moving along the curve at the time t_0.

With each point $(x(t), y(t))$ of the curve C we associate the vector

(2.16) $$\mathbf{v}(t) = \langle x'(t), y'(t) \rangle.$$

The condition (2.1) that the curve be regular, then takes the form

(2.17) $$\mathbf{v}(t) \neq \mathbf{0}.$$

---

[3] See [14], Section 7.3, in which this is proved for nonparametric curves $y = f(x)$, and see Exercise 2.18. Direct proofs may be found, for example, in [3], Vol. 1, Section 6.33, and in [27], Section 13.3. For a good review of parametric curves, and for further exercises, see Chapter 7 of [14].

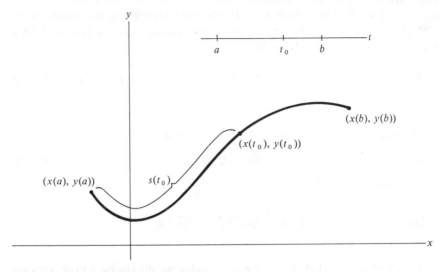

FIGURE 2.9 Arc-length function $s(t_0)$

When t represents time, $\mathbf{v}(t)$ is called the *velocity vector*. By Eq. (2.15), the magnitude of the velocity vector is the speed

$$(2.18) \qquad\qquad |\mathbf{v}(t)| = s'(t).$$

As for the direction of $\mathbf{v}(t)$, we note that by definition of the derivative,

$$x'(t_0) = \lim_{t \to t_0} \frac{x(t) - x(t_0)}{t - t_0}, \qquad y'(t_0) = \lim_{t \to t_0} \frac{y(t) - y(t_0)}{t - t_0}$$

so that[4]

$$\langle x'(t_0), y'(t_0) \rangle = \lim_{t \to t_0} \left\langle \frac{x(t) - x(t_0)}{t - t_0}, \frac{y(t) - y(t_0)}{t - t_0} \right\rangle$$

$$(2.19)$$

$$= \lim_{t \to t_0} \frac{1}{t - t_0} \langle x(t) - x(t_0), y(t) - y(t_0) \rangle.$$

The vectors on the right-hand side of Eq. (2.19) are in the direction of the displacement vector from $(x(t_0), y(t_0))$ to $(x(t), y(t))$. As t approaches t_0, the direction of these vectors approaches the direction of the tangent to the curve (Fig. 2.10).

---

[4] Here we are implicitly using the fact that limit operations on vectors may be performed by carrying out the limit process on each component.

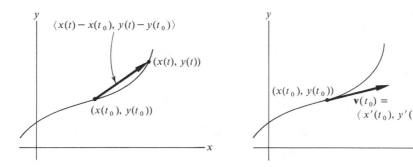

FIGURE 2.10 Velocity vector as a limit of displacement vectors

We may summarize as follows:

The vector $\mathbf{v}(t)$ *defined by Eq. (2.16) at each point of the curve C has a direction tangent to the curve at the point, and magnitude equal to the derivative of arc length with respect to the parameter t.*

The geometric interpretation of condition (2.1) is now clear. It implies that the vector $\mathbf{v}(t)$, being nonzero, has a well-defined direction, and hence the curve itself has a well-defined tangent direction at each point.

Definition 2.3 The vector **T** defined at each point of a regular curve C by

$$\mathbf{T} = \frac{\mathbf{v}(t)}{|\mathbf{v}(t)|}$$

is called the *unit tangent* to the curve at the point.

It is often convenient to use the arc length s as a parameter for the curve. Specifically, to each value of s we may assign the coordinate (x, y) of the point at a distance s along the curve from its beginning. Then by the chain rule,

$$\mathbf{v} = \left\langle \frac{dx}{dt}, \frac{dy}{dt} \right\rangle = \left\langle \frac{dx}{ds}\frac{ds}{dt}, \frac{dy}{ds}\frac{ds}{dt} \right\rangle = \frac{ds}{dt} \left\langle \frac{dx}{ds}, \frac{dy}{ds} \right\rangle$$

and using Eq. (2.18) we find that

(2.20) $$\mathbf{T} = \frac{\mathbf{v}}{|\mathbf{v}|} = \left\langle \frac{dx}{ds}, \frac{dy}{ds} \right\rangle.$$

We return to the examples given above, and see how these general considerations apply.

Example 2.1b

For the straight line of Eq. (2.5),

$$x = a + (c - a)t, \qquad y = b + (d - b)t, \qquad 0 \le t \le 1,$$

we have

$$\mathbf{v}(t) = \left\langle \frac{dx}{dt}, \frac{dy}{dt} \right\rangle = \langle c - a, d - b \rangle.$$

Thus $\mathbf{v}(t)$ is a constant vector. If we denote its magnitude by m, so that

$$m = |\mathbf{v}(t)| = \sqrt{(c - a)^2 + (d - b)^2},$$

then by Eq. (2.18)

$$\frac{ds}{dt} = m,$$

or

$$s = mt, \qquad t = \frac{s}{m}.$$

Substituting in the original equations of the line, we may express the coordinates explicitly in terms of arc length:

$$x = a + \frac{c - a}{m} s, \qquad y = b + \frac{d - b}{m} s, \qquad 0 \le s \le m.$$

Hence, using Eq. (2.20), we find the unit tangent $\mathbf{T}$ at each point is

$$\mathbf{T} = \left\langle \frac{c - a}{m}, \frac{d - b}{m} \right\rangle.$$

Example 2.2b

For the circle

$$x = c + r \cos t, \qquad y = d + r \sin t, \qquad 0 \le t \le 2\pi,$$

we find

$$\mathbf{v} = \langle -r \sin t, r \cos t \rangle,$$

and

$$\frac{ds}{dt} = |\mathbf{v}| = r.$$

Thus

$$s = rt, \qquad t = \frac{s}{r},$$

and we may write

$$x = c + r \cos \frac{s}{r}, \qquad y = d + r \sin \frac{s}{r}, \qquad 0 \le s \le 2\pi r.$$

The unit tangent **T** is then (Fig. 2.11)

$$\mathbf{T} = \left\langle \frac{dx}{ds}, \frac{dy}{ds} \right\rangle = \left\langle -\sin \frac{s}{r}, \cos \frac{s}{r} \right\rangle = \langle -\sin t, \cos t \rangle.$$

Note that it is customary to represent the velocity vector $\mathbf{v} = \langle x'(t), y'(t) \rangle$ and the corresponding unit tangent **T** as displacement vectors starting at $(x(t), y(t))$.

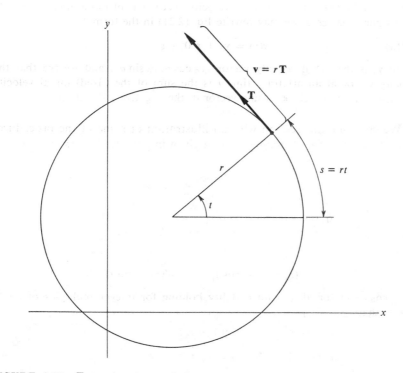

FIGURE 2.11 Tangent vector to circle

Example 2.3b

For the parabola

$$x = (v_0 \cos \alpha)t, \qquad y = (v_0 \sin \alpha)t - \tfrac{1}{2}gt^2,$$

we have

(2.21)
$$\mathbf{v}(t) = \langle v_0 \cos \alpha, v_0 \sin \alpha - gt \rangle.$$

Hence

$$\frac{ds}{dt} = |\mathbf{v}(t)| = \sqrt{v_0^2 - 2v_0(\sin \alpha)gt + g^2 t^2}.$$

This example is much more typical of the general case in which we cannot hope to represent x and y as explicit functions of the arc length s, as we were able to in the special cases of the straight line and the circle. We may, though, always express the unit tangent $\mathbf{T}$ in the form $\mathbf{v}/|\mathbf{v}|$, so that here

$$\mathbf{T} = \frac{\langle v_0 \cos \alpha, \, v_0(\sin \alpha) - gt \rangle}{\sqrt{v_0^2 - 2v_0(\sin \alpha)gt + g^2 t^2}}.$$

It should be noted that often the velocity vector $\mathbf{v}$ is of more interest than $\mathbf{T}$. In this particular case, we may rewrite Eq. (2.21) in the form

$$(2.21a) \qquad\qquad \mathbf{v}(t) = \mathbf{v}_0 + t\langle 0, -g \rangle,$$

where $\mathbf{v}_0$ is the initial velocity vector $\langle v_0 \cos \alpha, v_0 \sin \alpha \rangle$, and we see that the velocity vector at an arbitrary time t is the sum of the (fixed) initial velocity vector and a uniformly increasing vector in the negative y direction.

We conclude this section with an illustration of some of the procedures that may be used for sketching a curve given in parametric form.[5]

Example 2.5

Sketch the curve

$$(2.22) \qquad\qquad x = t - \sin t, \qquad y = 1 - \cos t.$$

We find

$$x'(t) = 1 - \cos t, \qquad y'(t) = \sin t.$$

Since $\cos t \le 1$ for all t, with equality holding for integer multiples of 2π, it follows that

$$x'(t) \ge 0 \qquad \text{for} \quad \text{all } t,$$

$$x'(t) = 0 \qquad \text{for} \quad t = 0, \pm 2\pi, \pm 4\pi, \ldots.$$

On the other hand,

$$y'(t) \ge 0 \qquad \text{for} \quad 0 \le t \le \pi, \quad 2\pi \le t \le 3\pi, \ldots$$

$$y'(t) \le 0 \qquad \text{for} \quad -\pi \le t \le 0, \quad \pi \le t \le 2\pi, \ldots$$

and

$$y'(t) = 0 \qquad \text{for} \quad t = 0, \pm\pi, \pm 2\pi, \ldots.$$

Thus $x'(t)$ and $y'(t)$ vanish simultaneously for $t = 0, \pm 2\pi, \pm 4\pi, \ldots$. In any interval excluding these values, condition (2.1) is satisfied, and we have a regular curve. The vector $\mathbf{v} = \langle x'(t), y'(t) \rangle$ always points to the right, since its first

---

[5] For a wealth of examples of plane curves, with illuminating discussions of each, see [24].

component is positive. The curve has a horizontal tangent at the points $t = \pm\pi, \pm3\pi, \ldots$, since at these points $y'(t) = 0$ and $x'(t) \neq 0$. But these are precisely the points at which $\cos t = -1$, and at these points y takes on its maximum value 2. On the other hand, $y \geq 0$ for all t, and $y = 0$ when $\cos t = 1$.

We can now form a clear qualitative picture of the entire curve. As t increases, the corresponding point $(x(t), y(t))$ on the curve moves to the right, since $x'(t) \geq 0$, while the y coordinate varies periodically between 0 and 2. The curve may therefore be described as a series of arches resting on the x axis at the points $x = 0, \pm2\pi, \ldots$. It lies below the horizontal line $y = 2$, and touches that line at the points $x = \pm\pi, \pm3\pi, \ldots$.

To obtain a more precise picture, we note that

$$x(t + 2\pi) = t + 2\pi - \sin(t + 2\pi) = t + 2\pi - \sin t = x(t) + 2\pi$$

$$y(t + 2\pi) = 1 - \cos(t + 2\pi) = 1 - \cos t = y(t).$$

This means that adding 2π to the parameter value shifts the corresponding point on the curve a distance 2π to the right. In other words, each arch of the curve is obtained by translating the previous arch a distance 2π to the right. We therefore get a complete description of the curve once we have plotted a single arch, corresponding, say, to the interval $0 \leq t \leq 2\pi$. For $0 < t < \pi$, y increases with t, while for $\pi < t < 2\pi$, y decreases. Furthermore, since

$$x(2\pi - t) = 2\pi - x(t), \qquad \text{and} \qquad y(2\pi - t) = y(t),$$

the right half of the arch is symmetric to the left half; it is sufficient to draw the part of the curve corresponding to $0 \leq t \leq \pi$, and then reflect across the line $x = \pi$. In Fig. 2.12, we have chosen a number of values of t between 0 and 2π, plotted the corresponding points $(x(t), y(t))$ on the curve, and drawn the vectors $\langle x'(t), y'(t) \rangle$ at each of these points. It is now a simple matter to sketch the complete curve.

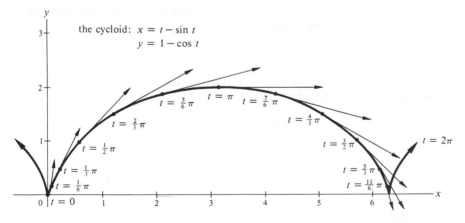

FIGURE 2.12 Cycloid

The only question remaining is the precise behavior at the points $0, \pm 2\pi, \ldots$. Since the vector $\langle x'(t), y'(t) \rangle$ vanishes at these points, we do not know if the curve has a well-defined direction. However, we may refer directly to the definition of the slope of a curve. We form the limit, as t tends to zero, of the slope $m(t)$ of the line joining the origin to an arbitrary point $(x(t), y(t))$. We have

$$m(t) = \frac{y(t) - y(0)}{x(t) - x(0)} = \frac{1 - \cos t}{t - \sin t}.$$

As t tends to zero, both the numerator and denominator on the right tend to zero. In order to determine whether or not a limit exists, we make a double application of l'Hôpital's rule,[6] which states that if the expression obtained by differentiating the numerator and denominator separately tends to a limit, then the original expression tends to the same limit; that is,

$$\lim_{t \to 0} m(t) = \lim_{t \to 0} \frac{1 - \cos t}{t - \sin t} = \lim_{t \to 0} \frac{\sin t}{1 - \cos t} = \lim_{t \to 0} \frac{\cos t}{\sin t} = \infty.$$

Thus the curve has a vertical tangent at the origin. It follows that at each of the values $t = 0, \pm 2\pi, \ldots$, where condition (2.1) fails, the curve has a cusp with vertical tangent.

The curve defined by Eqs. (2.22) is a *cycloid*. It is the locus of a point on the circumference of a circle of radius 1 as the circle is rolled along the x axis.

Exercises

2.1 Write parametric equations for the following straight-line segments:

  a. from $(1, 1)$ to $(4, 3)$
  b. from $(1, -2)$ to $(-3, 0)$
  c. from $(2, 3)$ to $(-1, 3)$
  d. starting at $(-1, 1)$ with tangent vector $\mathbf{T} = \langle \frac{3}{5}, \frac{4}{5} \rangle$ and length 10
  e. starting at the origin, having length 2 and meeting the line $y = x + 1$
at right angles

2.2 Sketch each of the following curves, and express y explicitly as a function of x by eliminating the parameter t. (Observe carefully the limitation imposed on each curve by the interval prescribed for the parameter t.)

  a. $x = t + 2,\quad y = 2t - 1,\qquad 0 \le t \le 1$
  b. $x = 2t + 1,\quad y = t^2 + t,\qquad -1 \le t \le 1$
  c. $x = t^2,\quad y = t^4,\qquad -1 \le t \le 1$
  d. $x = 1 + \sin t,\quad y = 3 - 2 \sin t,\qquad 0 \le t \le \frac{1}{2}\pi$
  e. $x = (1 - t^2)/(1 + t^2),\quad y = 2t/(1 + t^2),\qquad 0 \le t \le 1$

---

[6] See, for example [14], p. 222. In the form given there, the rule actually applies to the reciprocal quantity $1/m(t)$, which tends to zero as t tends to zero.

2.3   Show that the curve

$$x = \cos t, \qquad y = \cos 2t, \qquad\qquad 0 \le t \le \pi,$$

lies on a parabola. Sketch.

2.4   Show that the curve

$$x = a \cos t, \qquad y = b \sin t, \qquad\qquad 0 \le t \le 2\pi$$

is an ellipse, by obtaining an implicit equation of the form of Eq. (2.4).

2.5   Show that the curve

$$x = a \cosh t, \qquad y = b \sinh t, \qquad\qquad 0 \le t \le 1,$$

lies on a hyperbola. (Note that $\cosh t$ and $\sinh t$ are the hyperbolic functions—hyperbolic cosine and sine, respectively,

$$\cosh t = \tfrac{1}{2}(e^t + e^{-t}), \qquad \sinh t = \tfrac{1}{2}(e^t - e^{-t}).)$$

2.6   What conditions on the constants a, b, c, d are needed to guarantee that the line segment Eq. (2.5) of Example 2.1 be a regular curve? What do these conditions mean geometrically?

2.7   Repeat Ex. 2.6 for the constants c, d, r in Eq. (2.6) of Example 2.2.

*2.8   Repeat Ex. 2.6 for the constants v_0, α in Eq. (2.7) of Example 2.3.

2.9   Find the velocity vector **v** at an arbitrary point of each of the curves in Ex. 2.2.

2.10   Find the unit tangent **T** at an arbitrary point of each of the curves in Ex. 2.2.

2.11   Find the length of each of the following curves.

a. $x = 4t + 3$, $\quad y = 3t - 2$, $\qquad -3 \le t \le 1$
b. $x = 2 \cos t - 1$, $\quad y = 2 \sin t + 3$, $\qquad \tfrac{1}{6}\pi \le t \le \pi$
c. $x = 2t^2$, $\quad y = t^3$, $\qquad 0 \le t \le 1$
d. $x = t \cos t$, $\quad y = t \sin t$, $\qquad 0 \le t \le 1$
e. $x = t - \sin t$, $\quad y = 1 - \cos t$, $\quad$ (cycloid) $\qquad 0 \le t \le 2\pi$

*2.12   The *folium of Descartes* is defined by the equations

$$x = \frac{3t}{1 + t^3}, \qquad y = \frac{3t^2}{1 + t^3},$$

where t takes on all real values.

a. Write an equation for this curve in the implicit form of Eq. (2.4).
b. Sketch the curve.

(*Hint:* Note that the parameter t is equal to y/x, which is the slope of the line from the origin to the point (x, y) on the curve; x and y are both positive for t

positive, whereas for t negative, x and y have opposite signs, but $x + y = 3t/(1 - t + t^2) \geq -1$. It may also be helpful to sketch separately the graphs of x and y as functions of t.)

*2.13 The *strophoid* is given by the equation

$$x = \frac{1 - t^2}{1 + t^2}, \qquad y = t\frac{1 - t^2}{1 + t^2}.$$

    a. Write an equation for this curve in the implicit form of Eq. (2.4).
    b. Sketch the curve.

(*Hint:* Note that $t = y/x$, as in Ex. 2.12. Moreover $x(t) = x(-t)$, $y(t) = -y(-t)$ and $-1 \leq x(t) \leq 1$, for all parameter values t. Consider the cases $0 \leq t \leq 1$ and $1 \leq t < \infty$.)

Exercises 2.14 and 2.15 are designed to give further insight into the significance of condition (2.1) in the definition of a regular curve.

    2.14 Let C be the curve $x = t^3$, $y = t^6$, $-1 \leq t \leq 1$.

    a. Are x and y continuously differentiable functions of t?
    b. For which values of t is condition (2.1) satisfied?
    c. Find an explicit expression for the curve C by expressing y as a function of x, and sketch the curve.

(Note that this example shows that when the condition (2.1) fails to hold the curve *may* have an irregularity, such as a cusp or a corner, but that need not be the case. The irregularity may arise from a "poor choice of parameter.")

    *2.15 Let $f(t)$ be the function defined by

$$f(t) = \begin{cases} t^2 & \text{for} \quad t > 0 \\ 0 & \text{for} \quad t = 0 \\ -t^2 & \text{for} \quad t < 0. \end{cases}$$

Let C be the curve defined by $x = f(t)$, $y = t^2$, $-1 \leq t \leq 1$. Answer parts a, b, and c of Ex. 2.14.

    2.16 At each point $(x(t), y(t))$ of a curve C, the vector

$$\mathbf{a}(t) = \frac{d}{dt}\mathbf{v}(t) = \left\langle \frac{d^2x}{dt^2}, \frac{d^2y}{dt^2} \right\rangle$$

is called the *acceleration vector*. (It is defined only when the second derivatives of the functions $x(t)$ and $y(t)$ exist.)

    a. Find the acceleration vector at an arbitrary point of each of the three curves given as Examples 2.1, 2.2, and 2.3 in the text.

b. Note in Example 2.2 that even though the speed $|\mathbf{v}(t)|$ is constant, the acceleration vector is not zero. Explain why this is so.

*c. *Newton's second law of motion* states that a force **F** acting on a particle of mass m induces an acceleration **a** such that $\mathbf{F} = m\mathbf{a}$.

If a particle moves in the plane under the action of a force directed toward the origin, then the vector **F** at each point is a scalar multiple of the radius vector $\langle x, y \rangle$. (This situation is referred to as a *central force field.*) Show that under these conditions,

$$x \frac{d^2 y}{dt^2} - y \frac{d^2 x}{dt^2} \equiv 0$$

and

$$x \frac{dy}{dt} - y \frac{dx}{dt} \equiv \text{constant.}$$

*d. Show conversely, that if a particle moves in such a way that $x(dy/dt) - y(dx/dt)$ is constant, then at each point $(x(t), y(t))$ other than the origin, there is a scalar λ such that

$$\mathbf{a} = \left\langle \frac{d^2 x}{dt^2}, \frac{d^2 y}{dt^2} \right\rangle = \lambda \langle x, y \rangle.$$

Thus, by Newton's second law, the force inducing the motion is always directed toward the origin. (*Hint:* Use Ex. 1.9a.)

2.17 Curves of the form $x = a \cos pt$, $y = b \sin qt$, where a, b, p, q are arbitrary positive constants, are known as *Lissajous figures.* When p and q are equal they lie on an ellipse (see Ex. 2.4). When p and q are integers they are closed curves described infinitely often, since both x and y are periodic in t. If the ratio p/q is not equal to a ratio of integers, then the curve swept out never repeats itself, although it always remains in the rectangle $|x| \leq a$, $|y| \leq b$. Sketch the following Lissajous figures for $0 \leq t \leq 2\pi$.

    a. $x = \cos t$, $y = \sin 2t$
    b. $x = \cos 2t$, $y = \sin t$
  *c. $x = \cos 3t$, $y = \sin 2t$

(*Note.* There are a number of mechanical methods for drawing Lissajous figures, such as compound pendulums, which vibrate in transverse directions. There are also practical applications.[7] The variety of figures obtainable by simply varying the ratio p/q is astonishing. Some examples are shown in Fig. 2.13.)

---

[7] See [12], where it is shown how Lissajous figures are used to determine frequency and phase differences.

*2.18 Let a regular curve be defined by $x = \varphi(t)$, $y = \psi(t)$, $a \le t \le b$, where $\varphi'(t) > 0$ for $a \le t \le b$. Using results from the calculus of functions of one variable, show that

a. the function $x = \varphi(t)$, $a \le t \le b$ has an inverse $t = h(x)$, $A \le x \le B$, and substituting this inverse function in $y = \psi(t)$ gives a nonparametric form of the curve: $y = f(x)$, $A \le x \le B$;

b. $$\int_a^b \sqrt{x'(t)^2 + y'(t)^2} \, dt = \int_A^B \sqrt{1 + f'(x)^2} \, dx.$$

(*Note.* For an arbitrary regular curve, condition (2.1) implies that at each point either $x'(t) \ne 0$ or $y'(t) \ne 0$ (or both). As a consequence, it can be shown that the parameter interval for t can be divided into a finite number of intervals on each of which at least one of the conditions $x'(t) > 0$, $x'(t) < 0$, $y'(t) > 0$, or $y'(t) < 0$ holds throughout. It follows that on each of these intervals the curve can be expressed in nonparametric form, either as $y = f(x)$ or $x = g(y)$. Equation (2.13) is then a consequence of part b above and the arc-length formula for nonparametric curves.)

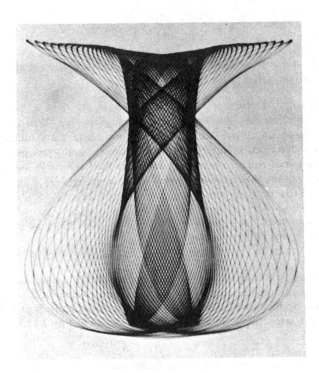

FIGURE 2.13 Lissajous figures [R. T. Lagemann.]

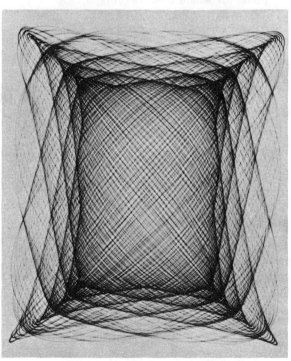

3 Functions of two variables

The following are some typical examples of functions of two variables.

1. $1 - 2x - y$ 4. $\sqrt{x^2 + y^2}$

2. $x^2 + y^2$ 5. $\sqrt{1 - x^2 - y^2}$

3. $2xy/(x^2 + y^2)$ 6. $\cos(xe^y)$.

The first two examples are *polynomials*. A *polynomial in two variables* is the sum of terms of the form $ax^m y^n$, where a is a constant, and m, n are nonnegative integers. The number $m + n$ is called the *degree* of this term. The *degree of a polynomial* is the highest degree of the terms it contains. A polynomial is called *homogeneous* if all terms have the same degree. Thus example 2 above is a homogeneous *quadratic* polynomial (degree 2), while the most general homogeneous cubic polynomial is

$$ax^3 + bx^2 y + cxy^2 + dy^3.$$

In studying functions of one variable, one can often gain a great deal of insight by associating with the function $f(x)$ its *graph*, consisting of those points in the x, y plane for which $y = f(x)$ (Fig. 3.1).

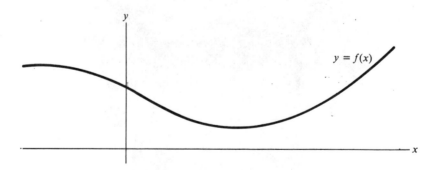

FIGURE 3.1 Graph of the function $f(x)$

An analogous procedure exists for visualizing a function $f(x, y)$ of two variables. Associated with this function is the surface $z = f(x, y)$ in 3-space. We illustrate this for each of the functions listed above.

Example 3.1

$$z = 1 - 2x - y.$$

This is a linear equation, and hence represents a plane. (In analytic geometry one would write it in the form $2x + y + z = 1$.) It intersects the z axis at the point where x and y are both zero, namely $z = 1$. Similarly, setting $y = z = 0$ yields $x = \frac{1}{2}$, and $x = z = 0$ gives $y = 1$. These three points uniquely determine the plane (Fig. 3.2).

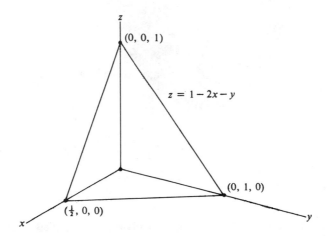

FIGURE 3.2   Plane

Example 3.2

$$z = x^2 + y^2.$$

This is the equation of a surface of revolution, since the value of z depends only on the distance $(x^2 + y^2)^{1/2}$ from the z axis. If one uses polar coordinates, $x = r \cos \theta, y = r \sin \theta$, in the x, y plane, then the surface is given by the equation $z = r^2$. This is a paraboloid of revolution (Fig. 3.3).

Example 3.3

$$z = \frac{2xy}{x^2 + y^2}.$$

An important difference between this example and the previous ones is that the first two functions were defined for all values of x and y, whereas this one is not. More generally, every polynomial $P(x, y)$ is defined for all x and y, whereas the quotient of two polynomials $P(x, y)/Q(x, y)$ is undefined at points where the denominator vanishes. In this example the denominator vanishes at a single point, the origin, and hence this function is defined everywhere in the plane except at the origin. At the origin the numerator also vanishes, and it is not immediately clear how this function behaves near the origin. As in the previous example, the

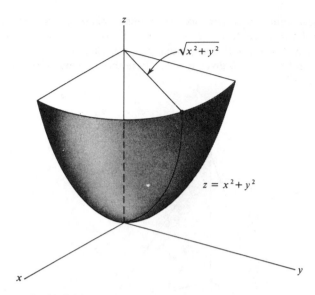

FIGURE 3.3 Paraboloid

introduction of polar coordinates in the x, y plane is helpful. Setting $x = r \cos \theta$, $y = r \sin \theta$, we find

$$z = \frac{2r^2 \cos \theta \sin \theta}{r^2(\cos^2 \theta + \sin^2 \theta)} = 2 \cos \theta \sin \theta = \sin 2\theta.$$

First we note that the value of z depends only on the polar angle θ and not on r. This means that z is constant on each ray $\theta = $ constant, or that the surface defined by $z = \sin 2\theta$ contains a horizontal ray over each of these rays in the x, y plane. Second, we observe that the value of z always stays between $+1$ and -1, so that the whole surface lies between the planes $z = 1$ and $z = -1$. Finally, we see that as θ goes from 0 to $\frac{1}{4}\pi$, z goes from 0 to 1, so that the surface is a kind of spiral ramp generated by starting with a ray along the positive x axis and moving it upwards, while holding it horizontally, as we rotate it through half of the first quadrant (Fig. 3.4).

If we continue to rotate the ray from $\theta = \frac{1}{4}\pi$ to $\theta = \frac{1}{2}\pi$, its height drops back down from 1 to 0, at which point it coincides with the positive y axis. In the second quadrant, for $\frac{1}{2}\pi < \theta < \pi$, we have $z = \sin 2\theta < 0$, and the line drops below the x, y plane, reaching its lowest point at $\theta = \frac{3}{4}\pi$, when $z = -1$, and then rising again until it coincides with the negative x axis. Finally, as θ goes from π to 2π, this pattern is repeated. The final surface is of undulating form, covering the whole x, y plane except the origin.

Example 3.4

$$z = \sqrt{x^2 + y^2}.$$

Once again it is easiest to use polar coordinates. We simply obtain $z = r$, which

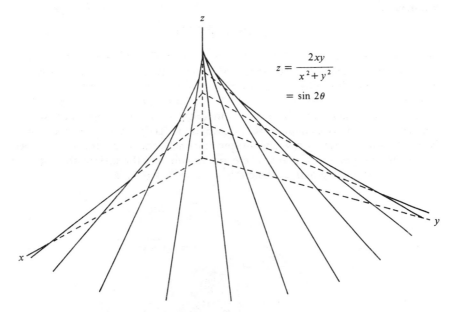

$$z = \frac{2xy}{x^2+y^2}$$
$$= \sin 2\theta$$

FIGURE 3.4 "Spiral ramp" surface

represents a cone with vertex at the origin (Fig. 3.5). In contrast to the previous example, this function is defined for all values of x and y, although the resulting surface is clearly not "smooth" at the origin.

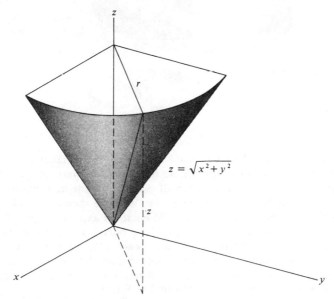

$$z = \sqrt{x^2+y^2}$$

FIGURE 3.5 Cone

Example 3.5

$$z = \sqrt{1 - x^2 - y^2}.$$

Squaring both sides, we find $z^2 = 1 - x^2 - y^2$ or $x^2 + y^2 + z^2 = 1$. Thus all points on this surface lie on the sphere of radius 1 about the origin. Conversely, every point on this sphere satisfies $z^2 = 1 - x^2 - y^2$ and hence either $z = (1 - x^2 - y^2)^{1/2}$ or $z = -(1 - x^2 - y^2)^{1/2}$. Thus our equation defines those points on the sphere for which $z \geq 0$; in other words, the upper hemisphere. Unlike the previous example, this surface is perfectly smooth at the origin. However, we now have a new phenomenon; namely, the function is only defined in a limited portion of the plane. There are no points of the surface above points in the x, y plane lying outside the circle $x^2 + y^2 = 1$ (Fig. 3.6).

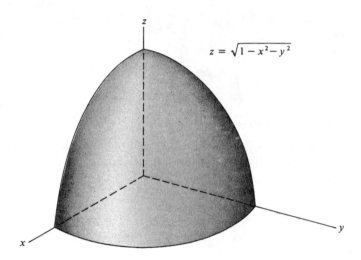

$$z = \sqrt{1 - x^2 - y^2}$$

FIGURE 3.6 Sphere

Example 3.6

$$z = \cos(xe^y).$$

As in Example 3.3, we should probably note first that z takes on only values between 1 and -1; and hence the whole surface lies between the corresponding horizontal planes. To obtain an over-all picture, we might examine how this function behaves on each straight line $y = c$. We have then $z = \cos kx$, where k is the positive constant e^c. In particular, along the x axis, where $y = 0$, we have $z = \cos x$. For $y < 0$ we have $k < 1$, and we obtain a "stretched out" cosine curve, while for $y > 0$, we have $k > 1$, and hence a "compressed" cosine curve. Thus the undulations become more and more rapid on each line $y = c$ as $c \to +\infty$. It may help also to note that on the y axis, where $x = 0$, we have $z = \cos 0 \equiv 1$. These features of the surface are illustrated in Fig. 3.7.

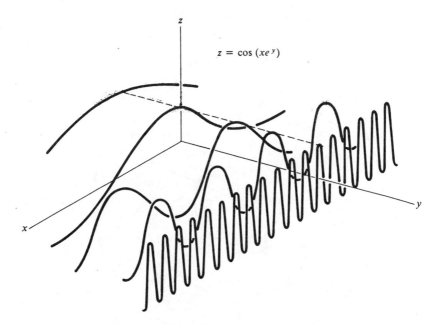

$$z = \cos (xe^y)$$

FIGURE 3.7 Cross sections of a surface

The following example, not included on our list at the beginning of this section, is of sufficient interest to warrant a separate discussion.

Example 3.7

$$z = \sqrt{1 - (\sqrt{x^2 + y^2} - 2)^2}.$$

We first try to determine the set of points in the x, y plane for which this function is defined. The quantity under the square root sign must be nonnegative, that is,

$$(\sqrt{x^2 + y^2} - 2)^2 \leq 1,$$

or

$$|\sqrt{x^2 + y^2} - 2| \leq 1,$$

or using polar coordinates in the x, y plane

$$|r - 2| \leq 1.$$

This says simply

$$1 \leq r \leq 3,$$

which describes all points between the circles of radius 1 and 3 about the origin (Fig. 3.8).

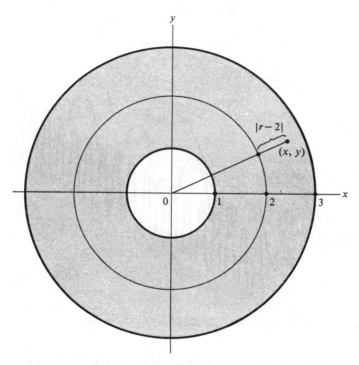

FIGURE 3.8 Set of points where a function is defined

We note next that z depends only on r, and not on θ, so that we have a surface of revolution whose intersection with each vertical plane through the z axis yields the same curve. Thus, in order to know the whole surface, it is sufficient to find its intersection with one half of the x, z plane, say $x \geq 0$, $y = 0$. We have then

$$z = \sqrt{1 - (x - 2)^2},$$

which is the upper half of the circle (Fig. 3.9)

$$z^2 = 1 - (x - 2)^2,$$

or

$$(x - 2)^2 + z^2 = 1.$$

Rotation about the z axis gives the upper half of a doughnut-shaped surface (Fig. 3.10). The entire surface is called a *torus*.

The above examples illustrate some of the methods by which we may obtain a geometric picture of a surface defined by a given function $f(x, y)$. One way, as in Example 3.6, was to choose various values for y, and find the

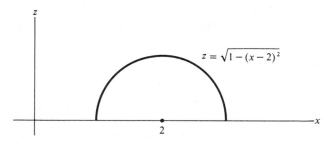

FIGURE 3.9   Cross section of a surface of revolution

curves on the surface that correspond to each of these values of y. In other words, find the intersection of the surface with vertical planes $y = c$ for various values of the constant c. Similarly, we can pick values for x and see how $f(x, y)$ behaves with respect to y, or we can introduce polar coordinates and study the function for values of r or θ. These are all special cases of the following general principle. If we are given a function $f(x, y)$ and if we choose a straight line or curve in the x, y plane, then setting $z = f(x, y)$ defines another curve lying over the given one. By choosing a suitable family of curves in the plane and constructing the corresponding curves on the surface, we may be able to visualize the entire surface.

There is a kind of inverse process to this one, which is also frequently of value. Instead of starting with curves in the x, y plane, we start with a fixed value of z and ask for those points in the x, y plane for which the function takes on the given value. In general, the set of points (x, y) for which $f(x, y) = c$, where c is any constant, is called a *level curve* of the function $f(x, y)$. By choosing various values of c, and constructing the corresponding level curves, we can often quickly obtain a picture of the entire function. It may also be helpful to shade in those portions of the plane where $f(x, y)$

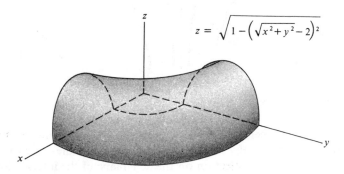

FIGURE 3.10   Torus

is positive. The corresponding surface $z = f(x, y)$ then lies above the x, y plane in the shaded portions and below it in the unshaded portions.

Example 3.8

$$f(x, y) = xy.$$

The level curves $xy = c$ are hyperbolas if $c \neq 0$. For c positive, they lie in the first and third quadrants, while for c negative, they lie in the second and fourth quadrants. The level curve $xy = 0$ consists of the x and y axes (Fig. 3.11). This gives us a reasonably good over-all description of the function. It is zero at the origin, increases as we move out in either direction along the line $y = x$, and decreases in both directions along the line $y = -x$. On the basis of Fig. 3.11 it is easy to visualize the surface $z = xy$ (Fig. 3.12). It is a "saddle-shaped" surface.

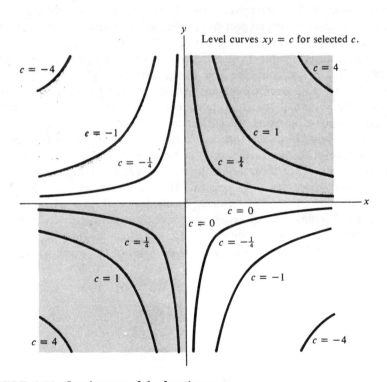

FIGURE 3.11 Level curves of the function xy

The relation between the above two methods of visualizing a function—by a surface in space or by level curves in the x, y plane—is clear; a level curve $f(x, y) = c$ is the projection on the x, y plane of the intersection of the surface $z = f(x, y)$ with a horizontal plane $z = c$ (Fig. 3.13). This relation

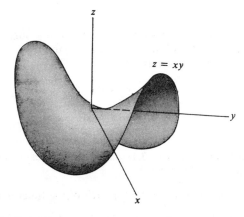

FIGURE 3.12   Surface $z = xy$

becomes more striking if we use the alternative expression "contour lines" for level curves. A contour line (or level curve) corresponds to the points on the surface $z = f(x, y)$, which are at a given height above the x, y plane. A "contour map," in the usual sense, gives a set of such curves when the function $f(x, y)$ is the altitude above sea level. We may picture each function $f(x, y)$ as defining a mountainous terrain, and our description of $f(x, y)$ by level curves is simply a contour map to help guide us through the terrain.

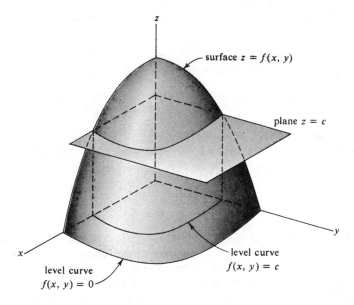

FIGURE 3.13   Level curves and horizontal cross sections

Exercises

3.1 Draw the intersection of each of the following surfaces with vertical planes $y = c$ for various values of the constant c, and use these, as in Example 3.6, to visualize and sketch the entire surface.

a. $z = xy$	**e.** $z = x/y$
b. $z = x^2 - y^2$	**f.** $z = \sin(x + y)$
c. $z = x^2 + 4y^2$	**g.** $z = \sin x \sin y$
d. $z = xy^2$	**h.** $z = ye^x$

3.2 Study the surfaces in Ex. 3.1, using their intersections with vertical planes $x = c$ for various values of c.

3.3 Draw a number of level curves for each of the functions in Ex. 3.1.

3.4 Using any or all of the devices in Exs. 3.1–3.3, try to visualize and sketch each of the following standard *quadric surfaces*. (*Note:* A, B, C are arbitrary positive constants. It may be easier in some cases to start by choosing two of these constants equal in order to obtain a surface of revolution.)

a. paraboloid
$$z = \frac{x^2}{A^2} + \frac{y^2}{B^2}$$

b. hyperbolic paraboloid
$$z = \frac{x^2}{A^2} - \frac{y^2}{B^2}$$

c. ellipsoid
$$\frac{x^2}{A^2} + \frac{y^2}{B^2} + \frac{z^2}{C^2} = 1$$

*d. hyperboloid of one sheet
$$\frac{x^2}{A^2} + \frac{y^2}{B^2} - \frac{z^2}{C^2} = 1$$

*e. hyperboloid of two sheets
$$-\frac{x^2}{A^2} - \frac{y^2}{B^2} + \frac{z^2}{C^2} = 1$$

3.5 For each of the following functions $f(x, y)$, draw the level curve $f(x, y) = 0$, shade in the region where $f(x, y) > 0$, and try to visualize the surface $z = f(x, y)$ near the origin.

a. $x + y^2$	**d.** $x^2 + 3xy + y^2$
b. $x^2 + 2xy + y^2$	**e.** $x^3 - 3xy^2$
c. $x^2 + xy + y^2$	**f.** $x^6 - yx^2 - yx^4 + y^2$

3.6 Let $\langle a, b \rangle$ be a unit vector, and let $x = at$, $y = bt$ be the line through the origin in the direction of that vector. For each of the functions $f(x, y)$ of Ex. 3.5, sketch the curve on the surface lying above the given line. (In other words, substitute $x = at$, $y = bt$ in the equation $z = f(x, y)$ and find z as a function of t.) Try to picture how these curves vary as the vector $\langle a, b \rangle$ is changed, and use them to help visualize the surface.

3.7   If a function $f(x, y)$ is of the form $f(x, y) = g(x) + h(y)$, then the surface $z = f(x, y)$ is called a *surface of translation*.

a. Explain this terminology by observing the cross sections of the surface $z = f(x, y)$ with an arbitrary plane $y = c$, and the way these cross sections vary as c varies.

b. Do the same for the cross sections with planes of the form $x = c$.

c. Find which of the surfaces in the previous exercises are surfaces of translation, and illustrate this fact with corresponding sketches.

d. If either g or h is constant, the surface is called *cylindrical*. Describe what this means geometrically, and give some examples.

3.8   Let $z = F(x)$, $0 < a \le x \le b$, define a curve in the x, z plane. The surface defined by revolving this curve about the z axis is a *surface of revolution*.

a. What is the equation of the surface obtained in this way? (That is, if the surface is $z = f(x, y)$, how is $f(x, y)$ expressed in terms of the function F?)

b. What is the set of points in the x, y plane for which the function $f(x, y)$ is defined?

c. What are the level curves of the function $f(x, y)$?

3.9   Describe geometrically how a surface $z = f(x, y)$ would have to be transformed in order to obtain each of the following surfaces $z = g(x, y)$, where $g(x, y)$ is:

a. $f(x, y) + 2$	**f.** $2f(x, y)$
b. $f(x + 2, y)$	**g.** $f(2x, y)$
c. $f(x, y - 2)$	**h.** $2f(\tfrac{1}{2}x, \tfrac{1}{2}y)$
d. $2 - f(x, y)$	**i.** $-f(-x, -y)$
e. $f(-x, y)$	**j.** $f(y, x)$

3.10   Let $h(t)$ be a strictly increasing function of t, and let $g(x, y) = h(f(x, y))$.

a. How are the level curves of $f(x, y)$ and $g(x, y)$ related?

b. How is the surface $z = g(x, y)$ related to the surface $z = f(x, y)$?

3.11   If the temperature at different points on the earth's surface is considered a function of position, then the level curves of this function are called *isotherms*. (Either the temperature at a given time or the average temperature over some period may be used.) Figure 3.14 shows such a set of level curves for July temperatures in North America. Find approximately those points on the map where the following conditions prevail:

a. the temperature is maximum

b. the temperature is maximum relative to nearby points

c. the temperature changes most rapidly with respect to nearby points.

3.12   Figure 3.15 is a portion of a contour map showing two mountain streams running vertically and a connecting stream between them.

a. Find approximately the point where the terrain is steepest.

*b. By examining the contour lines determine the direction of flow of each of the three streams.

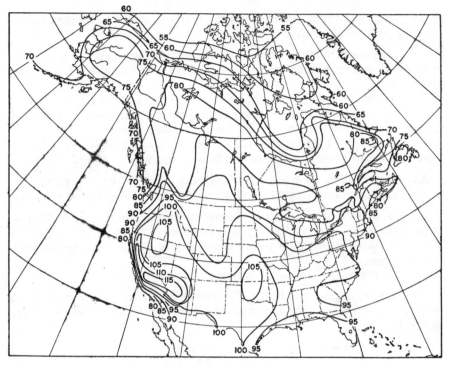

FIGURE 3.14 Isotherms for July temperatures

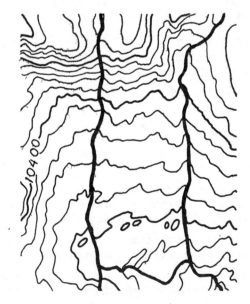

FIGURE 3.15 Portion of a contour map [U.S. Geological Survey topographic map.]

CHAPTER TWO

Differentiation

4 Partial derivatives and plane domains

The first observation to be made in the study of a function of two variables $f(x, y)$ is that when either x or y is held fixed the function is reduced to one of a single variable. The ordinary derivative with respect to that variable is called a *partial derivative* of the original function. To be precise we make the following definition.

Definition 4.1 The *partial derivative of $f(x, y)$ with respect to x at the point (x_0, y_0)* is denoted by $f_x(x_0, y_0)$ and is defined by

$$f_x(x_0, y_0) = \lim_{x \to x_0} \frac{f(x, y_0) - f(x_0, y_0)}{x - x_0}.$$

Similarly,

$$f_y(x_0, y_0) = \lim_{y \to y_0} \frac{f(x_0, y) - f(x_0, y_0)}{y - y_0}$$

is called the *partial derivative of $f(x, y)$ with respect to y at the point (x_0, y_0)*.

There are many different notations currently used for partial derivatives. Among the most common are

$$f_x = \frac{\partial f}{\partial x} = D_x f = D_1 f = f_1 \quad \text{and} \quad f_y = \frac{\partial f}{\partial y} = D_y f = D_2 f = f_2.$$

In this book we will generally use f_x and f_y, but occasionally also $\partial f/\partial x$, $\partial f/\partial y$. Also, when we set $z = f(x, y)$, we write z_x, z_y and $\partial z/\partial x$, $\partial z/\partial y$.

Returning to the definition, we see that in each case we have kept one of the variables fixed and formed the ordinary derivative with respect to the other variable. It follows immediately that all the basic laws concerning ordinary derivatives give rise to corresponding laws for partial derivatives. These include the rules for differentiating sums, products, and quotients. In particular, if $\partial f/\partial x$ and $\partial g/\partial x$ both exist at (x_0, y_0), then so do $\partial(fg)/\partial x$ and, providing $g(x_0, y_0) \neq 0$, $\partial(f/g)/\partial x$. Similarly for the partial derivative with respect to y. As a result, if f is a polynomial, then f_x and f_y exist at every point.

Example 4.1

If

$$f(x, y) = x^4 + y^3 + 5x^2y + 12,$$

then

$$f_x = 4x^3 + 10xy, \qquad f_y = 3y^2 + 5x^2.$$

Note that to take the partial derivative with respect to one variable we simply treat the other variable as a constant.

Example 4.2

If

$$f(x, y) = x \cos y,$$

then

$$f_x = \cos y, \qquad f_y = -x \sin y.$$

Example 4.3

If

$$f(x, y) = e^{xy},$$

then

$$f_x = ye^{xy}, \qquad f_y = xe^{xy}.$$

Note that here we have to use the chain rule. We have

$$f(x, y) = e^u, \qquad u = xy.$$

To take the partial derivative with respect to x, we consider y constant, and we multiply the derivative of f with respect to u by the derivative of u with respect to x.

Example 4.4

If

$$f(x, y) = (1 - x^2 - y^2)^{1/2},$$

then

$$f_x = \tfrac{1}{2}(1 - x^2 - y^2)^{-1/2} \cdot (-2x) = -x/\sqrt{1 - x^2 - y^2}.$$
$$f_y = -y/\sqrt{1 - x^2 - y^2}.$$

This last example is very instructive. Note that the function f is defined for $x^2 + y^2 \le 1$. The functions f_x and f_y are well-defined at all points inside this circle, that is, for $x^2 + y^2 < 1$. However, when $x^2 + y^2 = 1$, although f is defined, the expressions for f_x and f_y have a zero in the denominator. If we recall the definition of partial derivatives, we see that they are actually not defined at points (x_0, y_0) where $x_0^2 + y_0^2 = 1$; indeed, the expression whose limit we wish to take is not defined for all x sufficiently near x_0. It would be possible to consider one-sided limits and derivatives, but at certain points even these are not defined. For example, $f(x, 1)$ is defined only for the single value $x = 0$ (Fig. 4.1), and $f_x(0, 1)$ does not exist in *any* sense.

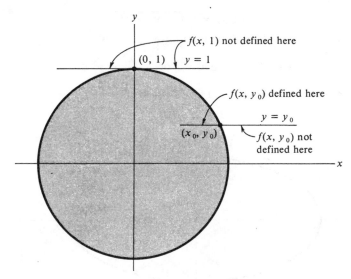

FIGURE 4.1 The domain of definition of $f(x, y) = \sqrt{1 - x^2 - y^2}$

Thus, before we proceed to compute partial derivatives blindly and obtain meaningless results, we should look a little more closely at the domain of definition of a function. In particular, we have to distinguish between those *interior points* where we may apply any differentiation rule from single-variable calculus without hesitation and the *boundary points* where trouble may arise.

We are confronted here with the first essential distinction between functions of one variable and functions of two variables. A function of one variable is defined on an interval or a set of intervals. A function of two variables may be defined on a very complicated domain in the plane. Even relatively simple functions, such as $f(x, y) = (P(x, y))^{1/2}$, where P is a polynomial, may lead to quite complicated geometrical figures. More generally, the function

(4.1) $$f(x, y) = \sqrt{P_1(x, y)} + \sqrt{P_2(x, y)} + \cdots + \sqrt{P_n(x, y)},$$

where $P_1, \ldots, P_n$ are polynomials, is defined on the set of points satisfying all of the inequalities

(4.2) $$P_1(x, y) \geq 0, P_2(x, y) \geq 0, \ldots, P_n(x, y) \geq 0.$$

We have already seen examples of such functions, including the one we have just discussed, where $P(x, y) = 1 - x^2 - y^2$. The following are some typical examples.

Example 4.5

$$f(x, y) = \sqrt{4 - x^2 - 4y^2}.$$

For this function to be defined, we must have

$$4 - x^2 - 4y^2 \geq 0, \quad \text{or} \quad \frac{x^2}{4} + y^2 \leq 1,$$

which describes a domain bounded by an ellipse (Fig. 4.2).

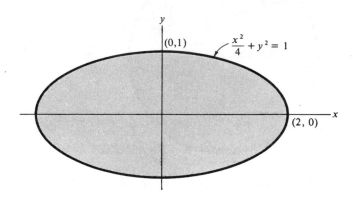

FIGURE 4.2 The set of points satisfying $(x^2/4) + y^2 \leq 1$

Example 4.6

$$f(x, y) = \sqrt{(4 - x^2 - y^2)(x^2 + y^2 - 1)}.$$

This function is defined, except when the two factors under the radical have opposite sign. Thus the function is defined only for those points (x, y) satisfying both

(4.3) $$4 - x^2 - y^2 \geq 0 \quad \text{and} \quad x^2 + y^2 - 1 \geq 0,$$

or both

(4.4) $$4 - x^2 - y^2 \leq 0 \quad \text{and} \quad x^2 + y^2 - 1 \leq 0.$$

The Eqs. (4.4) cannot be simultaneously satisfied, since they assert that $x^2 + y^2$

is a number at most equal to 1 and at least equal to 4. Thus the function is defined only for the set of points satisfying Eqs. (4.3); that is, the set of points bounded by the circles of radius 1 and 2 about the origin (Fig. 4.3). A set of points of this type (bounded by two concentric circles) is called an *annulus*.

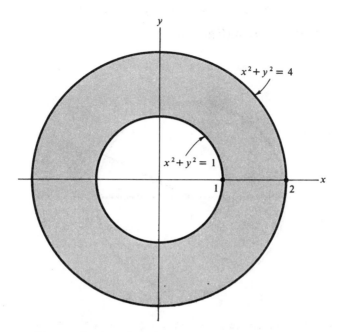

FIGURE 4.3 The annulus $1 \le x^2 + y^2 \le 4$

Example 4.7

$$f(x, y) = \sqrt{y - x^2} + \sqrt{1 - y}.$$

Here $y \ge x^2$ and $y \le 1$, which is a parabolic segment (Fig. 4.4).

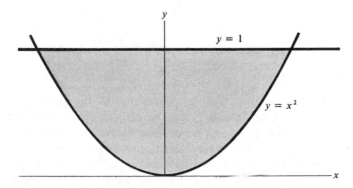

FIGURE 4.4 The parabolic segment $x^2 \le y \le 1$

Example 4.8

$$f(x, y) = \sqrt{4 - x^2 - y^2} + \sqrt{x - 2y} + \sqrt{y}.$$

We have

$$x^2 + y^2 \leq 4, \qquad y \leq \tfrac{1}{2}x, \qquad y \geq 0.$$

This defines a circular sector (Fig. 4.5).

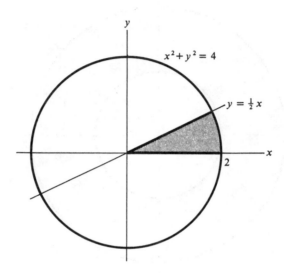

$x^2 + y^2 = 4$

$y = \tfrac{1}{2}x$

FIGURE 4.5 The circular sector $x^2 + y^2 \leq 4, 0 \leq y \leq x/2$

Example 4.9

$$f(x, y) = \sqrt{x - a} + \sqrt{b - x} + \sqrt{y - c} + \sqrt{d - y}, \qquad a < b, c < d.$$

This function is defined in the rectangle (Fig. 4.6):

$$a \leq x \leq b, \qquad c \leq y \leq d.$$

Example 4.10

$$f(x, y) = \sqrt{36 - 4x^2 - 9y^2} + \sqrt{(x + 1)^2 + y^2 - 1} + \sqrt{4(x - 2)^2 + 4y^2 - 1}.$$

See Fig. 4.7.

It is probably clear from these examples that all the figures one usually encounters in elementary geometry can be defined by a set of polynomial inequalities of the form (4.2), and hence these figures are the precise domains of definition of certain elementary functions (4.1). Clearly, one can also obtain far more complicated figures by such inequalities. For our purposes it makes little difference how simple or complicated a figure may be. The

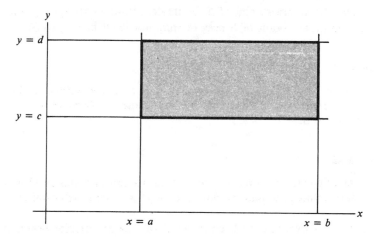

FIGURE 4.6   The rectangle $a \leq x \leq b, c \leq y \leq d$

important thing is to distinguish between its interior points and boundary points. We do this as follows.

Definition 4.2   Let S be any set of points in the plane. A point (x_0, y_0) is called a *boundary point* of S if

1.   There are points of S arbitrarily close to (x_0, y_0);
2.   There are points not in S that are arbitrarily close to (x_0, y_0).

A point of S that is not a boundary point is called an *interior point* of S.

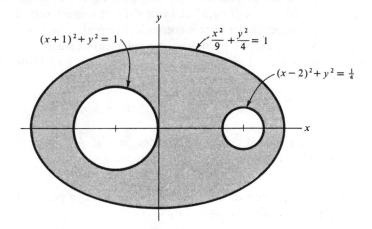

FIGURE 4.7   The set of points satisfying $4x^2 + 9y^2 \leq 36$, $(x + 1)^2 + y^2 \geq 1$ and $4(x - 2)^2 + 4y^2 \geq 1$

Remark An interior point of S, by its definition, is always a point of S, whereas a boundary point of S may or may not itself belong to S.

Example 4.11

If S is the set of points satisfying $x^2 + y^2 \leq 1$, then the points on the circle $x^2 + y^2 = 1$ are the boundary points of S, and those satisfying $x^2 + y^2 < 1$ are interior points of S.

Example 4.12

If S is the set of points satisfying $x^2 + y^2 < 1$, then again all points on $x^2 + y^2 = 1$ are boundary points of S, while all points $x^2 + y^2 < 1$ are interior points.

Note that in Example 4.12, every point of S is an interior point. As we mentioned earlier, we want to restrict our attention to sets having this property as the domains of definition of our functions, in order to be able to differentiate freely.

Definition 4.3 A set S of points is called *open* if every point of S is an interior point.

Examples of open sets are those sets of points (x, y) which satisfy a set of *strict* inequalities

(4.5) $$P_1(x, y) > 0, P_2(x, y) > 0, \ldots, P_n(x, y) > 0,$$

where $P_1, P_2, \ldots, P_n$ are polynomials. In fact, nothing is lost in this or the next chapter if functions are pictured as being defined on such sets. There is, however, one further restriction that is convenient to make. An open set may consist of several disconnected parts, and it is sufficient to study the function separately on each part. For example, the set consisting of those points (x, y) satisfying both

$$\frac{x^2}{9} + y^2 < 1 \quad \text{and} \quad x^2 + y^2 > 4$$

consists of the two shaded domains indicated in Fig. 4.8, lying inside an ellipse and outside a circle.

Definition 4.4 An open set S is *connected* if for any two points of S there is a curve lying in S joining the two points.

Definition 4.5 A *domain* is a connected open set.

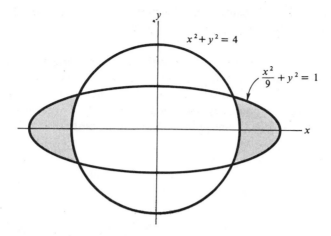

FIGURE 4.8 The set of points satisfying $(x^2/9) + y^2 < 1$, $x^2 + y^2 > 4$

We adhere to this terminology throughout the book. Whenever we speak of a function defined in a domain D, we mean that D is a set of points of this specific nature.

Remark The expression "the domain of a function" is often used to mean the set of points on which a function is defined. This set of points may or may not be "a domain" in our sense. In the following, we always refer to "a function defined in a domain D," with the understanding that the function may be defined at other points too, but that we are restricting our attention to the behavior of the function in D.

Example 4.13

Let D be the set of points (x, y) such that $x^2 + y^2 < 1$. The following functions are defined in D:

$$\log(1 - x^2 - y^2), \qquad 1/\sqrt{1 - x^2 - y^2},$$
$$(1 - x^2 - y^2)^{2/3}, \qquad \sqrt{4 - x^2 - y^2}.$$

The first two of these functions are defined precisely at points of D and nowhere else. The third is also defined at boundary points of D, but as we have seen, its partial derivatives are not. The last is actually defined in a larger domain, but we may wish to consider only the values that it takes on in D.

A basic ingredient of the above definitions is the notion of "all points sufficiently close to a given one." This notion can be formalized as follows.

Definition 4.6 A *neighborhood* of a point (x_0, y_0) is the set of all points (x, y) lying in some disk $(x - x_0)^2 + (y - y_0)^2 < r^2$.

This terminology leads to a convenient characterization of interior points and boundary points. Given a set S and a point (x_0, y_0), (x_0, y_0) is an interior point of S if and only if some neighborhood of (x_0, y_0) lies entirely in S, and (x_0, y_0) is a boundary point of S if and only if every neighborhood of (x_0, y_0) contains at least one point of S and one point not in S (see Ex. 4.16 and Fig. 4.9).

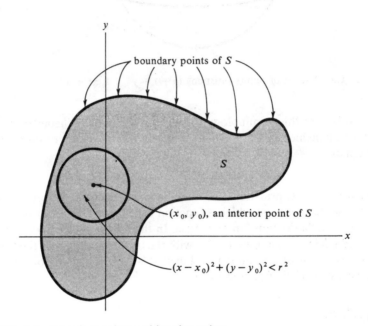

FIGURE 4.9 Boundary points and interior points

Exercises

4.1 Find the partial derivatives with respect to x and y of each of the following functions $f(x, y)$.

a. $x^4 + 3x^2y - 5xy^3$

b. $x^2 + 2xy + y^2$

c. $x^3 + 3x^2y + 3xy^2 + y^3$

d. $(x + y)^5$

e. $\sin(x + y)$

f. $\sin(x^2 + y^2)$

g. $\sqrt{x^2 + y^2}$

h. $(x^3 + y^3)^{1/3}$

i. $\arctan(y/x)$

j. $xe^{y/x}$

k. $xy/(x^2 + y^2)$

l. $xy \sin xy$

m. $\sec^3 x \log x + 5$

n. $\log \tan(x + y^2)$

4.2   Find the value of the following partial derivatives at the points indicated.

a. $f_x(0, 0)$ if $f(x, y) = \sin(xe^y)$
b. $f_y(\frac{1}{4}\pi, 0)$ if $f(x, y) = \log(\sin x + \sin y)$
c. $f_x(\frac{1}{4}\pi, 1)$ if $f(x, y) = e^{y\,\tan x}$
d. $f_y(0, 0)$ if $f(x, y) = \cosh x \sin y$

4.3   Compute the expression $xf_x + yf_y$ for each of the following functions $f(x, y)$.

a. $x^3 + y^3$

c. $\sqrt{x^6 + y^6}$

b. $(x + y)^3$

d. $\dfrac{x^4}{y} + xy^2$

4.4   Compute the partial derivatives of the following pairs of functions, and show that they satisfy the equations $f_x = g_y$, $f_y = -g_x$.

a. $f(x, y) = e^x \cos y$, $g(x, y) = e^x \sin y$

b. $f(x, y) = \log \sqrt{x^2 + y^2}$, $g(x, y) = \arctan \dfrac{y}{x}$

4.5   If $a(x, y)$, $b(x, y)$ are given functions, and if for each point (x, y) we consider the vector

$$v(x, y) = \langle a(x, y), b(x, y)\rangle$$

then the *partial derivatives* of $v(x, y)$ are defined by

$$v_x(x_0, y_0) = \lim_{x \to x_0} \frac{1}{x - x_0}\,(v(x, y_0) - v(x_0, y_0)),$$

$$v_y(x_0, y_0) = \lim_{y \to y_0} \frac{1}{y - y_0}\,(v(x_0, y) - v(x_0, y_0)).$$

Show that

$$v_x(x_0, y_0) = \langle a_x(x_0, y_0), b_x(x_0, y_0)\rangle,$$

$$v_y(x_0, y_0) = \langle a_y(x_0, y_0), b_y(x_0, y_0)\rangle.$$

4.6   Let $v = \langle a, b\rangle$, $w = \langle c, d\rangle$ be vectors, where a, b, c, and d are all functions of x and y. Suppose that λ is also a function of x and y. Show that

a. $\dfrac{\partial}{\partial x}(v + w) = v_x + w_x,$ $\dfrac{\partial}{\partial y}(v + w) = v_y + w_y$

b. $\dfrac{\partial}{\partial x}(\lambda v) = \lambda v_x + \lambda_x v,$ $\dfrac{\partial}{\partial y}(\lambda v) = \lambda v_y + \lambda_y v$

c. $\dfrac{\partial}{\partial x}(v \cdot w) = v \cdot w_x + v_x \cdot w,$ $\dfrac{\partial}{\partial y}(v \cdot w) = v \cdot w_y + v_y \cdot w.$

4.7 Using the notation ⋅

$$\begin{vmatrix} a & b \\ c & d \end{vmatrix} = ad - bc$$

for a 2×2 determinant, show that if a, b, c, d are functions of x and y, then

$$\frac{\partial}{\partial x} \begin{vmatrix} a & b \\ c & d \end{vmatrix} = \begin{vmatrix} a_x & b_x \\ c & d \end{vmatrix} + \begin{vmatrix} a & b \\ c_x & d_x \end{vmatrix}$$

$$= \begin{vmatrix} a_x & b \\ c_x & d \end{vmatrix} + \begin{vmatrix} a & b_x \\ c & d_x \end{vmatrix}$$

*4.8 State and prove a corresponding result for a 3×3 determinant.

4.9 Compute the determinant $\begin{vmatrix} f_x & f_y \\ g_x & g_y \end{vmatrix}$ for the following pairs of functions

a. $f(x, y) = x^2 - y^2$, $\qquad g(x, y) = 2xy$
b. $f(x, y) = e^x \cos y$, $\qquad g(x, y) = e^x \sin y$
c. $f(x, y) = x^2 + y^2 + 2xy + 2$, $\qquad g(x, y) = e^x e^y$

d. $f(x, y) = x - 2 \log y$, $\qquad g(x, y) = \frac{1}{y^2} e^x$

4.10 Let $f(x, y) = \sin xy$.

a. Show that $xf_x - yf_y \equiv 0$.
b. Let $F(t)$ be an arbitrary differentiable function of one variable, and let $g(x, y) = F(f(x, y))$. Show that also $xg_x - yg_y \equiv 0$.

4.11 Show that Def. 4.1 that we have given for partial derivatives is equivalent to the following:

$$f_x(x_0, y_0) = \lim_{h \to 0} \frac{f(x_0 + h, y_0) - f(x_0, y_0)}{h},$$

$$f_y(x_0, y_0) = \lim_{k \to 0} \frac{f(x_0, y_0 + k) - f(x_0, y_0)}{k}.$$

4.12 By referring directly to the definition of partial derivatives, compute $f_x(0, 0)$ and $f_y(0, 0)$, where

a. $f(x, y) = x - y^2$
b. $f(x, y) = 1 + x \cos y$

*c. $f(x, y) = \begin{cases} \dfrac{x^3 + y^4}{x^2 + y^2} & \text{for} \quad (x, y) \neq (0, 0) \\ 0 & \text{for} \quad (x, y) = (0, 0) \end{cases}$

*d. $f(x, y) = \begin{cases} \dfrac{(\cos x - 1)^2}{x^2 + y^2} & \text{for} \quad (x, y) \neq (0, 0) \\ 0 & \text{for} \quad (x, y) = (0, 0) \end{cases}$

4.13 Find the set of points where each of the following functions is defined, and sketch.

    a. $\sqrt{1 - x^2 - y^2} + \sqrt{x - y}$

    b. $\sqrt{x^2 - y^2} + \sqrt{1 - x^2}$

    c. $\log x + 1/\sqrt{y} + (1 - x - y)^{-1/4}$

    d. $\arcsin x + \arccos y$

4.14 Describe the boundary points of each of the following sets of points (x, y). Illustrate with a sketch.

    a. $x^2 + 4y^2 < 4$

    b. $x^2 + 4y^2 > 4$

    c. $x^2 + 4y^2 \geq 4$ and $x^2 + 4y^2 \leq 16$

    d. $x \geq 2y$ and $y \geq 2x$

    e. $x^2 \leq 1$ and $y^2 < 4$

    f. $x^2 + y^2 = 1$

4.15 *a.* Which of the sets in Ex. 4.14 are open?

    b. For each set in Ex. 4.14 which is not open, name a point of the set that is not an interior point.

4.16 Show that the following characterizations of boundary points and interior points of a set S are equivalent to the definitions in the text. A point (x_0, y_0) is a *boundary point* of S if for every positive number r, the disk $(x - x_0)^2 + (y - y_0)^2 < r^2$ contains at least one point in S and at least one point not in S. A point (x_0, y_0) is an *interior point* of S if for some positive r (sufficiently small), all points of the disk $(x - x_0)^2 + (y - y_0)^2 < r^2$ are in S (Fig. 4.9).

4.17 A set S is called *closed* if S contains all its boundary points. Which sets in Ex. 4.14 are closed? (Note that a set S may be *neither* open *nor* closed. This happens when S contains some of its boundary points, but not others.)

4.18 A set S is called *convex* if for any two points of S, all points on the line segment joining them also lie in S.

    a. Which sets in Ex. 4.14 are convex? (Do not try to justify your answers. This is one place where geometric intuition is almost infallible. Actual proofs are not difficult, but they require some practice in juggling inequalities.)

    b. Show that an open convex set is necessarily a domain.

4.19 Prove carefully, by referring directly to the definition, that each of the following sets is a domain

    a. the whole plane

    b. the half-plane $y > 0$

    c. the disk $x^2 + y^2 < 1$

4.20 Show that for the following functions $f(x, y)$ the partial derivatives do *not* exist at the points indicated

    a. $f(x, y) = \sqrt{x^2 + y^2}$ at $(0, 0)$

    b. $f(x, y) = (x + y - 3)^{1/3}$ at $(1, 2)$

(Note that both functions are defined throughout the plane.)

5 Continuity

The function $f(x, y) = 2xy/(x^2 + y^2)$ is defined everywhere except at the origin. Since it is the quotient of two polynomials, its partial derivatives f_x and f_y exist wherever the denominator is nonzero, which is everywhere except at the origin. Since the function is not defined at the origin, we may ask whether we cannot extend the function by suitably defining it at the origin so that its partial derivatives exist there too. We note that along the x and y axes, where $y = 0$ or $x = 0$, $f(x, y) = 0$. Hence we make the definition

$$(5.1) \qquad f(x, y) = \begin{cases} \dfrac{2xy}{x^2 + y^2} & \text{if } (x, y) \neq (0, 0) \\ 0 & \text{if } (x, y) = (0, 0). \end{cases}$$

Then $f(x, 0) \equiv 0$ for all x, and hence $f_x(x, 0) \equiv 0$. In particular, $f_x(0, 0)$ exists and is zero. Similarly $f_y(0, 0) = 0$. Thus the function defined by (5.1) has both partial derivatives f_x and f_y at every point.

For a function of one variable, the existence of a derivative at a point implies that the function is continuous at that point. On the other hand, for a function of two variables both partial derivatives may exist at every point, and the function may still not be continuous. In particular, the function defined by (5.1) is not continuous at the origin, since approaching the origin along the line $y = x$, we have $f(x, y) \equiv 1$ for $(x, y) \neq (0, 0)$, but at the origin $f(x, y)$ suddenly jumps from 1 to 0 (see Fig. 3.4).

In order to express precisely our intuitive notions of continuity, we introduce the following notation:

$$(5.2) \qquad \lim_{n \to \infty} (x_n, y_n) = (x_0, y_0) \qquad \text{if} \quad \lim_{n \to \infty} \sqrt{(x_n - x_0)^2 + (y_n - y_0)^2} = 0.$$

Note that for each n, the quantity

$$d_n = \sqrt{(x_n - x_0)^2 + (y_n - y_0)^2}$$

is a fixed number, representing the distance of the point (x_n, y_n) from the point (x_0, y_0). Thus condition (5.2) simply says that the ordinary sequence of numbers d_n tends to zero. Geometrically, the points (x_n, y_n) are tending to (x_0, y_0) although they may not be approaching from any fixed direction. Several examples are pictured in Fig. 5.1.

Note also that we have

$$\lim_{n \to \infty} \sqrt{(x_n - x_0)^2 + (y_n - y_0)^2} = 0 \Leftrightarrow \lim_{n \to \infty} x_n = x_0 \quad \text{and} \quad \lim_{n \to \infty} y = y_0.$$

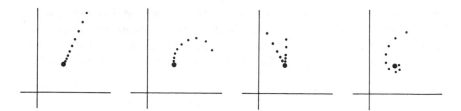

FIGURE 5.1 Illustrations of "$\lim_{n \to \infty} (x_n, y_n) = (x_0, y_0)$"

Definition 5.1 *$f(x, y)$ tends to the limit L as (x, y) tends to (x_0, y_0) if*

$$\lim_{n \to \infty} f(x_n, y_n) = L$$

whenever

$$\lim_{n \to \infty} (x_n, y_n) = (x_0, y_0).$$

We adopt the notation

$$\lim_{(x,y) \to (x_0,y_0)} f(x, y) = L$$

or

$$f(x, y) \to L \quad \text{as} \quad (x, y) \to (x_0, y_0).$$

Definition 5.2 *$f(x, y)$ is continuous at (x_0, y_0) if*

$$\lim_{(x,y) \to (x_0,y_0)} f(x, y) = f(x_0, y_0).$$

Thus the idea of continuity is essentially the same for functions of two variables as for one variable. We simply require that as a variable point approaches a fixed point, the value of the function should approach its value at that point.

Remark Although we only use the property of continuity given in the definition, there is an alternative characterization that may be helpful to those who are more accustomed to "ϵ, δ" definitions of continuity. Namely, $f(x, y)$ is continuous at (x_0, y_0) if and only if for every $\epsilon > 0$ there is a $\delta > 0$ such that $|f(x, y) - f(x_0, y_0)| < \epsilon$ whenever $\sqrt{(x - x_0)^2 + (y - y_0)^2} < \delta$.

It is important to note that it is not the same thing for a function $f(x, y)$ to be continuous as a function of two variables, and to be continuous in each variable when the other is held fixed. It is an immediate consequence of our definition of continuity that if $f(x, y)$ is continuous at (x_0, y_0), then $f(x, y_0)$ is a continuous function of x at $x = x_0$, and $f(x_0, y)$ is a continuous function of y at $y = y_0$. However, the converse is not true. For example, the function defined by Eq. (5.1) is continuous in each variable separately at the origin, since $f(x, 0) \equiv 0$ and $f(0, y) \equiv 0$; however, it is not continuous at the origin, since if we consider, for example, the sequence of points $(1/n, 1/n)$, then

$$\left(\frac{1}{n}, \frac{1}{n}\right) \to (0, 0) \quad \text{as} \quad n \to \infty$$

but

$$f\left(\frac{1}{n}, \frac{1}{n}\right) = 1 \quad \text{for all } n;$$

hence

$$\lim_{n \to \infty} f\left(\frac{1}{n}, \frac{1}{n}\right) = 1 \neq f(0, 0).$$

So far, we have discussed only the notion of continuity at a point. In general, we are interested in continuity of a function in a domain.

Definition 5.3 A function $f(x, y)$ defined in a domain D is *continuous in D* if it is continuous at each point of D.

Suppose now that $f(x, y)$ is defined in D and that both partial derivatives f_x and f_y exist at each point of D. Then f_x and f_y are themselves functions defined throughout D, and we may ask whether they are continuous.

Definition 5.4 $f(x, y)$ is called *continuously differentiable* in D if f_x and f_y exist at each point and if they are continuous functions in D.

By adding the hypothesis that f_x and f_y not only exist but are continuous, we rule out all "pathological" functions of the type (5.1). This is a consequence of the following lemma, which plays a key role in the theory of functions of two variables.

Lemma 5.1 Fundamental Lemma *Let $f(x, y)$ be continuously differentiable in a domain D, and let (x_0, y_0) be a point in D. Then there is a function $h(x, y)$, defined in D, satisfying*

(5.3) $f(x, y) = f(x_0, y_0) + (x - x_0)f_x(x_0, y_0) + (y - y_0)f_y(x_0, y_0) + h(x, y)$

and

(5.4) $$\lim_{(x,y) \to (x_0, y_0)} \frac{h(x, y)}{d(x, y)} = 0,$$

where

$$d(x, y) = \sqrt{(x - x_0)^2 + (y - y_0)^2}.$$

Remark The function $h(x, y)$ in Eq. (5.3) should be thought of as a remainder term. It is actually defined as the difference of the value of the function $f(x, y)$ and the sum of the first three terms on the right-hand side of Eq. (5.3). The whole content of the lemma is given by Eq. (5.4) which says that as $(x, y) \to (x_0, y_0)$, the remainder term $h(x, y)$ tends to zero *faster* than the distance $d(x, y)$.

PROOF. Since Eq. (5.4) is concerned with the limit as $(x, y) \to (x_0, y_0)$, it is sufficient to consider only values of (x, y) in some circle

$$(5.5) \qquad (x - x_0)^2 + (y - y_0)^2 < r^2.$$

By our definition of a domain, (x_0, y_0) is an interior point of D, and hence if r is chosen sufficiently small all points of the circle (5.5) lie in D.

The proof is based on two observations. First, to evaluate the change in f as we move from (x_0, y_0) to (x, y), we may proceed in two steps, first fixing y and letting x vary and then fixing x and letting y vary. Second, we may apply to each step the mean-value theorem for a function of a single variable.[1] Thus, fixing $y = y_0$, the mean-value theorem asserts that there exists a value x_1 between x_0 and x such that

$$(5.6) \qquad f(x, y_0) - f(x_0, y_0) = (x - x_0)f_x(x_1, y_0).$$

(See Fig. 5.2.)

Similarly, we have

$$(5.7) \qquad f(x, y) - f(x, y_0) = (y - y_0)f_y(x, y_1),$$

where y_1 lies between y_0 and y.

Adding the above equations, we find that

$$(5.8) \quad f(x, y) - f(x_0, y_0) = (x - x_0)f_x(x_1, y_0) + (y - y_0)f_y(x, y_1).$$

Substituting Eq. (5.8) into Eq. (5.3), we obtain

$$(5.9) \qquad \begin{aligned} h(x, y) &= (x - x_0)[f_x(x_1, y_0) - f_x(x_0, y_0)] \\ &\quad + (y - y_0)[f_y(x, y_1) - f_y(x_0, y_0)]. \end{aligned}$$

But as $x \to x_0$ we have $x_1 \to x_0$, and as $y \to y_0$ we have $y_1 \to y_0$. Thus, by the continuity of f_x and f_y at (x_0, y_0), we have

$$\lim_{(x,y) \to (x_0,y_0)} f_x(x_1, y_0) = f_x(x_0, y_0),$$

$$\lim_{(x,y) \to (x_0,y_0)} f_y(x, y_1) = f_y(x_0, y_0).$$

---

[1] See [14], Section 9.4, or [3], Vol. I, Chapters 7 and 8, concerning the mean-value theorem for one variable.

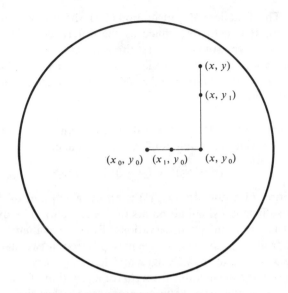

FIGURE 5.2 "Two-step" method of moving from (x_0, y_0) to (x, y)

Thus each of the expressions in the square brackets in Eq. (5.9) tends to zero as $(x, y) \to (x_0, y_0)$. On the other hand, it is evident that for all (x, y)

$$\frac{|x - x_0|}{\sqrt{(x - x_0)^2 + (y - y_0)^2}} \leq 1, \qquad \frac{|y - y_0|}{\sqrt{(x - x_0)^2 + (y - y_0)^2}} \leq 1.$$

Thus, dividing through Eq. (5.9) by $\sqrt{(x - x_0)^2 + (y - y_0)^2}$, we obtain the sum of two terms, each of which tends to zero as $(x, y) \to (x_0, y_0)$. This shows that Eq. (5.4) holds, and the lemma is proved. ◆

In the next section we shall give a number of applications of this basic lemma. For now, we shall restrict ourselves to one immediate consequence and to a geometrical interpretation.

Corollary *A continuously differentiable function is continuous.*

PROOF. Using the lemma, at any point (x_0, y_0) we can express the function $f(x, y)$ by Eq. (5.3). Then by Eq. (5.4) we have certainly

$$\lim_{(x,y) \to (x_0, y_0)} h(x, y) = 0,$$

and hence from Eq. (5.3),

$$\lim_{(x,y) \to (x_0, y_0)} f(x, y) = f(x_0, y_0).$$ ◆

We turn now to the geometrical interpretation of Lemma 5.1. Given the function $f(x, y)$ and a fixed point (x_0, y_0) we construct the function

(5.10) $L(x, y) = f(x_0, y_0) + (x - x_0)f_x(x_0, y_0) + (y - y_0)f_y(x_0, y_0).$

$L(x, y)$ is simply a linear function of the form $Ax + By + C$. Hence the equation $z = L(x, y)$ defines a plane. Further, $L(x_0, y_0) = f(x_0, y_0)$, so that this plane touches the surface $z = f(x, y)$ above the point (x_0, y_0). Equation (5.3) can now be written in the form

$$f(x, y) = L(x, y) + h(x, y).$$

Thus $h(x, y)$ represents the difference between the height of the surface $z = f(x, y)$ above the x, y plane, and the height of the plane $z = L(x, y)$ (Fig. 5.3). For any plane passing through the same point, say

$$z = l(x, y), \qquad \text{where} \quad l(x_0, y_0) = f(x_0, y_0),$$

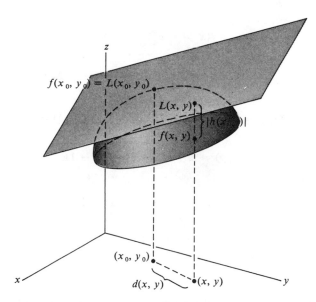

FIGURE 5.3   Geometric interpretation of the fundamental lemma: the tangent plane

the difference $f(x, y) - l(x, y)$ would tend to zero as $(x, y) \to (x_0, y_0)$. The significance of Eq. (5.4) is that for the particular plane defined by Eq. (5.10), the difference $h(x, y)$ tends to zero *to a higher order*, so that even the ratio

$$\frac{h(x, y)}{d(x, y)} \to 0 \qquad \text{as} \quad (x, y) \to (x_0, y_0).$$

One can show that given a function $f(x, y)$ and a point (x_0, y_0), there cannot be more than one plane $z = l(x, y)$ for which

$$\frac{f(x, y) - l(x, y)}{d(x, y)} \to 0 \quad \text{as} \quad (x, y) \to (x_0, y_0).$$

(See Ex. 5.4 below.) If there is a plane satisfying this condition, it is called the *tangent plane* to the surface $z = f(x, y)$ at the point. Thus, the geometric content of Lemma 5.1 is the following: *if $f(x, y)$ is continuously differentiable, then it has a tangent plane at each point, and the equation of the tangent plane is $z = L(x, y)$, where $L(x, y)$ is defined by Eq. (5.10).*

Example 5.1

Find the equation of the tangent plane to the surface

$$z = e^{x-y} + xy^2$$

at the point (1, 1).
We set

$$f(x, y) = e^{x-y} + xy^2;$$

then

$$f_x = e^{x-y} + y^2, \qquad f_y = -e^{x-y} + 2xy$$

and

$$f(1, 1) = 2, \qquad f_x(1, 1) = 2, \qquad f_y(1, 1) = 1.$$

Hence,

$$L(x, y) = 2 + 2(x - 1) + (y - 1),$$

and the tangent plane is

$$z = 2(x - 1) + (y - 1) + 2,$$

or

$$z = 2x + y - 1.$$

Exercises

5.1 Find the equation of the tangent plane to each of the following surfaces at the points indicated.

a. $z = x^4 - 3x^2y^3 + 2xy^2 + 3x - 4y + 5$ at (0, 0)
b. $z = 3x - 4y + 5$ at (-1, 2)
c. $z = \sqrt{4 - x^2 - y^2}$ at (0, 0)
d. $z = \sqrt{x^2 + y^2}$ at (3, -4)
e. $z = 2xy/(x^2 + y^2)$ at (2, -1)
f. $z = ye^{x^2+y}$ at (0, 0)

5.2   Show that for the quadric surface $ax^2 + by^2 + cz^2 = 1$, where $c > 0$, the tangent plane at the point (x_0, y_0, z_0), where

$$z_0 = \sqrt{1 - ax_0^2 - by_0^2} / \sqrt{c},$$

has the equation

$$ax_0 x + by_0 y + cz_0 z = 1.$$

5.3   **a.** Find the equation of the tangent plane to the surface $x^{1/2} + y^{1/2} + z^{1/2} = 1$ at an arbitrary point (x_0, y_0, z_0).
   b. Find the intersection of the tangent plane with each of the axes.
   c. Show that the sum of the intercepts does not depend on the choice of the point (x_0, y_0, z_0).

5.4   Fix a point (x_0, y_0) and let

$$d(x, y) = \sqrt{(x - x_0)^2 + (y - y_0)^2}.$$

Show that the following statements are true.

   a. If $g(x, y)$ is defined in some domain containing (x_0, y_0) and if

$$\lim_{(x,y) \to (x_0, y_0)} \frac{g(x, y)}{d(x, y)} = 0,$$

then

$$\lim_{(x,y) \to (x_0, y_0)} g(x, y) = 0.$$

(*Hint:* $g(x, y) = d(x, y)[g(x, y)/d(x, y)]$.)

   b. If $L(x, y) = Ax + By + C$, and if

$$\lim_{(x,y) \to (x_0, y_0)} \frac{L(x, y)}{d(x, y)} = 0,$$

then

$$A = B = C = 0.$$

(*Hint:* First use part a, and then show that $A = 0$ and $B = 0$ by forming the limit first with a sequence such as $(x_0 + 1/n, y_0)$ and then with the sequence $(x_0, y_0 + 1/n)$.)

   c. Let

$$l_1(x, y) = a_1 x + b_1 y + c_1, \qquad l_2(x, y) = a_2 x + b_2 y + c_2.$$

If

$$\lim_{(x,y) \to (x_0, y_0)} \frac{l_1(x, y) - l_2(x, y)}{d(x, y)} = 0,$$

then

$$l_1(x, y) = l_2(x, y).$$

d. Suppose $f(x, y)$ is defined in a domain D containing (x_0, y_0). If

$$\lim_{(x,y)\to(x_0,y_0)} \frac{f(x, y) - l_1(x, y)}{d(x, y)} = 0$$

and

$$\lim_{(x,y)\to(x_0,y_0)} \frac{f(x, y) - l_2(x, y)}{d(x, y)} = 0$$

then

$$l_1(x, y) = l_2(x, y).$$

(*Hint:* $|l_1 - l_2| \leq |l_1 - f| + |f - l_2|$.)

Note that Ex. 5.4d is the statement in the text that there cannot be more than one plane $z = l(x, y)$ such that

$$\lim_{(x,y)\to(x_0,y_0)} \frac{f(x, y) - l(x, y)}{d(x, y)} = 0.$$

If such a plane exists, it is called the tangent plane to the surface $z = f(x, y)$ at the point. The following terminology is also frequently used. A function $f(x, y)$ defined in a domain D is called *differentiable* at a point (x_0, y_0) in D if $f(x, y)$ is continuous at (x_0, y_0) and if the surface $z = f(x, y)$ has a tangent plane at the point. For functions of one variable the corresponding property (that is, existence of a tangent line) is equivalent to the existence of the derivative at the point. However, the situation in two variables is far more complicated. Since we make no use of this concept, we leave it to the interested reader to explore its basic properties in Exs. 5.5–5.8. These properties may be summarized as follows.

1. If $f(x, y)$ is continuously differentiable in a domain D, then it is differentiable at each point of D (Lemma 5.1).
2. If $f(x, y)$ is differentiable at a point (x_0, y_0), then $f_x(x_0, y_0)$ and $f_y(x_0, y_0)$ exist, and the equation of the tangent plane is

$$z = f_x(x_0, y_0)(x - x_0) + f_y(x_0, y_0)(y - y_0) + f(x_0, y_0)$$

(Ex. 5.5).
3. A function may have both partial derivatives at a point but not be differentiable (Exs. 5.6 and 5.7).
4. A function may be differentiable, but not continuously differentiable (Ex. 5.8).

Thus differentiability implies more than mere existence of partial derivatives and less than existence and continuity of the partial derivatives. By property 2 above, differentiability is *exactly* the conclusion of Lemma 5.1, that is, that Eqs. (5.3) and (5.4) hold.

*5.5 Using the notation of Ex. 5.4, show that if

$$\lim_{(x,y)\to(x_0,y_0)} \frac{f(x, y) - L(x, y)}{d(x, y)} = 0,$$

and if $f(x, y)$ is continuous at (x_0, y_0), then $f_x(x_0, y_0)$ and $f_y(x_0, y_0)$ exist, and

$$L(x, y) = f_x(x_0, y_0)(x - x_0) + f_y(x_0, y_0)(y - y_0) + f(x_0, y_0).$$

(*Hint:* first apply the statement in Ex. 5.4a to the function $g(x, y) = f(x, y) - L(x, y)$ to deduce $L(x_0, y_0) = f(x_0, y_0)$. Then take the above limit, first fixing $y = y_0$ and then $x = x_0$.)

*5.6 Show that the function defined by Eq. (5.1) does not have a tangent plane at the origin, even though both partial derivatives exist at every point. (*Hint:* if there were a tangent plane, and if we took limits first along $y = 0$ and then along $x = 0$, the tangent plane would have to be the x, y plane. But taking limits along $y = x$ gives a contradiction.)

*5.7 Show that the function $f(x, y) = (x^{1/3} + y^{1/3})^3$ is continuous at $(0, 0)$ and that both partial derivatives $f_x(0, 0)$ and $f_y(0, 0)$ exist, but that $f(x, y)$ is not differentiable at $(0, 0)$. (*Hint:* use the same reasoning as Ex. 5.6.)

5.8 Let $f(x, y) = x^3 \sin(1/x) + y^2$ for $x \neq 0$, and $f(0, y) = y^2$.

     a. Find the partial derivatives of f at any point (x_0, y_0) where $x_0 \neq 0$.
     b. Find the partial derivatives of f at the point $(0, y_0)$ for any y_0. (Use the definition of $f_x(0, y_0)$.)
     c. Show that $f(x, y)$ is continuous at $(0, 0)$.
     *d. Show that $f_x(x, y)$ is *not* continuous at $(0, 0)$.
     *e. Show that $f(x, y)$ is differentiable at $(0, 0)$. In fact, show that the plane $z = 0$ is the tangent plane to the surface $z = f(x, y)$ at the origin.

5.9 Show that the cone $z = \sqrt{x^2 + y^2}$, does not have a tangent plane at the origin. (*Hint:* assuming that a tangent plane $z = L(x, y)$, where $L(x, y) = Ax + By + C$, does exist, then applying Ex. 5.4a to the condition

$$\lim_{(x,y) \to (0,0)} \frac{\sqrt{x^2 + y^2} - L(x, y)}{d(x, y)} = 0$$

gives $C = 0$. Forming this limit with the sequence $(1/n, 0)$ gives $A = 1$ and using $(0, 1/n)$ gives $B = 1$. But using $(1/n, 1/n)$ gives $A + B = \sqrt{2}$, a contradiction.)

5.10 Let $f(x, y)$ be the function defined in Eq. (5.1).

     a. Compute the partial derivatives of f with respect to x and y at any point other than $(0, 0)$.
     b. Show that f_x and f_y are not continuous at $(0, 0)$.

5.11 Let $f(x, y)$ be defined in a domain D containing the point (x_0, y_0). Let $x = x_0 + at$, $y = y_0 + bt$, define an arbitrary straight line through the point (x_0, y_0). Show that the function $g(t) = f(x_0 + at, y_0 + bt)$ is a continuous function of t at $t = 0$.

Note. It was mentioned earlier that a continuous function of two variables reduces to a continuous function of each variable separately if the other variable is held fixed. Geometrically, this may be interpreted as saying that a continuous function of two variables is continuous along every horizontal and vertical line. What Ex. 5.11 states is that a continuous function of two variables is in fact continuous along *every* straight line. It may come as a surprise that the converse is not true.

The following exercise gives an example of a function that is continuous when restricted to any straight line through the origin, but is not continuous at the origin when considered as a function of two variables.

5.12 Let

$$f(x, y) = \begin{cases} \dfrac{2x^2y}{x^4 + y^2} & \text{if } (x, y) \neq (0, 0) \\ 0 & \text{if } (x, y) = (0, 0). \end{cases}$$

a. Show that if $x = at$, $y = bt$ is any straight line through the origin, where a and b are not both zero, then

$$\lim_{t \to 0} f(at, bt) = f(0, 0) = 0.$$

b. Show that at every point $(1/n, 1/n^2)$, $f(x, y)$ is equal to 1. Conclude that $f(x, y)$ is not continuous at $(0, 0)$.

*c. Sketch the surface $z = f(x, y)$ by observing that the parabolas $y = ax^2$ are all level curves of the function, and by seeing how the value of the function on each of these curves depends on the constant a. Try to visualize how the function can be continuous along every straight line through the origin, but still fail to be continuous if one approaches the origin along a parabola $y = ax^2$.

5.13 Let $x(t)$, $y(t)$ be continuous functions for $a \leq t \leq b$. Let $x_0 = x(t_0)$ and $y_0 = y(t_0)$. Let $f(x, y)$ be defined in a domain containing the point (x_0, y_0) and let $f(x, y)$ be continuous at (x_0, y_0). Show that the function $g(t) = f(x(t), y(t))$ is continuous at $t = t_0$.

Note. The property stated in Ex. 5.13 may be reformulated roughly in the following geometric form: a continuous function of two variables is continuous along every curve.

5.14 Show that if $f(x, y)$ is continuous in a domain D, and if $f(x, y)$ is never zero, then either $f(x, y) > 0$ throughout D, or else $f(x, y) < 0$ throughout D. (*Hint:* prove the contrapositive; if $f(x_1, y_1) > 0$ and $f(x_2, y_2) < 0$, then by connectedness (x_1, y_1) can be joined to (x_2, y_2) by a curve in D, and applying the intermediate-value theorem for functions of one variable to the function $g(t)$ of Ex. 5.13 gives a point in D where $f(x, y) = 0$.)

*5.15 Prove the equivalence of Def. 5.2 of continuity and the ϵ, δ definition mentioned in the Remark following Def. 5.2.

5.16 Let $f(x, y)$ be continuous in a domain D.

a. Prove that if $f(x_0, y_0) = \eta > 0$, then there exists a positive number r such that $f(x, y) > \eta/2$ for all points (x, y) satisfying $(x - x_0)^2 + (y - y_0)^2 < r^2$. (*Hint:* either use Ex. 5.15, or else prove by contradiction, choosing a sequence of circles whose radii tend to zero, in each of which there is a point (x_n, y_n), where $f(x_n, y_n) \leq \eta/2$.)

b. Show that the set of points (x, y) in D, where $f(x, y) > 0$, is an open set.

6   Directional derivatives and gradient

The partial derivatives f_x and f_y at a point measure the rate of change of the function f in the x and y directions, respectively. They may be considered to be special cases of the *directional derivative*, which measures the rate of change of f in an arbitrary direction, and which we define below.

We use the following notation throughout this section. Let

(6.1)
$$\mathbf{T}_\alpha = \langle \cos \alpha, \sin \alpha \rangle$$

be the unit vector making an angle α with the positive x direction.

Given any point (x_0, y_0), we wish to measure the rate of change of f in the direction of $\mathbf{T}_\alpha$. We simply take the change in the function as we move a distance Δs in this direction, divide by Δs, and take the limit as Δs tends to zero (Fig. 6.1). We use the notation $\nabla_\alpha f(x_0, y_0)$ for this quantity.

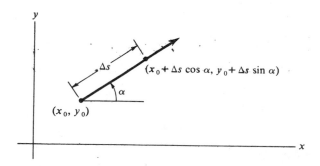

FIGURE 6.1   Directional derivative ∇_α

Definition 6.1   The *directional derivative* of a function $f(x, y)$ at a point (x_0, y_0) in the direction of the vector $\langle \cos \alpha, \sin \alpha \rangle$ is

(6.2)
$$\nabla_\alpha f(x_0, y_0) = \lim_{\Delta s \to 0} \frac{\Delta f}{\Delta s}$$

$$= \lim_{\Delta s \to 0} \frac{f(x_0 + \Delta s \cos \alpha,\ y_0 + \Delta s \sin \alpha) - f(x_0, y_0)}{\Delta s}.$$

Note that the partial derivatives f_x and f_y are special cases:

(6.3) $\nabla_0 f(x_0, y_0) = f_x(x_0, y_0);$ $\nabla_{\pi/2} f(x_0, y_0) = f_y(x_0, y_0).$

In general, a function may have directional derivatives in some directions, but not in others. Again we may refer to the example defined by Eq. (5.1).

For that function, the directional derivatives at the origin exist in the x and y directions, but in no other direction. This illustrates again the inadequacy of partial derivatives for describing the way a function varies in an arbitrary direction. The surprising fact is that if we add to the assumption that the partial derivatives exist at each point the further assumption that they are continuous, then we are able to draw two conclusions.

1. The directional derivatives must exist in all directions.
2. The value of the directional derivative at any point in a given direction depends only on the direction and on the partial derivatives f_x, f_y at the point.

The proof of these facts is based on the fundamental Lemma 5.1. We have the following theorem.

Theorem 6.1 *Let* $f(x, y)$ *be continuously differentiable in a domain D. Then at any point* (x_0, y_0), *the directional derivative* $\nabla_\alpha f(x_0, y_0)$ *exists for all* α *and is given by*

$$(6.4) \qquad \nabla_\alpha f(x_0, y_0) = f_x(x_0, y_0) \cos \alpha + f_y(x_0, y_0) \sin \alpha.$$

PROOF. Setting $x = x_0 + \Delta s \cos \alpha$, $y = y_0 + \Delta s \sin \alpha$ in Eq. (5.3), we find

$$f(x_0 + \Delta s \cos \alpha, y_0 + \Delta s \sin \alpha) - f(x_0, y_0)$$
$$= \Delta s \cos \alpha \, f_x(x_0, y_0) + \Delta s \sin \alpha \, f_y(x_0, y_0) + h(x, y).$$

Dividing by Δs yields

$$(6.5) \qquad \frac{\Delta f}{\Delta s} = \cos \alpha \, f_x(x_0, y_0) + \sin \alpha \, f_y(x_0, y_0) + \frac{h(x, y)}{\Delta s}.$$

But $\Delta s = \sqrt{(x - x_0)^2 + (y - y_0)^2}$, and the fundamental Lemma 5.1 asserts precisely that

$$\lim_{\Delta s \to 0} \frac{h(x, y)}{\Delta s} = 0.$$

Thus, applying Def. 6.1 to Eq. (6.5) we obtain Eq. (6.4). ◆

Example 6.1

Let

$$f(x, y) = \sqrt{25 - x^2 - y^2},$$

and

$$(x_0, y_0) = (3, 2), \qquad \alpha = \tfrac{1}{3}\pi;$$

then

$$f_x = \frac{-x}{\sqrt{25 - x^2 - y^2}}, \qquad f_y = \frac{-y}{\sqrt{25 - x^2 - y^2}};$$

$$f_x(3, 2) = \frac{-3}{\sqrt{12}} = \frac{-\sqrt{3}}{2}, \qquad f_y(3, 2) = \frac{-2}{\sqrt{12}} = \frac{-1}{\sqrt{3}};$$

$$\cos \alpha = \frac{1}{2}, \qquad \sin \alpha = \frac{\sqrt{3}}{2};$$

$$\nabla_{\pi/3}f(3, 2) = \frac{-\sqrt{3}}{2} \cdot \frac{1}{2} + \frac{-1}{\sqrt{3}} \cdot \frac{\sqrt{3}}{2} = -\frac{\sqrt{3} + 2}{4}.$$

Note that the fact that the directional derivative is *negative* means that the function is decreasing in this direction (Fig. 6.2a).

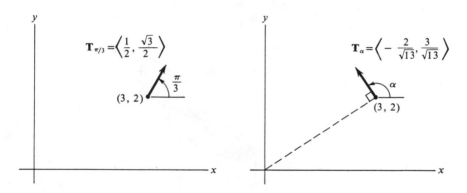

FIGURE 6.2   Directions of derivatives in Examples 6.1 and 6.2

Example 6.2

Using the same function and same point as in Example 6.1, let

$$T_\alpha = \left\langle \frac{-2}{\sqrt{13}}, \frac{3}{\sqrt{13}} \right\rangle;$$

then

$$\cos \alpha = \frac{-2}{\sqrt{13}}, \qquad \sin \alpha = \frac{3}{\sqrt{13}},$$

and

$$\nabla_\alpha f(3, 2) = \frac{-\sqrt{3}}{2} \cdot \frac{-2}{\sqrt{13}} + \frac{-1}{\sqrt{3}} \cdot \frac{3}{\sqrt{13}} = \frac{\sqrt{3}}{\sqrt{13}} - \frac{\sqrt{3}}{\sqrt{13}} = 0.$$

Note that in this case T_α is a unit vector in a direction perpendicular to the radius vector $\langle 3, 2 \rangle$. The rate of change of the function in this direction is zero (Fig. 6.2b).

There are two physical interpretations that are most helpful in under-standing the directional derivative. The first is obtained by picturing the domain D to be a thin plate and the function $f(x, y)$ the temperature at each point of the plate. Then the directional derivative at any point tells us whether, as we move in a given direction, we are getting hotter or colder, and how fast.

The other interpretation concerns the surface $z = f(x, y)$. We draw a line through (x_0, y_0) in the direction of $\mathbf{T}_\alpha$ and consider the vertical plane through this line. The intersection of that plane with the surface defines a curve whose slope is precisely $\nabla_\alpha f(x_0, y_0)$ (Fig. 6.3). If we return to our description of the surface as a mountainous terrain, where $f(x, y)$ is the altitude, then the directional derivative is positive or negative according to whether we are ascending or descending when walking in the given direction. The magnitude of the directional derivative simply measures how steep a path would be in that direction.

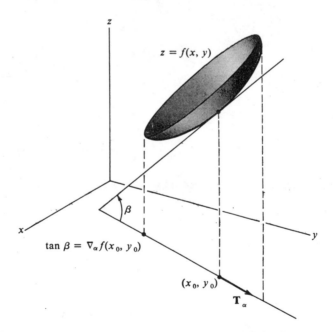

FIGURE 6.3 Geometric interpretation of the directional derivative

We return now to Eq. (6.4), and we observe that the right-hand side has the form of a dot product of two vectors, one of them being $\mathbf{T}_\alpha$. The other one would have components f_x, f_y at the point. This vector is of basic importance for functions of two variables. It is called the *gradient vector* of f and is denoted by ∇f.

Definition 6.2   Let $f(x, y)$ be defined in a domain D. At any point (x_0, y_0) where the partial derivatives f_x and f_y exist, we define the *gradient of f* to be the vector whose components are these partial derivatives:

(6.6) $$\nabla f(x_0, y_0) = \langle f_x(x_0, y_0), f_y(x_0, y_0) \rangle.$$

Remark on Notation   One often sees the notation grad f, for the gradient, or **grad** f, where the boldface type indicates a vector quantity. The upside-down delta used in Eq. (6.6) is called a *nabla*, but when used to operate on a function, as in Eq. (6.6), it is usually referred to as *del*. Thus one reads "∇f" as "del f." The boldface ∇ is used to emphasize the fact that the gradient of a function defines a vector at each point. In the notation $\nabla_\alpha f$, we do not use boldface, since the directional derivative is a number, rather than a vector.

We may now rewrite Eq. (6.4) in vector notation, using Eqs. (6.1) and (6.6):

(6.7) $$\nabla_\alpha f(x_0, y_0) = \nabla f(x_0, y_0) \cdot \mathbf{T}_\alpha.$$

This is at first merely a different notation for the same thing. However, it leads to important new information. Recalling that the dot product of two vectors is the product of their lengths times the cosine of the angle in between and observing that $\mathbf{T}_\alpha$ is a vector of length 1, we have

(6.8) $$\nabla_\alpha f(x_0, y_0) = |\nabla f(x_0, y_0)| \cos \theta$$

where θ is the angle between $\nabla f(x_0, y_0)$ and $\mathbf{T}_\alpha$ (Fig. 6.4).

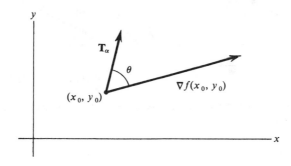

FIGURE 6.4   Gradient vector and directional derivative

From Eq. (6.8) we deduce immediately the following basic facts.

Let $f(x, y)$ be continuously differentiable in D. Then at each point (x_0, y_0) of D we have a fixed vector

$$\nabla f(x_0, y_0) = \langle f_x(x_0, y_0), f_y(x_0, y_0) \rangle.$$

1. If $|\nabla f(x_0, y_0)| = 0$ (i.e., if both partial derivatives at the point are zero), then *all* directional derivatives at the point are zero.

2. If $|\nabla f(x_0, y_0)| \neq 0$, then

$$\nabla_\alpha f(x_0, y_0) = 0 \Leftrightarrow \cos \theta = 0 \Leftrightarrow \theta = \tfrac{1}{2}\pi$$

$\Leftrightarrow \mathbf{T}_\alpha$ is one of the two directions perpendicular to $\nabla f(x_0, y_0)$ (Fig. 6.5).

3. Since $|\cos \theta| \leq 1$, the value of the directional derivative always varies between $\pm|\nabla f(x_0, y_0)|$. If $|\nabla f(x_0, y_0)| \neq 0$, then as α varies the directional derivative takes on its maximum value, $|\nabla f(x_0, y_0)|$, precisely when $\theta = 0$, i.e., when $\mathbf{T}_\alpha$ is in the same direction as $\nabla f(x_0, y_0)$. It takes on its minimum value, $-|\nabla f(x_0, y_0)|$, in the opposite direction, when $\theta = \pi$ (Fig. 6.5).

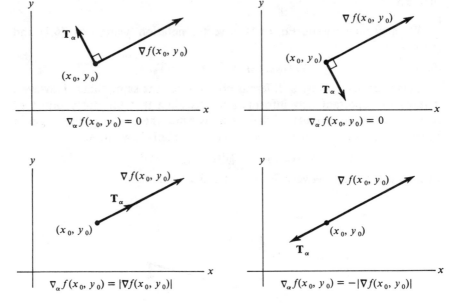

FIGURE 6.5 Directional derivatives in directions parallel to and perpendicular to the gradient vector

Example 6.3

In Example 6.2 above, we have

$$\nabla f(3, 2) = \langle f_x(3, 2), f_y(3, 2) \rangle = \left\langle \frac{-\sqrt{3}}{2}, \frac{-\sqrt{3}}{3} \right\rangle = \frac{\sqrt{3}}{6} \langle -3, -2 \rangle.$$

This is a vector directed toward the origin. The vector $\mathbf{T}_\alpha = (1/\sqrt{13})\langle -2, 3 \rangle$ is perpendicular to it, and hence the directional derivative $\nabla_\alpha f(3, 2) = 0$ (Fig. 6.6).

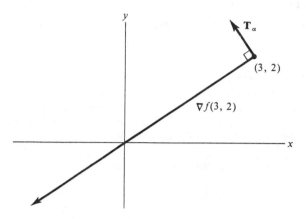

FIGURE 6.6 $\nabla f(3, 2)$, where $f(x, y) = \sqrt{25 - x^2 - y^2}$

Note that in terms of the surface $z = (25 - x^2 - y^2)^{1/2}$, which is a her‌ sphere, the gradient points in the direction of steepest ascent, whereas the vec‌ $\mathbf{T}_\alpha$ points in the direction of the level curve $x^2 + y^2 = 13$ (Fig. 6.7).

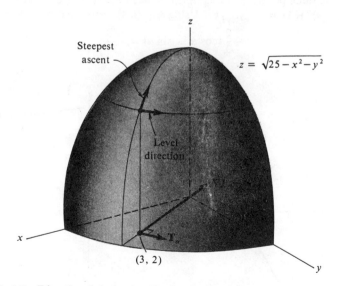

FIGURE 6.7   Directional derivatives for hemisphere

Example 6.4

A road is to be built to the top of a mountain whose equation is

$$z = 1 - \frac{x^2}{2} - \frac{y^2}{3}$$

(in miles). The shortest possible road is desired with a maximum grade of 30°. In which direction should the road be built at the point $(0, -1)$?

Since we want the shortest road to the top, we want the maximum increase in $f(x, y) = 1 - x^2/2 - y^2/3$, subject to the restriction. Thus, we would make the grade precisely 30°, and hence choose α so that

$$\nabla_\alpha f(0, -1) = \tan 30° = \frac{\sqrt{3}}{3}.$$

But $f_y = -x$, $f_y = -2y/3$, and $\nabla f(0, -1) = \langle 0, \frac{2}{3} \rangle$; hence,

$$\nabla_\alpha f(0, -1) = \tfrac{2}{3} \sin \alpha \quad \text{and} \quad \tfrac{2}{3} \sin \alpha = \frac{\sqrt{3}}{3} \Leftrightarrow \sin \alpha = \frac{\sqrt{3}}{2}$$

$$\Leftrightarrow \alpha = \tfrac{1}{3}\pi, \tfrac{2}{3}\pi.$$

Thus we proceed in a direction that makes an angle of 60° with the positive or negative x axis (Fig. 6.8a).

Example 6.5

Consider the same problem as Example 6.4, but start at the point $(\frac{2}{3}, 1)$.

We have $\nabla f(\frac{2}{3}, 1) = \langle -\frac{2}{3}, -\frac{2}{3} \rangle$. Using Eq. (6.4), we would have to solve

$$-\frac{2}{3}(\cos \alpha + \sin \alpha) = \frac{\sqrt{3}}{3}.$$

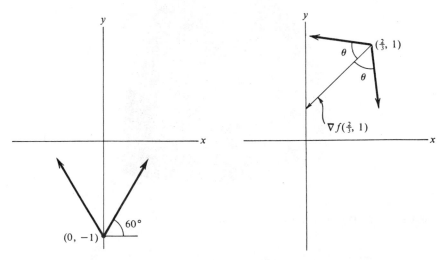

FIGURE 6.8 Finding α so that $\nabla_\alpha f(x_0, y_0) = \sqrt{3}/3$ where $f(x, y) = 1 - x^2/3 - y^2/3$

It is easier to apply Eq. (6.8). We have $|\nabla f(\frac{2}{3}, 1)| = \frac{2}{3}\sqrt{2}$, and we must choose θ so that

$$\frac{2}{3} \sqrt{2} \cos \theta = \frac{\sqrt{3}}{3} \quad \text{or} \quad \cos \theta = \frac{\sqrt{3}}{2\sqrt{2}} = \frac{\sqrt{6}}{4} \sim 0.612.$$

Referring to trigonometric tables, we find that θ is approximately $52°$. Thus the gradient vector at $(\frac{2}{3}, 1)$ points downward and to the left at a $45°$ angle, and we must proceed in a direction making an angle of $52°$ with the gradient (Fig. 6.8b).

Remark Equation (6.8) and its consequences listed above are of basic importance in working with functions of more than one variable. They describe how to determine all directional derivatives from a knowledge of the gradient vector. Equally important is the observation that when the above facts are viewed from the opposite direction, they allow us to describe the gradient vector in terms of the totality of directional derivatives.

Theorem 6.2 *If $f(x, y)$ is continuously differentiable in a domain D, then the gradient vector of f at a point (x_0, y_0) in D is the vector whose magnitude equals the maximum directional derivative of f at (x_0, y_0) and whose direction is the direction in which this maximum occurs.*

NOTE This description of the gradient vector may be expressed more compactly as

$$\nabla f(x_0, y_0) = M\mathbf{T}_{\alpha_0},$$

where

$$M = |\nabla f(x_0, y_0)| = \nabla_{\alpha_0} f(x_0, y_0) = \max_{0 \le \alpha \le 2\pi} \nabla_\alpha f(x_0, y_0).$$

PROOF. (See also Ex. 6.8 below.) This is essentially a restatement of the facts listed after Eq. (6.8). If all directional derivatives at (x_0, y_0) are zero, then the gradient of f is the zero vector. If some directional derivative is different from zero, then by Eq. (6.8)

$$|\nabla f(x_0, y_0)| = \max_{0 \le \alpha \le 2\pi} \nabla_\alpha f(x_0, y_0) \ne 0.$$

Furthermore, there is a *unique* angle α_0 such that $\nabla_{\alpha_0} f(x_0, y_0) = |\nabla f(x_0, y_0)|$; namely, α_0 must be chosen so that $\cos \theta = 1$ in Eq. (6.8). In other words, the angle θ between $\nabla f(x_0, y_0)$ and $\mathbf{T}_{\alpha_0}$ must be zero. This means that $\mathbf{T}_{\alpha_0}$ is the unit vector in the direction of $\nabla f(x_0, y_0)$, or that

$$\mathbf{T}_{\alpha_0} = \frac{\nabla f(x_0, y_0)}{|\nabla f(x_0, y_0)|}.$$

This proves the theorem. ◆

We conclude this section with a brief discussion of maxima and minima for functions of two variables. We shall return to this subject in Sect. 12.

Definition 6.3   A function $f(x, y)$ defined in a domain D is said to have a *local maximum* (or *relative maximum*) at a point (x_0, y_0) in D if there is some circle about (x_0, y_0) such that $f(x, y) \le f(x_0, y_0)$ for (x, y) inside that circle. Similarly for a *local minimum* (or *relative minimum*).

Definition 6.4   A function $f(x, y)$ defined in a domain D is said to have an *absolute maximum* (or to *attain its maximum*) at a point (x_0, y_0) in D if $f(x, y) \le f(x_0, y_0)$ for all points (x, y) in D. Similarly, for an *absolute minimum*.

Note that a local maximum of $f(x, y)$ may correspond to a point on the surface $z = f(x, y)$ that looks like a mountain peak, a point on a ridge, or a point on a plateau (Fig. 6.9).

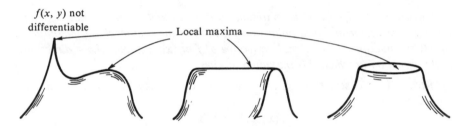

$f(x, y)$ not differentiable — Local maxima —

FIGURE 6.9   Different types of local maxima

Lemma 6.1   *If $f(x, y)$ has a local maximum or minimum at a point (x_0, y_0) in a domain D, and if the gradient of f exists at (x_0, y_0), then*

$$(6.9) \qquad\qquad \nabla f(x_0, y_0) = \mathbf{0}.$$

PROOF.   If $f(x, y)$ has a local maximum at (x_0, y_0), then the function $f(x, y_0)$ of the single variable x has a local maximum at $x = x_0$, and hence if its derivative exists, it must vanish; that is, $f_x(x_0, y_0) = 0$. Similarly, $f(x_0, y)$ has a local maximum at $y = y_0$, and hence $f_y(x_0, y_0) = 0$. Thus Eq. (6.9) holds. The same argument applies if $f(x, y)$ has a local minimum at (x_0, y_0). ◆

Example 6.6

At which points does the function

$$f(x, y) = x^2 + y^2 + 2x - 4y$$

have a local maximum or minimum?

We have in this case

$$f_x = 2x + 2 = 2(x + 1), \qquad f_y = 2y - 4 = 2(y - 2).$$

Hence

$$f_x = 0 \Leftrightarrow x = -1, \qquad f_y = 0 \Leftrightarrow y = 2.$$

Thus the gradient exists for all x, y, and it vanishes only at the point $(-1, 2)$. This point is therefore the only possible local maximum or minimum.

We may note that in this case we can determine purely algebraically the behavior of $f(x, y)$ near $(-1, 2)$. Namely, by completing the squares:

$$x^2 + 2x = x^2 + 2x + 1 - 1 = (x + 1)^2 - 1, \qquad y^2 - 4y = (y - 2)^2 - 4,$$

we may write

$$f(x, y) = (x + 1)^2 + (y - 2)^2 - 5,$$

and this function clearly has a local minimum (in fact an absolute minimum) at $(-1, 2)$.

Example 6.7

At which points does the function $f(x, y) = \cos(xe^y)$ have a local maximum or minimum?

Here we have

$$f_x = -e^y \sin(xe^y), \qquad f_y = -xe^y \sin(xe^y).$$

Since e^y never vanishes, we have

$$f_x = 0 \Leftrightarrow \sin(xe^y) = 0 \Leftrightarrow xe^y = n\pi,$$

where n is an arbitrary integer. Thus the gradient can only vanish at points where $xe^y = n\pi$, and it does in fact vanish at all these points, since f_y vanishes there too. When $n = 0$, we obtain the entire y axis or $x = 0$. For every other integer n we obtain a curve $x = n\pi e^{-y}$. Each of these curves corresponds to a "ridge" on the surface $z = f(x, y)$. Note that along each curve $xe^y = n\pi$,

$$f(x, y) = \cos(xe^y) = \cos n\pi = \pm 1,$$

the sign being positive or negative depending on whether n is even or odd. Since $|f(x, y)| \leq 1$ for all x, y, it follows that for n even, all points on the curve $x = n\pi e^{-y}$ are local maxima; while for n odd, all points are local minima. (See the sketch of this surface in Fig. 3.7.)

In both of these examples we could tell whether a given point where the gradient vanished was a local maximum or minimum by using special properties of the functions considered. In general it may be difficult to decide if a point where the gradient vanishes is a local maximum, a local minimum, or neither. One method for doing this will be given in Th. 12.2.

Exercises

6.1   For each of the following functions, find the gradient vector at the points indicated.

  a. $f(x, y) = 3x - 7y + 2$ at $(1, -5)$
  b. $f(x, y) = 5x^3 - 6x^2y + 7y^2$ at $(0, 0)$
  c. $f(x, y) = 5x^3 - 6x^2y + 7y^2 + 3x - 7y + 2$ at $(0, 0)$
  d. $f(x, y) = x^5 + e^{\sin x}$ at $(0, \pi)$

6.2   For each of the following functions $f(x, y)$ find the directional derivative $\nabla_\alpha f$ at the point indicated and in the direction prescribed.

  a. $f(x, y) = \sinh xy$ at $(2, 0)$; $\alpha = \frac{1}{6}\pi$
  b. $f(x, y) = \arctan (y/x)$ at $(1, 2)$; $\alpha = \frac{1}{4}\pi$
  c. $f(x, y) = (x^2 - y^2)/(x^2 + y^2)$ at $(1, 1)$; $\alpha = \frac{1}{2}\pi$
  d. $f(x, y) = (x^2 + y^2)^2$ at $(\frac{1}{2}, -\frac{1}{2})$; $\alpha = \frac{2}{3}\pi$

6.3   For each of the following functions, find the gradient at the point $(1, 1)$, and then compute the directional derivatives at that point in each of the directions $\alpha = 0, \frac{1}{4}\pi, \frac{1}{2}\pi, \frac{3}{4}\pi, \pi, \frac{5}{4}\pi, \frac{3}{2}\pi, \frac{7}{4}\pi$.

  a. $f(x, y) = x^2 + (y - 1)^2$ **c.** $f(x, y) = \log (x^2 + y^2)$
  b. $f(x, y) = (1/x) \log (1/y)$ **d.** $f(x, y) = e^{y/x}$

6.4   For each of the following functions, find the maximal directional derivative at the point indicated, and find the direction for which this maximum is attained.

  a. $f(x, y) = x \sin y$ at $(1, 0)$
  b. $f(x, y) = x \sin y$ at $(1, \frac{1}{2}\pi)$
  c. $f(x, y) = \sqrt{x^2 + y^2}$ at $(3, 4)$
  d. $f(x, y) = \log (y/x)$ at $(3, 4)$

6.5   Let $F(t)$ be a continuously differentiable function of t, $F'(t) \neq 0$, and let $f(x, y) = F(x^2 + y^2)$. Find ∇f at any arbitrary point, and show that it is directed along the line joining that point to the origin.

6.6   Let $G(t)$ be a continuously differentiable function of t, $G'(t) \neq 0$, and let $g(x, y) = G(y/x)$. Find ∇g at any point where $x \neq 0$ and show that it is perpendicular to the line joining that point to the origin.

6.7   Let $f(x, y)$ and $g(x, y)$ be continuously differentiable in a domain D, and suppose that they satisfy the equations $f_x = g_y$, $f_y = -g_x$ at each point of D. (These equations are called the *Cauchy-Riemann equations*. Examples of such pairs of functions are given in Ex. 4.4.) Show that at any point (x_0, y_0) of D, the equation

$$\nabla_\alpha f(x_0, y_0) = \nabla_{\alpha + \pi/2} g(x_0, y_0)$$

holds for all α.

6.8   Let $g(\alpha) = A \cos \alpha + B \sin \alpha$, where A and B are constants, not both zero.

  a. Show that the derivative $g'(\alpha)$ vanishes for those values of α for which

$$\langle \cos \alpha, \sin \alpha \rangle = \pm \frac{\langle A, B \rangle}{\sqrt{A^2 + B^2}}.$$

b. Show that the maximum value of $g(\alpha)$ is $\sqrt{A^2 + B^2}$ and is attained for the unique value of α satisfying

$$\cos \alpha = \frac{A}{\sqrt{A^2 + B^2}}, \qquad \sin \alpha = \frac{B}{\sqrt{A^2 + B^2}}.$$

c. Use parts a and b in conjunction with Th. 6.1 to give an alternative proof of Th. 6.2.

6.9   Let $f(x, y)$ be continuously differentiable, and suppose that the maximal directional derivative of f at the origin is equal to 5 and is attained in the direction of the vector from the origin toward the point $(-1, 2)$. Find $\nabla f(0, 0)$.

6.10   Let $f(x, y)$ be continuously differentiable, and let $z = L(x, y)$ be the equation of the tangent plane to the surface $z = f(x, y)$ at (x_0, y_0). Show that

$$\nabla_\alpha L(x_0, y_0) = \nabla_\alpha f(x_0, y_0)$$

for all α.

6.11   If $f(x, y)$ is continuously differentiable at (x_0, y_0), show that for any α and β the following are true.

a. $\nabla_{\alpha+\beta} f(x_0, y_0) = \cos \beta \, \nabla_\alpha f(x_0, y_0) + \sin \beta \, \nabla_{\alpha+\pi/2} f(x_0, y_0)$

b. $|\nabla f(x_0, y_0)| = \sqrt{(\nabla_\alpha f(x_0, y_0))^2 + (\nabla_{\alpha+\pi/2} f(x_0, y_0))^2}$

6.12   Using Lemma 6.1, determine the points where the following functions may have a maximum or minimum. Then try to verify directly whether the points in question are local maxima or minima.

a. $f(x, y) = 1 - x^2 - 2y^2 + x + 4y$
b. $f(x, y) = (\cos x - y)^2$
c. $f(x, y) = \log \log (x^2 + y^2 + 2)$
d. $f(x, y) = e^{1 - x^2 - y^2}$

6.13   Let $f(x, y)$ be a general quadratic polynomial

$$f(x, y) = Ax^2 + 2Bxy + Cy^2 + 2Dx + 2Ey + F.$$

Show that if $f(x, y)$ has a local maximum or minimum at a point, then the coordinates (x, y) of that point must satisfy the simultaneous linear equations

$$Ax + By + D = 0,$$
$$Bx + Cy + E = 0.$$

6.14   Applying Ex. 6.13 to each of the following quadratic polynomials, find the point or points where a maximum or minimum may occur.

a. $x^2 + 4xy + 5y^2 - 2y + 1$
b. $8x^2 + 2xy - y^2 - 14x + 5y$
c. $4x^2 - 12xy + 9y^2 + 4x - 6y + 3$
d. $-5x^2 + 6xy - 2y^2 - 10x + 6y + 2$

6.15 *a.* Show that if the coefficients of $f(x, y)$ in Ex. 6.13 satisfy the condition $AC - B^2 \neq 0$, then there is at most one point at which $f(x, y)$ can have a local maximum or minimum.

b. Find the coordinates of that point in terms of the coefficients of $f(x, y)$.

c. Test the functions in Ex. 6.14 to see which satisfy the condition of part a. (Note that the coefficient of the xy term was denoted by $2B$ rather than B.)

d. Show that the function $f(x, y) = (ax + by + c)^2$ does *not* satisfy the condition of part a, and that this function has a local minimum at more than one point.

6.16 A situation that arises in many different contexts is the following. Two observable quantities x and y are known to be connected approximately by a linear relation of the form $y = ax + b$. The values of the coefficients a and b are not known, and the problem is to determine them by a number of observations. Using different values $x_1, \ldots, x_n$, one obtains a series of values $y_1, \ldots, y_n$. If there exist constants a, b such that $y_k = ax_k + b$ holds exactly for each $k = 1, \ldots, n$, the problem is solved. In general, for any choice of a and b there are certain "residuals" $r_k = ax_k + b - y_k$. The problem is to find the most favorable choice of a and b; geometrically, the problem is to find the straight line that "most nearly" passes through the points (x_n, y_n). From many points of view the optimal choice is that provided by the *method of least squares*. This method consists of choosing a and b by minimizing the sums of the squares of the residuals.

a. Let

$$f(a, b) = r_1^2 + \cdots + r_n^2 = (ax_1 + b - y_1)^2 + \cdots + (ax_n + b - y_n)^2.$$

Show that $f(a, b)$ is a quadratic polynomial in the unknowns a and b.

b. Use Ex. 6.13 to show that the values of a and b obtained by the method of least squares must satisfy a pair of simultaneous linear equations whose coefficients depend on $x_1, \ldots, x_n$ and $y_1, \ldots, y_n$.

c. Show that if a and b are chosen by the method of least squares, then the positive and negative residuals cancel out in the sense that the sum of the residuals equals zero. (*Hint:* apply Lemma 6.1 to the function $f(a, b)$.)

6.17 Apply the method of least squares to find the best linear relation $y = ax + b$ corresponding to the pairs of observed values $(x_1, y_1) = (1, 5)$, $(x_2, y_2) = (2, 6)$, $(x_3, y_3) = (3, 8)$. After finding a and b, compute the residuals and show that they satisfy the condition in Ex. 6.16c.

6.18 Let $(x_1, y_1), \ldots, (x_n, y_n)$ be any n fixed points in the plane. For any point (x, y), let

$$f(x, y) = \sum_{k=1}^{n} [(x - x_k)^2 + (y - y_k)^2]$$

be the sum of the squares of the distances from (x, y) to the given points. Show that if $f(x, y)$ has a minimum at (x_0, y_0), then

$$x_0 = \frac{x_1 + \cdots + x_n}{n}, \qquad y_0 = \frac{y_1 + \cdots + y_n}{n}.$$

6.19   In most problems encountered in applications it is impossible to find the maximum or minimum of a function $f(x, y)$ by purely analytic methods. One of the numerical procedures that may be used is the *gradient method* or *method of steepest descent*. This is based on the fact that the function increases most rapidly in the direction of the gradient vector. Suppose for example, that we are trying to minimize the function $f(x, y)$. Starting at any point (x_1, y_1), choose a positive constant h and let

$$x_2 = x_1 - hf_x(x_1, y_1), \qquad y_2 = y_1 - hf_y(x_1, y_1).$$

Since the displacement vector

$$\langle x_2 - x_1, y_2 - y_1 \rangle = -h\langle f_x(x_1, y_1), f_y(x_1, y_1)\rangle$$

is in the *opposite direction* from the gradient vector, the function $f(x, y)$ will have a *smaller* value at (x_2, y_2) than at (x_1, y_1), provided that h is sufficiently small. Repeating the process, let

$$x_3 = x_2 - hf_x(x_2, y_2), \qquad y_3 = -y_2 - hf_y(x_2, y_2).$$

By continuing in this manner, we obtain a sequence of points (x_k, y_k), and it can be shown in many cases that when h is sufficiently small, this sequence will converge to a point where $f(x, y)$ has a minimum. (See Sects. 4.41 and 6.45 of [34] for further discussion, and for variations on this method.)

Carry out several steps of this procedure for the function

$$f(x, y) = 2x^2 - 2xy + y^2 - 2y + 2,$$

starting at the origin $(x_1, y_1) = (0, 0)$, and using the value $h = \frac{1}{4}$. Note how the points obtained gradually approach the actual minimum, which can be obtained analytically in this case by the method of Ex. 6.13.

7   The chain rule

The following situation arises in many different guises. We have a function $f(x, y)$ defined in a domain D, and x and y are themselves functions of a third variable t. Our problem is, how does f behave as a function of t? This is often referred to as the problem of "composite functions," or in certain contexts, of "related rates." One way to visualize the situation, is to consider the functions $x(t)$, $y(t)$ as defining a curve C in D, and to interpret the problem as one of describing the behavior of f along this curve.

In most cases we are concerned with regular curves. There are, however, situations in which it is desirable to allow more freedom. The word *curve*, without qualifications, is used whenever the defining functions $x(t)$, $y(t)$ are continuous. A *differentiable curve* is a curve for which $x(t)$ and $y(t)$ are continuously differentiable.

We adopt the following notation. Let $f(x, y)$ be defined in D. Let

$$C: x(t), y(t), \qquad a \le t \le b$$

be a curve in D. Then we set

(7.1) $$f_C(t) = f(x(t), y(t)), \qquad a \le t \le b.$$

Example 7.1

$$f(x, y) = x^2 + 2xy,$$

and

$$C: x(t) = \cos t, y(t) = \sin t,$$

then

$$f_C(t) = \cos^2 t + 2 \sin t \cos t.$$

Thus $f_C(t)$ is simply the function of t obtained by substituting for x and y their expressions in terms of t. It is a function of a single variable, and we may ask when it is differentiable, and what is the value of its derivative.

Theorem 7.1 *If $f(x, y)$ is continuously differentiable in D, and if C is a differentiable curve in D defined by $x(t), y(t), a \le t \le b$, then $f_C(t)$ is differentiable, and we have*

(7.2) $$f'_C(t_0) = f_x(x_0, y_0)x'(t_0) + f_y(x_0, y_0)y'(t_0),$$

where $x_0 = x(t_0), y_0 = y(t_0)$.

Remark Equation (7.2) is known as the *chain rule*. It is often written more briefly in the form

$$\frac{df}{dt} = \frac{\partial f}{\partial x}\frac{dx}{dt} + \frac{\partial f}{\partial y}\frac{dy}{dt}.$$

Here it is understood that on the left-hand side we have substituted in f the expressions for x and y as functions of t, and that each factor on the right-hand side is evaluated at a particular point of the t axis or the corresponding point in the x, y plane.

PROOF.   We must again use the fundamental Lemma 5.1. Setting $x = x(t)$, $y = y(t)$ and $x_0 = x(t_0), y_0 = y(t_0)$, we have from Eq. (5.3)

$$\frac{f_C(t) - f_C(t_0)}{t - t_0} = \frac{f(x, y) - f(x_0, y_0)}{t - t_0}$$

$$= \frac{x - x_0}{t - t_0} f_x(x_0, y_0) + \frac{y - y_0}{t - t_0} f_y(x_0, y_0) + \frac{h(x, y)}{t - t_0}.$$

Then as $t \to t_0$, $(x - x_0)/(t - t_0) \to x'(t_0)$, $(y - y_0)/(t - t_0) \to y'(t_0)$, and to prove Eq. (7.2) we must only show that $h(x, y)/(t - t_0) \to 0$. But

$$\frac{h(x, y)}{t - t_0} = \frac{h(x, y)}{\sqrt{(x - x_0)^2 + (y - y_0)^2}} \cdot \frac{\sqrt{(x - x_0)^2 + (y - y_0)^2}}{t - t_0}$$

$$= \frac{h(x, y)}{d(x, y)} \cdot \sqrt{\left(\frac{x - x_0}{t - t_0}\right)^2 + \left(\frac{y - y_0}{t - t_0}\right)^2},$$

the second factor on the right-hand side tends to the limit $\sqrt{x'(t_0)^2 + y'(t_0)^2}$, whereas the first factor, by Lemma 5.1 tends to zero.   ◆

Example 7.2

In Example 7.1 we have

$$f(x, y) = x^2 + 2xy, \qquad f_x = 2x + 2y, \qquad f_y = 2x.$$

Further

$$x'(t) = -\sin t, \qquad y'(t) = \cos t.$$

Thus, if $t_0 = \tfrac{1}{4}\pi$, we have $x_0 = \sqrt{2}/2$, $y_0 = \sqrt{2}/2$,

$$f_x(x_0, y_0) = 2\sqrt{2}, \qquad f_y(x_0, y_0) = \sqrt{2}.$$

Hence,

$$f'_C(\tfrac{1}{4}\pi) = 2\sqrt{2} \cdot \frac{-\sqrt{2}}{2} + \sqrt{2}\frac{\sqrt{2}}{2} = -1.$$

In this case we could also calculate $f'_C(t)$ directly, obtaining

$$f'_C(t) = -2 \cos t \sin t + 2 \cos^2 t - 2 \sin^2 t.$$

Setting $t = \tfrac{1}{4}\pi$, we find $f'_C(\tfrac{1}{4}\pi) = -1$.

Before we go on to applications of the chain rule, we note some important special cases.

First, consider the case where the curve C is a straight line, and the parameter t is the distance from (x_0, y_0) a variable point on the line. Then C is given by $x = x_0 + t \cos \alpha$, $y = y_0 + t \sin \alpha$, where $\langle \cos \alpha, \sin \alpha \rangle$ is a unit vector in the direction of the line. Then $x'(t) = \cos \alpha$, $y'(t) = \sin \alpha$, and

$$f'_C(0) = f_x(x_0, y_0) \cos \alpha + f_y(x_0, y_0) \sin \alpha = \nabla f_\alpha(x_0, y_0).$$

Thus the directional derivative appears as a special case of (7.2) in which the curve C is a straight line.

Consider next the case where C is an arbitrary regular curve, but the parameter t represents arc length along C. Then $\langle x'(t_0), y'(t_0) \rangle$ represents a

unit vector tangent to C. We have denoted this vector earlier by **T**. If we set
T $= \langle \cos \alpha, \sin \alpha \rangle$, then Eq. (7.2) may be written

(7.3) $$f_C'(t_0) = \nabla f(x_0, y_0) \cdot \mathbf{T} = \nabla_\alpha f(x_0, y_0),$$

where t is arc length parameter on C.

We may describe Eq. (7.3) as follows. *The rate of change of f with respect to arc length along any regular curve C through a point (x_0, y_0), depends only on the unit tangent* **T** *to C at (x_0, y_0) and equals the directional derivative of f in the direction of* **T***.*

Example 7.3

Find the directional derivative $\nabla_\alpha f(1, 1)$, where $f(x, y) = x^2 + y^2$ and $\alpha = \frac{3}{4}\pi$.

We note that the vector **T** $= \langle \cos \alpha, \sin \alpha \rangle = \langle -\sqrt{2}/2, \sqrt{2}/2 \rangle$ is tangent to the circle $x^2 + y^2 = 2$ at the point $(1, 1)$. But this circle is a regular curve $C: x = \sqrt{2} \cos t, y = \sqrt{2} \sin t$, along which the function f is constant (Fig. 7.1). Namely, $f_C(t) \equiv 2$. Hence $f_C'(t) \equiv 0$, and by Eq. (7.3), $\nabla_\alpha f(1, 1) = 0$.

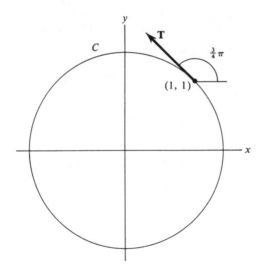

FIGURE 7.1 Directional derivative in tangent direction to a level curve

Example 7.3 is a special case of the situation in which we apply Eq. (7.3) to a curve C which is a level curve of the function $f(x, y)$. We discuss the general case in detail in Sect. 8.

By a *pass* through a mountain range we mean a path across the mountains, whose highest point is as low as possible. The highest point of the path is called the *summit* of the pass. We use the following precise characterization of such a point.

Definition 7.1   The *summit of a pass* is a point having the property that there are two paths through the point in different directions, such that it is the highest point along one of the paths and the lowest point along the other (Fig. 7.2).

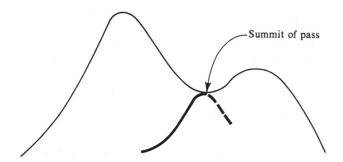

—Summit of pass

FIGURE 7.2   Summit of a pass

As another application of Eq. (7.3), we can state the following:

At a summit of a pass, the gradient is zero.

Proof.   We assume, as usual, that the mountain range is defined by $z = f(x, y)$, and we let (x_0, y_0) correspond to the summit of the pass. Let C be one of the two paths through the summit referred to in Def. 7.1. Then $f_C(t)$ has a local maximum or minimum as a function of t at the summit, and hence $f_C'(t) = 0$ at this point. By Eq. (7.3) the directional derivative of f is zero in the direction of each of these paths. But then the gradient of f must be zero, since if it were different from zero, the directional derivative could be zero only in the direction perpendicular to the gradient vector, as we observed in consequence of Eq. (6.8).   ♦

The following are two important applications of the chain rule.

Theorem 7.2   *Let $f(x, y)$ be continuously differentiable in a domain D. Then f is constant if and only if $\nabla f \equiv 0$ in D.*

PROOF.   If f is constant, then $f_x \equiv 0$, $f_y \equiv 0$, and hence $\nabla f \equiv 0$ in D. To prove the converse, assume $\nabla f \equiv 0$, and let (x_1, y_1) be any point of D. Let $k = f(x_1, y_1)$. We shall show that $f(x, y) \equiv k$ in D. For any point (x_2, y_2) in D, let C be a regular curve in D joining (x_1, y_1) to (x_2, y_2); let C be given by $x(t)$, $y(t)$, $a \le t \le b$. Then the mean-value theorem applied to the function $f_C(t)$ states that

(7.4)
$$f_C(b) - f_C(a) = (b - a)f_C'(t_0),$$

where t_0 is some value of t between a and b. If we let $x_0 = x(t_0)$, $y_0 = y(t_0)$, then substituting the definition of $f_C(t)$ in Eq. (7.4), and applying Eq. (7.2), we have

$$f(x_2, y_2) - f(x_1, y_1) = (b - a)[f_x(x_0, y_0)x'(t_0) + f_y(x_0, y_0)y'(t_0)] = 0,$$

since by assumption $f_x \equiv 0$ and $f_y \equiv 0$. Hence $f(x_2, y_2) = f(x_1, y_1) = k$, and since (x_2, y_2) is an arbitrary point of D, we have $f(x, y) = k$ for all (x, y) in D. ◆

Corollary *If $\nabla g \equiv \nabla h$ in D, then $g(x, y) = h(x, y) + k$, where k is a constant.*

PROOF. Let $f(x, y) = g(x, y) - h(x, y)$. Then $\nabla f \equiv 0$ in D, and applying Th. 7.2, $f(x, y) \equiv k$. ◆

Remark Theorem 7.2 and its Corollary would be false if D were not connected. We made important use of the fact that any two points in D could be joined by a curve in D.

Example 7.4

Let $ax + by + c = 0$ be the equation of a fixed line in the plane. Find the distance to this line from an arbitrary point (X, Y).

Let us denote the distance in question by $f(X, Y)$. If we consider the distance to be a function of the variables X, Y, then we can determine the gradient of this function by our geometric characterization of the gradient vector in terms of the direction and magnitude of the greatest change in the function (Theorem 6.2). Starting at any point, the rate of change of distance to the fixed line is greatest if we move directly away from the line. Since the vector $\langle a, b \rangle$ is perpendicular to the line $ax + by + c = 0$, the direction of ∇f must be the same as (or opposite to) that of $\langle a, b \rangle$. Furthermore, if a point moves directly away from the line, its distance to the line increases by exactly the distance it traverses, so that the rate of change of $f(X, Y)$ in this direction is equal to 1. Hence $|\nabla f| = 1$, and we have

$$\nabla f = \pm \frac{\langle a, b \rangle}{\sqrt{a^2 + b^2}}.$$

Now if

$$g(X, Y) = \frac{1}{\sqrt{a^2 + b^2}} (aX + bY),$$

then

$$g_X = \frac{a}{\sqrt{a^2 + b^2}}, \qquad g_Y = \frac{b}{\sqrt{a^2 + b^2}}, \qquad \text{and} \qquad \nabla g = \frac{\langle a, b \rangle}{\sqrt{a^2 + b^2}}.$$

Applying the above Corollary we have

$$f(X, Y) = \pm g(X, Y) + k = \pm \frac{aX + bY}{\sqrt{a^2 + b^2}} + k.$$

To find the value of k, we note that for any point (X_0, Y_0) on the given line, we have $aX_0 + bY_0 + c = 0$, and $f(X_0, Y_0) = 0$. Hence,

$$0 = f(X_0, Y_0) = \pm \frac{aX_0 + bY_0}{\sqrt{a^2 + b^2}} + k = \pm \frac{-c}{\sqrt{a^2 + b^2}} + k,$$

and

$$k = \pm \frac{c}{\sqrt{a^2 + b^2}}.$$

Thus,

$$f(X, Y) = \pm \frac{aX + bY + c}{\sqrt{a^2 + b^2}},$$

and since $f(X, Y) \geq 0$ everywhere, we must have

$$f(X, Y) = \frac{|aX + bY + c|}{\sqrt{a^2 + b^2}}.$$

This is precisely the formula for the distance from the point (X, Y) to the line $ax + by + c = 0$.

Theorem 7.3   Mean-Value Theorem   *Let $f(x, y)$ be continuously differentiable in a domain D, and let (x_1, y_1) and (x_2, y_2) be points of D such that the entire line segment L from (x_1, y_1) to (x_2, y_2) lies in D (Fig. 7.3). Then there is a point (x_0, y_0) on L such that*

(7.5) $f(x_2, y_2) - f(x_1, y_1) = (x_2 - x_1)f_x(x_0, y_0) + (y_2 - y_1)f_y(x_0, y_0).$

PROOF.   We consider the line segment L to be a regular curve

$$C: x = x_1 + t(x_2 - x_1), \; y = y_1 + t(y_2, y_1), \qquad 0 \leq t \leq 1.$$

Applying the mean-value theorem to the function $f_C(t)$, we obtain

(7.6) $$f_C(1) - f_C(0) = f_C'(t_0), \qquad\qquad 0 < t_0 < 1.$$

If we let $x_0 = x(t_0)$, $y_0 = y(t_0)$, and if we note that $x'(t) = x_2 - x_1$, $y'(t) = y_2 - y_1$, then applying Eq. (7.2), Eq. (7.6) takes the form

$$f(x_2, y_2) - f(x_1, y_1) = f_x(x_0, y_0)x'(t_0) + f_y(x_0, y_0)y'(t)$$
$$= f_x(x_0, y_0)(x_2 - x_1) + f_y(y_0, y_0)(y_2 - y_1). \qquad \blacklozenge$$

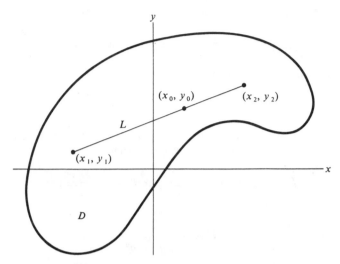

FIGURE 7.3 Mean-value theorem for a function of two variables

Remark It is instructive to compare the statement of the mean-value theorem, Eq. (7.5), with the fundamental Lemma 5.1, Eq. (5.3). Both express the change in a function f as we move from one point to another, essentially as the change in x times the x derivative of f plus the change in y times the y derivative of f. In Lemma 5.1, the partial derivatives f_x and f_y are evaluated at (x_0, y_0), and a remainder term $h(x, y)$ must be included. In the mean-value theorem, we obtain a precise expression with no remainder term, but f_x and f_y are evaluated at an unknown point, about which we can say only that it lies between the two given points.

Example 7.5

Find approximately the distance of the point $(4.2, 3.1)$ to the origin.

We consider the function $f(x, y) = (x^2 + y^2)^{1/2}$, representing the distance of any point (x, y) to the origin. Let $(x_1, y_1) = (4, 3)$, and $(x_2, y_2) = (4.2, 3.1)$. We have $f_x = x/(x^2 + y^2)^{1/2}$ and $f_y = y/(x^2 + y^2)^{1/2}$. We apply the fundamental Lemma 5.1 and obtain

$$f(4.2, 3.1) = f(x_1, y_1) + (x_2 - x_1)f_x(x_1, y_1) + (y_2 - y_1)f_y(x_1, y_1) + h(x_2, y_2)$$
$$= 5 + 0.2(\tfrac{4}{5}) + 0.1(\tfrac{3}{5}) + h(x_2, y_2)$$
$$= 5.22 + h(x_2, y_2).$$

Since $h(x, y)$ is a remainder term, we have an approximate value,

$$\sqrt{(4.2)^2 + (3.1)^2} \sim 5.22.$$

This method is excellent for obtaining a quick estimate for a function near a point where we know its value. The disadvantage of this method is that it gives no indication of the accuracy of the estimate. One way to test the accuracy is to use Th. 7.3. We have a point (x_0, y_0) on the line segment from $(4, 3)$ to $(4.2, 3.1)$. Thus, $4 < x_0 < 4.2$, and $3 < y_0 < 3.1$. Hence, $(x_0^2 + y_0^2)^{1/2} > 5$,

$$f_x(x_0, y_0) = \frac{x_0}{\sqrt{x_0^2 + y_0^2}} < \frac{4.2}{5} = 0.84, \qquad f_y(x_0, y_0) < \frac{3.1}{5} = 0.62.$$

$$f(4.2, 3.1) = f(4, 3) + 0.2f_x(x_0, y_0) + 0.1f_y(x_0, y_0)$$
$$< 5 + 0.168 + 0.062 = 5.23.$$

Thus,

(7.7) $$\sqrt{(4.2)^2 + (3.1)^2} < 5.23 = 5.22 + 0.01.$$

This gives an upper estimate, and we can use it in turn to obtain a lower estimate. We have

$$\frac{1}{\sqrt{x_0^2 + y_0^2}} > \frac{1}{\sqrt{(4.2)^2 + (3.1)^2}} > \frac{1}{5.23} > 0.19,$$

and, consequently,

$$f_x(x_0, y_0) = \frac{x_0}{\sqrt{x_0^2 + y_0^2}} > 4(0.19) = 0.76, \qquad f_y(x_0, y_0) > 3(0.19) = 0.57,$$

and

(7.8) $$f(4.2, 3.1) > 5 + 0.2(0.76) + 0.1(0.57) = 5.209.$$

Combining Eqs. (7.7) and (7.8), we have

(7.9) $$5.209 < \sqrt{(4.2)^2 + (3.1)^2} < 5.23,$$

or

$$\sqrt{(4.2)^2 + (3.1)^2} = 5.22 \pm 0.011.$$

Thus our original estimate was correct to within essentially one-hundredth. Equation (7.9) gives precise bounds.

We conclude this section with two rather special, but useful formulas, which may be derived from the chain rule. The first involves the notion of *homogeneity*, abstracted from the basic property of homogeneous polynomials.

Definition 7.2 A function $f(x, y)$ is *homogeneous of degree k* if

(7.10) $$f(\lambda x, \lambda y) = \lambda^k f(x, y) \qquad \text{for all } \lambda > 0.$$

Example 7.6

Consider a homogeneous polynomial $P(x, y)$ of degree k. Each of its terms is of the form $cx^m y^n$, where $m + n = k$. Hence,

$$c(\lambda x)^m (\lambda y)^n = c\lambda^{m+n} x^m y^n = \lambda^k (cx^m y^n),$$

and λ^k factors out from any sum of such terms. Thus $P(\lambda x, \lambda y) = \lambda^k P(x, y)$, and $P(x, y)$ satisfies the definition of a function homogeneous of degree k.

Example 7.7

$f(x, y) = 1/(x^2 + y^2)$. Then $f(\lambda x, \lambda y) = 1/(\lambda^2(x^2 + y^2)) = \lambda^{-2} f(x, y)$. Here $k = -2$.

Example 7.8

$f(x, y) = xy/(x^2 + y^2)$. Then $f(\lambda x, \lambda y) = f(x, y) = \lambda^0 f(x, y)$. Hence $k = 0$.

Example 7.9

$f(x, y) = \sqrt{x^3 - y^3}$. Here $f(\lambda x, \lambda y) = \lambda^{3/2} f(x, y)$, and $k = \frac{3}{2}$.

Note that, in general, k need not be a positive integer, as in the case of polynomials, but may have any real value, positive, negative, or zero.

Lemma 7.1 *Let $f(t)$ be defined and differentiable for all $t > 0$. Then $f(t) = ct^k$ for some constant $c \Leftrightarrow$*

$$(7.11) \qquad\qquad tf'(t) \equiv kf(t).$$

PROOF. Let $g(t) = f(t)/t^k$. Then

$$g'(t) = \frac{t^k f'(t) - kt^{k-1} f(t)}{t^{2k}} = \frac{tf'(t) - kf(t)}{t^{k+1}}.$$

Hence condition (7.11) holds $\Leftrightarrow g'(t) \equiv 0 \Leftrightarrow g(t) \equiv c$. ◆

Theorem 7.4 Euler's Theorem *A continuously differentiable function $f(x,y)$ is homogeneous of degree k if and only if it satisfies*

$$(7.12) \qquad\qquad xf_x(x, y) + yf_y(x, y) = kf(x, y).$$

PROOF. Let (x_0, y_0) be any fixed point, and let C be the curve $x = tx_0$, $y = ty_0$, $t > 0$. Then $x'(t) = x_0$, $y'(t) = y_0$, and by the chain rule, we have for the function $f_C(t) = f(tx_0, ty_0)$ that

$$f_C'(t) = f_x(tx_0, ty_0) \cdot x_0 + f_y(tx_0, ty_0) \cdot y_0,$$

and hence

$$tf_C'(t) = f_x(tx_0, ty_0) \cdot (tx_0) + f_y(tx_0, ty_0) \cdot (ty_0).$$

Thus, if we set $x = tx_0$, $y = ty_0$, then in our present notation, Eq. (7.12) takes the form

$$(7.13) \qquad\qquad tf_C'(t) = kf_C(t).$$

By Lemma 7.1, this is equivalent to $f_C(t) = ct^k$. The constant c is determined by setting $t = 1$. We find $c = f_C(1) = f(x_0, y_0)$. Thus

$$f_C(t) = f(tx_0, ty_0) = f(x_0, y_0)t^k$$

for all $t > 0$, which is precisely the condition (Eq. (7.10)) for homogeneity of degree k.
$\blacklozenge$

Remarks

1. Geometrically, the meaning of homogeneity is that the surface $z = f(x, y)$ consists of curves $z = cs^k$ over each ray in the x, y plane through the origin, where s is distance to the origin, and the coefficient c varies from ray to ray, but the exponent k is fixed (Fig. 7.4). Thus a homogeneous function of degree k is completely known if it is known at one point on each ray, since that is sufficient to determine the value of the coefficient c.

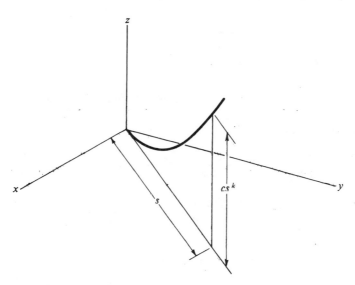

FIGURE 7.4   Geometric meaning of homogeneity

2. Although the domain of $f(x, y)$ was not explicitly stated, it is generally the whole plane with the possible exception of the origin. In some cases it is a sector, or a collection of sectors, with vertex at the origin. What is needed in the proof is to know that if f is defined at some point, then it is defined on the entire ray through that point.

We conclude with an application of the chain rule to the following situation. Let $f(x, u)$ be continuous in some square $a \leq x \leq b$, $a \leq u \leq b$. Define the function $g(x)$ for $a \leq x \leq b$ by

$$g(x) = \int_a^x f(x, u) \, du.$$

How do we find $g'(x)$?

This is a typical example of a problem that can be answered more easily if stated in an apparently more difficult form. Consider the function

(7.14) $$h(x, y) = \int_a^y f(x, u) \, du,$$

where y may be an arbitrary function of x. We start with the important special case when y is a constant.

Lemma 7.2 **Differentiation under the Integral Sign** *Let $f(x, u)$ and $f_x(x, u)$ be continuous for $x_0 - \delta \leq x \leq x_0 + \delta$, $a \leq u \leq b$. Let*

$$\varphi(x) = \int_a^b f(x, u) \, du.$$

Then $\varphi(x)$ is differentiable and

$$\varphi'(x_0) = \int_a^b f_x(x_0, u) \, du.$$

PROOF. We form

$$\frac{\varphi(x) - \varphi(x_0)}{x - x_0} = \frac{1}{x - x_0} \left[\int_a^b f(x, u) \, du - \int_a^b f(x_0, u) \, du \right]$$

$$= \int_a^b \frac{f(x, u) - f(x_0, u)}{x - x_0} \, du$$

$$= \int_a^b f_x(x_1, u) \, du,$$

where x_1 lies between x and x_0, by the mean-value theorem for functions of one variable. Thus, by the continuity of f_x, we have

$$\varphi'(x_0) = \lim_{x \to x_0} \frac{\varphi(x) - \varphi(x_0)}{x - x_0} = \lim_{x \to x_0} \int_a^b f_x(x_1, u) \, du$$

$$= \int_a^b \lim_{x \to x_0} f_x(x_1, u) \, du = \int_a^b f_x(x_0, u) \, du. \qquad \blacklozenge$$

Remark A close examination of the last step, in which we interchanged the order of integration and limit as $x \to x_0$, shows that one actually needs *uniform continuity*[2] of the function $f_x(x, u)$. However, that is a consequence of ordinary continuity on the closed rectangle $x_0 - \delta \le x \le x_0 + \delta$, $a \le u \le b$.

We return now to the function $h(x, y)$ defined in Eq. (7.14). If we fix x and let y vary, then by the fundamental theorem of calculus

$$h_y(x, y) = f(x, y).$$

On the other hand, if y is fixed and x varies, then by Lemma 7.2,

$$h_x(x, y) = \int_a^y f_x(x, u)\, du.$$

Thus, if f and f_x are both continuous, h is continuously differentiable, and we may apply the chain rule, where x and y are given as arbitrary differentiable functions of our auxiliary variable t:

$$\frac{d}{dt} h(x, y) = h_x(x, y) \frac{dx}{dt} + h_y(x, y) \frac{dy}{dt}$$

$$= \frac{dx}{dt} \int_a^y f_x(x, u)\, du + \frac{dy}{dt} f(x, y).$$

If we now specialize to the case where y is a given function of x, $y = F(x)$, then we may choose $x \equiv t$, and we have

(7.15) $\qquad \dfrac{d}{dx} \displaystyle\int_a^{F(x)} f(x, u)\, du = \int_a^{F(x)} f_x(x, u)\, du + F'(x) f(x, F(x)).$

Equation (7.15) contains both the fundamental theorem of calculus and Lemma 7.2 as special cases. It also provides the answer to our original question concerning the function $g(x)$ obtained by setting $F(x) \equiv x$. We find

$$\frac{d}{dx} \int_a^x f(x, u)\, du = \int_a^x f_x(x, u)\, du + f(x, x).$$

---

[2] A function $g(x, y)$ is called *uniformly continuous* on a set E, if for every $\epsilon > 0$ there exists a $\delta > 0$ such that for every pair of points (x_1, y_1), (x_2, y_2) in E whose distance apart is less than δ, $|g(x_1, y_1) - g(x_2, y_2)| < \epsilon$. The notion of uniform continuity, although important in certain connections, is not central to the subject matter of this book. A discussion of uniform continuity for functions of two variables may be found in [6], Section 2.3. Theorem 6 of that section is the result quoted in the above remark.

Exercises

7.1 For each of the following functions $f(x, y)$ and curves $C:x(t), y(t)$ find $f_C(t) = f(x(t), y(t))$.

    a. $f(x, y) = x^2 - y^2$; $x = 2 \cos t$, $y = 2 \sin t$
    b. $f(x, y) = xy$; $x = 2 \cos t$, $y = 2 \sin t$
    c. $f(x, y) = x^2 - y^2$; $x = 2 \cosh t$, $y = 2 \sinh t$
    d. $f(x, y) = e^{xy}$; $x = t$, $y = e^t$
    e. $f(x, y) = x^3 + 2x^2y + xy^2$; $x = 2t$, $y = 3t$
    f. $f(x, y) = \log(1 + x^2 + y^2) - 2 \arctan y$; $x = \log(1 + t^2)$, $y = e^t$

7.2 For each function and curve in Ex. 7.1, find the point on the curve corresponding to $t = 0$, find the gradient of the function $f(x, y)$ at that point, and compute $f_C'(0)$ by the chain rule. Then check your answer by differentiating $f_C(t)$ and setting $t = 0$.

7.3 Given the differentiation formulas for functions of one variable:

$$\frac{d}{dx} x^c = cx^{c-1}, \qquad \frac{d}{dx} c^x = c^x \log c, \qquad c \text{ constant,}$$

find a formula for $(d/dx)x^x$, by applying the chain rule to $f(x, y) = x^y$, where $x = t, y = t$.

Note: in order to compute $f_C'(0)$, as in Ex. 7.2, in some cases it is easier to find the function $f_C(t)$ explicitly and then differentiate, while in other cases it is simpler to compute the gradient of f at the point corresponding to $t = 0$ and then apply the chain rule. One example of the latter case is when the gradient turns out to be zero. There is also an important situation in which *only* the chain rule can be applied. Namely, if the curve C is not known explicitly, but if the velocity vector is known at a given point, then $f_C'(t)$ can be found at that point. This is illustrated in Exs. 7.4–7.6.

7.4 A rectangular sheet of rubber is being stretched in such a way that at a certain moment when its length is 10 inches and its width is 4 inches, its length is increasing at the rate of 2 inches per second and its width is decreasing at the rate of 1 inch per second. Is its area increasing or decreasing at that moment, and by how much?

7.5 The pressure P of a gas is related to the volume V and to the absolute temperature T by the relation $P = cT/V$, where c is a constant. If hot air is being pumped into a balloon so that at a given instant $V = 1200$, $T = 360$, $P = 30$, $dV/dt = 8$, and $dT/dt = 4$, what is dP/dt? Is the pressure increasing or decreasing?

7.6 On a part of a map, the height (in miles) above sea level at the point (x, y) is equal to $1 - x^2/64 - y^2/60$. If a car going 50 miles per hour is headed directly away from the origin as it passes the point $(4, 3)$, how fast is its height above sea level changing at that point? Is it increasing or decreasing?

7.7 Suppose that $f(x, y)$ is homogeneous of degree $k \neq 0$, and that $\nabla f \equiv 0$. Show that $f(x, y) \equiv 0$.

7.8 Determine which of the following functions are homogeneous, and for the homogeneous ones find the degree.

a. $5x^6y - x^2y^5$ *e.* $x^2y - 2x^{3/2}y^{3/2}$

b. $8x^7y + 7x^6y^2 + 3x^2y^5$ *f.* $x^{1/2}y - 2y^{3/2}e^{y/x}$

c. $\dfrac{x^2 - y}{x + y^2}$ *g.* $\dfrac{\sqrt{x} - \sqrt{y}}{x + y}$

d. $x^{1/3}y^{-1/3} \log \dfrac{(x + y)^2}{xy}$ *h.* $\dfrac{y^2}{x^2} + 2\dfrac{y}{x} \sin \dfrac{y}{x}$

7.9 Verify Euler's theorem for each of the homogeneous functions in Ex. 7.8.

7.10 Let $f(x, y)$ be homogeneous of degree k and let $g(x, y)$ be homogeneous of degree l. For each of the following functions $h(x, y)$, determine whether $h(x, y)$ is homogeneous, and if so, determine the degree.

a. $h(x, y) = f(x, y) + g(x, y)$ *d.* $h(x, y) = f(x, y)^3$

b. $h(x, y) = f(x, y)g(x, y)$ *e.* $h(x, y) = f(x^2, y^2)$

c. $h(x, y) = f(x, y)/g(x, y)$ *f.* $h(x, y) = f(g(x, y), g(x, y))$

7.11 Show that $f(x, y)$ is homogeneous of degree zero if and only if there exists a function $g(t)$ of one variable such that

$$f(x, y) = g(y/x).$$

7.12 Show that $f(x, y)$ is homogeneous of degree k if and only if there exists a function $g(t)$ of one variable such that

$$f(x, y) = x^k g(y/x).$$

7.13 Give a geometric interpretation of Ex. 7.11 in terms of the surface $z = f(x, y)$, where $f(x, y)$ is homogeneous of degree zero.

7.14 Derive Euler's Eq. (7.12) from Ex. 7.12.

7.15 Use Ex. 7.12 to show that if $f(x, y)$ is homogeneous of degree k, then f_x and f_y are homogeneous of degree $k - 1$.

7.16 For each of the following functions $f(x, y)$ find an approximate value at the point (x_2, y_2) by applying Lemma 5.1 at the point (x_1, y_1) and neglecting the remainder term.

a. $f(x, y) = x^2 + xy + y^2$, $(x_2, y_2) = (2.1, 3.2)$, $(x_1, y_1) = (2, 3)$

b. $f(x, y) = \sqrt[3]{x^2 + y^2}$, $(x_2, y_2) = (0.9, 0.1)$, $(x_1, y_1) = (1, 0)$

c. $f(x, y) = \sqrt{\dfrac{3 + x}{2 + y}}$, $(x_2, y_2) = (1.2, 2.4)$, $(x_1, y_1) = (1, 2)$

d. $f(x, y) = \log(x - \sin y)$, $(x_2, y_2) = (1.2, 0.3)$, $(x_1, y_1) = (1, 0)$

7.17 Use Th. 7.3 to obtain precise bounds in Ex. 7.16a.

7.18 Find $g'(x)$, where

a. $g(x) = \int_0^\pi \dfrac{\sin xt}{t}\, dt$ **d.** $g(x) = \int_\pi^{\pi/x} \dfrac{\cos xt}{t}\, dt$

b. $g(x) = \int_1^2 \dfrac{e^{xt}}{t}\, dt$ **e.** $g(x) = \int_1^x \dfrac{e^{-xt}}{t}\, dt$

c. $g(x) = \int_2^3 \dfrac{t^{x-1}}{\log t}\, dt$

7.19 **a.** Setting $g(x) = \int_0^\pi \cos xt\, dt$, we have $g'(x) = -\int_0^\pi t \sin xt\, dt$, and $g'(1) = -\int_0^\pi t \sin t\, dt$. Evaluate $g(x)$ explicitly, differentiate, and set $x = 1$ in order to compute $\int_0^\pi t \sin t\, dt$. Check your answer by integrating by parts.

    b. Use a similar procedure to evaluate $\int_0^1 te^{2t}\, dt$.

7.20 Show that if

$$g(x) = \int_{F(x)}^{G(x)} f(x, t)\, dt,$$

then

$$g'(x) = \int_{F(x)}^{G(x)} f_x(x, t)\, dt + G'(x)f(x, G(x)) - F'(x)f(x, F(x)).$$

(*Hint:* Write $g(x) = \int_a^{G(x)} f(x, t)\, dt - \int_a^{F(x)} f(x, t)\, dt$, for some constant a.)

7.21 Use Ex. 7.20 to find $g'(x)$, where

a. $g(x) = \int_{\sqrt{x}}^{x^2} \sin(t^2 - x^4)\, dt$

b. $g(x) = \int_x^{e^x} \log(x + t)\, dt$

7.22 Find the answer to Ex. 7.21b by computing the integral in order to obtain $g(x)$ explicitly, and then differentiating.

7.23 **a.** Show that the function $g(x) = \int_0^x (x - t)h(t)\, dt$ satisfies $g(0) = g'(0) = 0$, $g''(x) = h(x)$.

    b. Show that if

$$g_n(x) = \frac{1}{n!} \int_0^x (x - t)^n h(t)\, dt,$$

then

$$g_n(0) = 0, \qquad g_n'(x) = g_{n-1}(x),$$

and hence

$$g_n(0) = g_n'(0) = g_n''(0) = \cdots = g_n^{(n)}(0) = 0, \qquad g_n^{(n+1)}(x) = h(x).$$

Note: Ex. 7.23b can be used to derive *Taylor's formula* with integral remainder term:

$$f(x) = f(0) + f'(0)x + \frac{f''(0)}{2} x^2 + \cdots + \frac{f^{(n)}(0)}{n!} x^n + \frac{1}{n!} \int_0^x (x - t)^n f^{(n+1)}(t)\, dt,$$

since both sides of this equation have the same $(n + 1)$st derivative, while the first n derivatives agree at the origin.

7.24   *a.* Use Eq. (7.15) to show that if $g(x) = \int_a^{F(x)} h(u)\, du$, then

$$g'(x) = h(F(x))F'(x).$$

   b. If a curve is given in polar coordinates by $r = r(\theta)$, then the area A swept out by the radius vector for $a \leq \theta \leq b$ is given by

$$A = \frac{1}{2} \int_a^b r^2\, d\theta.$$

If r and θ are given parametrically as functions of t, and if $A(t)$ is the area swept out between a fixed t_0 and t, show that

$$\frac{dA}{dt} = \frac{1}{2} r^2 \frac{d\theta}{dt}.$$

7.25   Let a curve C be given by $x(t)$, $y(t)$.

   a. Let $r = \sqrt{x^2 + y^2}$, $\theta = \arctan(y/x)$. Use the chain rule to find dr/dt, $d\theta/dt$.

   b. Using Ex. 7.24b, show that

$$\frac{dA}{dt} = \frac{1}{2} \left(x \frac{dy}{dt} - y \frac{dx}{dt} \right).$$

   c. Show that dA/dt is constant if and only if the acceleration vector is a scalar multiple of the radius vector $\langle x, y \rangle$. (*Hint:* see Ex. 2.16c, d.)

Note: Kepler's first law of planetary motion states that each planet moves in an ellipse with the sun at one focus. *Kepler's second law* states that planetary motion is such that the radius vector from the sun to a planet sweeps out equal areas in equal times; in other words dA/dt is constant. Exercise 7.25 shows that Kepler's second law does not depend on any special properties of gravitational force, other than the fact that it is a *central force*, directed along the line from the planet to the sun.

7.26   Prove the following theorems.

   a. If $f(x, y)$ is continuously differentiable in the whole plane, and if $f_y \equiv 0$, then $f(x, y) = g(x)$, where g is a function of one variable. (*Hint:* let $g(x) = f(x, 0)$ and apply the mean value theorem to $f(x, y) - f(x, 0)$.)
   b. If $f(x, y)$ is continuously differentiable in the whole plane, and if $f_x \equiv 0$, then $f(x, y) = h(y)$, where h is a function of one variable.

7.27 Show that the theorems in Ex. 7.26 remain valid if $f(x, y)$ is continuously differentiable in a domain D consisting of the upper half-plane, the right half-plane, a rectangle, or a disk.

*7.28 Formulate a general condition on a domain D such that the first theorem in Ex. 7.26 will be valid for an arbitrary continuously differentiable function $f(x, y)$ in D.

*7.29 Show that the theorems in Ex. 7.26 are *not* valid in an arbitrary domain D, by verifying the details in the following example.

    a. Sketch the domain D defined by $x > 0$, $1 < x^2 + y^2 < 4$.

    b. Let $f(x, y)$ be defined in the domain D of part a by

$$f(x, y) = \begin{cases} 0 & \text{if } x \geq 1 \\ (x - 1)^2 & \text{if } x < 1, y > 0 \\ -(x - 1)^2 & \text{if } x < 1, y < 0. \end{cases}$$

Show that $f(x, y)$ is continuously differentiable in D, and $f_y \equiv 0$.

    c. Show that, for example, $f(\frac{1}{2}, \frac{3}{2}) = \frac{1}{4}$, $f(\frac{1}{2}, -\frac{3}{2}) = -\frac{1}{4}$, and hence $f(x, y)$ can not be written as $g(x)$, a function of x alone.

8 Level curves and the implicit function theorem

Let $f(x, y)$ be continuously differentiable in a domain D and let (x_0, y_0) be any point in D. The equation $f(x, y) = f(x_0, y_0)$ defines a level curve through the point (x_0, y_0). Let us assume for the moment that this level curve is the implicit form of a regular curve C, at least near (x_0, y_0). In other words, suppose we have a regular curve $C: x(t), y(t)$, passing through (x_0, y_0), such that $f(x, y)$ is constant on C. Using our earlier notation, this means

$$f(x(t), y(t)) = f_C(t) \equiv k$$

for some constant k. Then

$$f_C'(t) \equiv 0,$$

and by the chain rule, at the point t_0 corresponding to (x_0, y_0), we have

$$(8.1) \qquad f_x(x_0, y_0)x'(t_0) + f_y(x_0, y_0)y'(t_0) = 0.$$

This equation may be interpreted in various ways. First, if the unit tangent to C at (x_0, y_0) is $\mathbf{T}_\alpha = \langle \cos \alpha, \sin \alpha \rangle$, and if s is the parameter of arc length along C, we have

$$\frac{ds}{dt} = |\langle x'(t), y'(t) \rangle| \neq 0,$$

and

$$\langle x'(t_0), y'(t_0) \rangle = s'(t_0)\mathbf{T}_\alpha.$$

Thus we may write Eq. (8.1) as

(8.2) $\qquad s'(t_0)[f_x(x_0, y_0) \cos \alpha + f_y(x_0, y_0) \sin \alpha] = 0,$

and since $s'(t_0) \neq 0$, we have

(8.3) $\qquad \nabla f(x_0, y_0) \cdot \mathbf{T}_\alpha = 0.$

But the left-hand side of Eq. (8.3) is precisely $\nabla_\alpha f(x_0, y_0)$. We are led to the following theorem.

Theorem 8.1 *The directional derivative of a function at a point in the direction of a level curve through the point is always zero. The gradient of a function at a point is always orthogonal to the level curve through the point. (Both statements are made under the assumption that the level curve is a regular curve.)*

PROOF. Both statements are immediate consequences of Eq. (8.3). ◆

Example 8.1

$$f(x, y) = x^2 + y^2.$$

Here the level curves $x^2 + y^2 = k$ are circles about the origin; $\nabla f(x, y) = \langle 2x, 2y \rangle$ is a vector in the direction of the radius to these circles at each point and is therefore orthogonal to the circle (Fig. 8.1).

The directional derivative of $x^2 + y^2$ in a direction tangent to the circle $x^2 + y^2 = k$ is always zero.

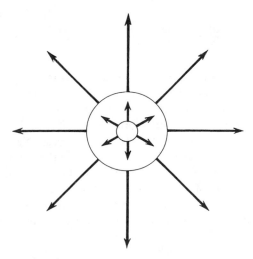

FIGURE 8.1 Level curves and gradient vectors for $f(x, y) = x^2 + y^2$

Theorem 8.1 is based on the assumption that the equation $f(x, y) = k$ actually represents a regular curve C, at least near a given point. Our main objective now is to prove that this is the typical case. In order to arrive at a general theorem, we consider first a special case. We ask when the equation $f(x, y) = k$ can be solved in the form $y = g(x)$, where $g(x)$ is a differentiable function of x. If it can be solved, then we have

$$f(x, g(x)) \equiv k,$$

and by the chain rule

(8.4) $$\frac{\partial f}{\partial x} + \frac{\partial f}{\partial y} g'(x) = 0,$$

which is nothing but a special case of Eq. (8.1), where the parameter t is equal to x. From Eq. (8.4) it follows that if $f_y \neq 0$,

(8.5) $$\frac{dy}{dx} = g'(x) = -\frac{f_x}{f_y}.$$

Example 8.2

Find the slope of the curve

(8.6) $$(x + y)^5 - xy = 1$$

at the point $(1, -1)$.

We let

$$f(x, y) = (x + y)^5 - xy.$$

Then

$$f_x = 5(x + y)^4 - y, \qquad f_y = 5(x + y)^4 - x.$$

Thus,

$$\frac{dy}{dx} = -\frac{5(x + y)^4 - y}{5(x + y)^4 - x}\bigg|_{(1, -1)} = 1.$$

The significance of Eq. (8.5) is that it allows us to compute the derivative of the function $g(x)$ without actually knowing the function itself. In fact, there are extremely few cases where an equation $f(x, y) = k$ can be solved explicitly in the form $y = g(x)$. In Example 8.2, solving Eq. (8.6) for y as a function of x would be complicated, to say the least. Nevertheless, we can find the derivative of this function at any point on the curve.

One may well ask, what do we mean by "the function obtained by solving Eq. (8.6) for y in terms of x," when we cannot in any practical sense solve Eq. (8.6) for y in terms of x. The answer to this question requires an understanding of the answer to the more basic question "what do we mean by a function?" For example, what do we mean by $y = \log_{10} x$? The answer is that we mean the number y such that $10^y = x$. No explicit way for finding this number y need be given, but we must prove that for each $x > 0$ there is one and only one y such that $10^y = x$ in order to conclude that we have defined y as a function of x.

In the same way, if we wish to assert that Eq. (8.6) defines y as a function of x, at least near some point such as $(1, -1)$, we have to show that for all values of x in some interval about $x = 1$, there is a corresponding value of y satisfying Eq. (8.6). In this case it is easy to show that for any fixed value x_0, the equation

$$(x_0 + y)^5 - x_0 y = 1$$

is satisfied by at least one value of y. (Namely, the left-hand side of this equation is a polynomial in y of odd degree, hence tends to $+\infty$ as $y \to +\infty$ and tends to $-\infty$ as $y \to -\infty$, and hence must be equal to 1 at some point.) The trouble here is that there may be several values of y satisfying this equation, and in order to obtain a well-defined function we must also restrict the interval in which y is allowed to lie (Fig. 8.2).

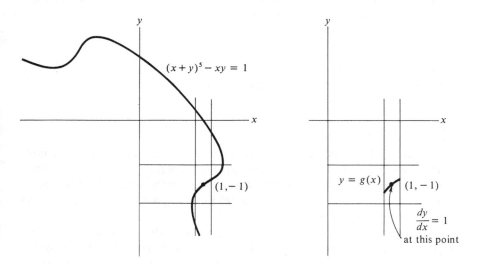

FIGURE 8.2 "Solving" $(x + y)^5 - xy = 1$ in the form $y = g(x)$ near $(1, -1)$

We may now make our earlier statements precise. When we say that

"Equation (8.6) can be solved in the form $y = g(x)$ near the point $(1, -1)$,"

we mean that

"there is some interval about $x = 1$, such that if we restrict y to an interval about $y = -1$, then Eq. (8.6) has one and only one solution for each fixed x."

Alternatively, we may say that

"there exist positive numbers a, b such that for each x satisfying $|x - 1| < a$, there is exactly one solution of Eq. (8.6) satisfying $|y + 1| < b$."

A final, more geometric, wording of the same statement, is that

"all points satisfying Eq. (8.6) and lying inside some rectangle $|x - 1| < a$, $|y + 1| < b$ about the point $(1, -1)$ lie on a curve $y = g(x)$."

We now state the general theorem which allows us to conclude that an equation $f(x, y) = k$ can be solved in the form $y = g(x)$. It is called the *implicit function theorem*, and an approximate form is: *if $f_y(x_0, y_0) \neq 0$, then the equation of the level curve $f(x, y) = k$ passing through (x_0, y_0) can be solved in the form $y = g(x)$ near (x_0, y_0).*

Theorem 8.2   Implicit Function Theorem   *Let $f(x, y)$ be continuously differentiable in D. Let (x_0, y_0) be any point in D such that*

$$(8.7) \qquad\qquad f_y(x_0, y_0) \neq 0.$$

Then there exist numbers $\delta > 0$ and $\epsilon > 0$, and there exists a continuously differentiable function $g(x)$ defined for $|x - x_0| \leq \delta$ such that if $|x - x_0| \leq \delta$ and $|y - y_0| \leq \epsilon$, then

$$(8.8) \qquad\qquad f(x, y) = f(x_0, y_0) \Leftrightarrow y = g(x).$$

PROOF.   The proof falls into two completely separate parts. The first part shows that there exists a function $g(x)$ satisfying Eq. (8.8). By its definition, this function need not even be continuous. In the second part we show that *any* function $g(x)$ satisfying Eq. (8.8) must in fact be continuously differentiable if $f(x, y)$ is.

Part I.   Existence of $g(x)$ Let $f_y(x_0, y_0) = \eta$. Our basic assumption is that $\eta \neq 0$. We may assume $\eta > 0$. (If $\eta < 0$ apply the reasoning below to the function $-f$.) Then, by the continuity of f_y, we have for r sufficiently small,

$$(8.9) \qquad\qquad f_y(x, y) \geq \frac{\eta}{2} > 0$$

if

$$(x - x_0)^2 + (y - y_0)^2 < r^2.$$

This means that inside this circle f is a strictly increasing function of y for each fixed x. In particular, if we choose any positive number $\epsilon < r$, we have

$$(8.10) \qquad\qquad f(x_0, y_0 - \epsilon) < f(x_0, y_0) < f(x_0, y_0 + \epsilon).$$

By the continuity of $f(x, y)$, we can vary x_0 slightly on the left- and right-hand sides of (8.10), and the inequality still holds. More precisely, we can choose $\delta > 0$ so that

(8.11) $\qquad f(x, y_0 - \epsilon) < f(x_0, y_0) < f(x, y_0 + \epsilon) \qquad$ for $\quad |x - x_0| < \delta$.

If δ is sufficiently small, then the rectangle

(8.12) $\qquad\qquad |x - x_0| \le \delta, \qquad |y - y_0| \le \epsilon$

lies inside the circle, so that Eq. (8.9) still holds in this rectangle (Fig. 8.3). Equation (8.11) asserts that along the bottom of this rectangle $f(x, y) < f(x_0, y_0)$, while on the top $f(x, y) > f(x_0, y_0)$. Thus on each vertical line segment $x = x_1$, the function $f(x_1, y)$ must pass through the value $f(x_0, y_0)$ at some point (x_1, y_1) (Fig. 8.4). Further, there is only one such point on each vertical segment, since by (8.9), $f(x_1, y)$ is a strictly increasing function of y. We set $g(x_1)$ equal to this unique number y_1. This defines the function $g(x)$ for $|x - x_0| < \delta$ in such a way that Eq. (8.8) holds.

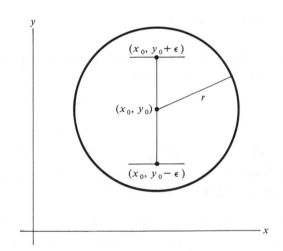

FIGURE 8.3 A neighborhood of (x_0, y_0) in which $f_y(x, y) \ge \eta/2 > 0$

Part II. Smoothness of $g(x)$ The fact that the function $g(x)$ is continuously differentiable does not depend on the particular way in which we defined it, but is a general property, which may be stated roughly as follows: *the level curves of a smooth function are smooth curves.* We give this statement precise form in the following "regularity lemma."

Lemma 8.1 *If $f(x, y)$ is continuously differentiable in D and satisfies* (8.7), *then any function $g(x)$ satisfying* (8.8) *must be continuously differentiable.*

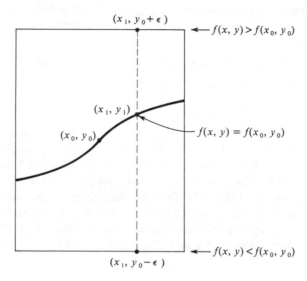

FIGURE 8.4 Existence of y_1 such that $f(x_1, y_1) = f(x_0, y_0)$

PROOF. Define the numbers δ and ϵ as in part I of the proof of Th. 8.2. We consider any two values x_1, x_2 in $|x - x_0| < \delta$, and let

(8.13) $$y_1 = g(x_1), \qquad y_2 = g(x_2).$$

Then, by (8.8),

(8.14) $$f(x_1, y_1) = f(x_0, y_0), \qquad f(x_2, y_2) = f(x_0, y_0).$$

The mean-value theorem (Th. 7.3) tells us that there is a point (x_3, y_3) on the line segment from (x_1, y_1) to (x_2, y_2) such that

(8.15) $$f(x_1, y_1) - f(x_2, y_2) = (x_2 - x_1)f_x(x_3, y_3) + (y_2 - y_1)f_y(x_3, y_3).$$

But by (8.14), the left-hand side of Eq. (8.15) is zero, and hence

(8.16) $$(y_2 - y_1)f_y(x_3, y_3) = -(x_2 - x_1)f_x(x_3, y_3),$$

whence

$$|y_2 - y_1| = |x_2 - x_1| \frac{|f_x(x_3, y_3)|}{|f_y(x_3, y_3)|} \leq |x_2 - x_1| \frac{2|f_x(x_3, y_3)|}{\eta}$$

by (8.9). But f_x is continuous in the rectangle defined by (8.12), and hence $|f_x|$ has a maximum M there. Thus

$$|y_2 - y_1| \leq \frac{2M}{\eta} |x_2 - x_1|.$$

Hence $x_2 \to x_1 \Rightarrow y_2 \to y_1 \Rightarrow g(x_2) \to g(x_1)$, by (8.13). This means precisely that $g(x)$ is continuous at x_1. Since (x_3, y_3) is on the line segment between (x_1, y_1) and (x_2, y_2), we have further that $x_2 \to x_1 \Rightarrow x_3 \to x_1$ and $y_3 \to y_1$. From Eq. (8.16), we conclude that

$$\lim_{x_2 \to x_1} \frac{y_2 - y_1}{x_2 - x_1} = \lim_{x_2 \to x_1} -\frac{f_x(x_3, y_3)}{f_y(x_3, y_3)} = -\frac{f_x(x_1, y_1)}{f_y(x_1, y_1)}.$$

Thus the limit on the left-hand side exists, and it is by definition (using Eq. (8.13)), $g'(x_1)$. Explicitly,

$$g'(x_1) = -\frac{f_x(x_1, y_1)}{f_y(x_1, y_1)},$$

where $y_1 = g(x_1)$. Since x_1 was any value of x in $|x - x_0| < \delta$, we have

(8.17)
$$g'(x) = -\frac{f_x(x, g(x))}{f_y(x, g(x))}, \qquad \text{in } |x - x_0| < \delta.$$

Thus $g(x)$ is differentiable in $|x - x_0| < \delta$, and since the right-hand side of Eq. (8.17) is a continuous function of x, $g'(x)$ is continuous in $|x - x_0| < \delta$. This concludes the proof of the lemma, and hence of the theorem.   ◆

Remark.   In the course of this proof we have used several basic properties of continuous functions. They are the following.

1.   If $f(x)$ is continuous for $a \le x \le b$, and if $f(a) = c$, $f(b) = d$, then given any number y_0 between c and d, there is some x_0 between a and b such that $f(x_0) = y_0$. Geometrically, this means that the graph of a continuous function intersects every horizontal line between its initial and terminal values (Fig. 8.5). This is called the "intermediate-value property" of continuous functions, and is often used (even if not explicitly stated) in calculus. To prove it requires a careful study of the real number system.[3]

2.   If $f(x, y)$ is continuous at (x_0, y_0), and if $f(x_0, y_0) = \eta > 0$, then for some $r > 0$, $f(x, y) \ge \eta/2$ in the whole circle $(x - x_0)^2 + (y - y_0)^2 < r^2$. This is an elementary fact that follows directly from the definition of continuity (see Ex. 5.16a).

3.   If $f(x, y)$ is continuous in a rectangle $|x - x_0| \le \delta$, $|y - y_0| \le \epsilon$ then it is bounded; that is, for some fixed M, $|f(x, y)| \le M$ in the whole rectangle. The proof of this again entails a fundamental study of the real number system. We actually stated a somewhat stronger result; namely that $f(x, y)$ has a maximum in the rectangle. This means that $f(x, y) \le M$ in the

---

[3] See, for example, [3] Vol. I, Section 2.30.

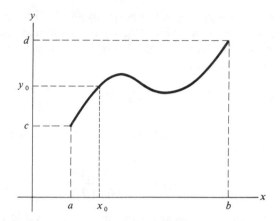

FIGURE 8.5 Intermediate-value property of continuous functions

whole rectangle, and further that $f(x_1, y_1) = M$ for some point (x_1, y_1) in the rectangle.

For the sake of completeness we state (without proof) a general theorem of which the statements in paragraph 3 are special cases. We need first two definitions.

Definition 8.1 A set S of points in the plane is *bounded* if it lies inside some circle $x^2 + y^2 < r^2$.

Definition 8.2 A set S of points in the plane is *closed* if it contains all its boundary points.

Example 8.3

If $P_1(x, y), \ldots, P_r(x, y)$ are polynomials (or any functions continuous in the whole plane), then the set of points satisfying all of the inequalities

$$P_1(x, y) \le 0, \ldots, P_r(x, y) \le 0$$

is a closed set. Thus, a disk $x^2 + y^2 \le 1$ and a rectangle $|x - x_0| \le \delta$, $|y - y_0| \le \epsilon$ are examples of closed, bounded sets.

Theorem 8.3 *If $f(x, y)$ is continuous on a closed, bounded set S, then it has a maximum M at some point of S.*[4]

---

[4] For proofs of this and related theorems, see [6], Section 2.4.

Returning to Th. 8.2, we see that if the roles of x and y are reversed and if $f_x(x_0, y_0) \neq 0$, then the equation $f(x, y) = f(x_0, y_0)$ can be solved near (x_0, y_0) in the form $x = h(y)$.

Corollary   *Let $f(x, y)$ be continuously differentiable in D and suppose $\nabla f(x_0, y_0) \neq 0$. Then near (x_0, y_0), the level curve $f(x, y) = f(x_0, y_0)$ defines a regular curve $x(t), y(t)$.*

PROOF.   Since $\nabla f(x_0, y_0) \neq 0$,   we   have   either $f_y(x_0, y_0) \neq 0$ or $f_x(x_0, y_0) \neq 0$ (or both). In the first case we can solve $f(x, y) = f(x_0, y_0)$ in the form $y = g(x)$, which gives a regular curve with $x = t$. In the second case we may write $x = h(y)$, which is a regular curve with $y = t$.   ◆

Geometrically, we may picture the situation as follows. A point where $\nabla f \neq 0$, defines a point on the side of the mountain $z = f(x, y)$. At such a point, the intersection of $z = f(x, y)$ with a horizontal plane $z = f(x_0, y_0)$, yields a smooth contour line (Fig. 8.6). Somewhat less picturesquely, the condition $\nabla f \neq 0$ means precisely that the tangent plane to the surface $z = f(x, y)$ at the point is not horizontal.

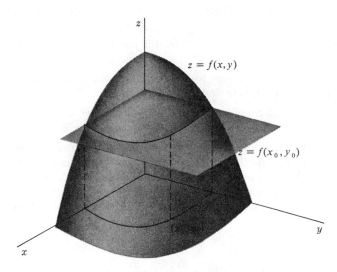

FIGURE 8.6   Geometric interpretation of the implicit-function theorem

Example 8.4

$$f(x, y) = x^2 + y^2, \qquad \nabla f = \langle 2x, 2y \rangle.$$

Hence $\nabla f = 0$ only at the origin. The level curve through any other point is a regular curve. The "level curve" $x^2 + y^2 = 0$ consists of the single point $(0, 0)$.

Example 8.5

$$f(x, y) = xy; \qquad \nabla f = \langle y, x \rangle.$$

Again $\nabla f = 0$ only at the origin. The "level curve" $xy = 0$ through the origin consists of the entire x and y axes. Since these cross at the origin, they clearly do not form a regular curve in any circle about the origin.

Example 8.6

$$f(x, y) = (x^4 + x^2 + 2)y + \sin y.$$

Here,

$$\nabla f = \langle (4x^3 + 2x)y, x^4 + x^2 + 2 + \cos y \rangle.$$

Since $\cos y \geq -1$ for every y, we have

$$x^4 + x^2 + 2 + \cos y \geq x^4 + x^2 + 1 \geq 1$$

at every point. Hence ∇f is never zero, and the level curves of this function are regular curves near every point. In fact, $f_y > 0$ everywhere, and hence the equation $f(x, y) = c$ can be solved in the form $y = g(x)$ near each point.

Exercises

8.1 Find the slope of the curve $(x + y)^5 - xy = 1$ at the following points:

    a. $(1, 0)$ *d.* $(-2, 1)$
    b. $(0, 1)$ *e.* the point on the curve where $y = x$
    c. $(1, -2)$

8.2 Assume that each of the following equations can be solved in the form $y = g(x)$, and use Eq. (8.5) to find dy/dx.

    a. $x \sin y - y \cos x = 0$ *c.* $x^3 - 2xy + y^2 = 5$
    b. $x + 2y^2 + \sin xy = 1$ *d.* $xe^y - y^2 + x = 0$

Note: in practice, it is often easier to use the method that led to Eq. (8.5) rather than Eq. (8.5) itself. In other words, given an equation that defines y implicitly as a function of x, consider y replaced by that function so that the equation becomes an identity in x, and then differentiate with respect to x, using the rule for composite functions of one variable. For example, differentiating

$$(x + y)^5 - xy = 1$$

yields

$$5(x + y)^4 \left(1 + \frac{dy}{dx} \right) - x \frac{dy}{dx} - y = 0,$$

and solving for dy/dx, we obtain

$$\frac{dy}{dx} = -\frac{5(x + y)^4 - y}{5(x + y)^4 - x}.$$

8.3 Find dy/dx for each of the equations in Ex. 8.2 by the method just indicated.

8.4   For each of the following equations, find dy/dx by the implicit function method of Exs. 8.2 or 8.3. Then solve explicitly for y as a function of x and differentiate. Check that your answer is the same in both cases.

a. $\dfrac{x^2}{a^2} + \dfrac{y^2}{b^2} = 1$ c. $e^x + e^y + e^{x+y} = 1$

b. $ye^x - x^2 + y = 0$ d. $x + xy + y = 1$

Note: an advantage of the implicit function method for finding the derivative dy/dx, even in the cases where it is possible to solve explicitly for y as a function of x, is that the computations are often simpler. A disadvantage is that the answer is expressed in terms of both x and y. It is still possible to take higher derivatives by using the same method. The original equation can often be used at various stages along the way to simplify the expressions obtained. For example, if $x^2 + y^2 = 1$, then

$$2x + 2y\frac{dy}{dx} = 0, \qquad \text{or} \qquad \frac{dy}{dx} = -\frac{x}{y}.$$

Hence,

$$\frac{d^2y}{dx^2} = -\frac{y - x(dy/dx)}{y^2} = -\frac{y - x(-x/y)}{y^2} = -\frac{y^2 + x^2}{y^3} = -\frac{1}{y^3}.$$

8.5   Use the above method of implicit differentiation to find d^2y/dx^2 for each of the equations in Ex. 8.4.

8.6 Show that if $\nabla f(x_0, y_0) \neq \mathbf{0}$, then the equation of the tangent line to the level curve $f(x, y) = f(x_0, y_0)$ at the point (x_0, y_0) is

$$f_x(x_0, y_0)(x - x_0) + f_y(x_0, y_0)(y - y_0) = 0.$$

*8.7   Show that all level curves of the function

$$f(x, y) = \tfrac{1}{2} \log (x^2 + y^2) - \arctan (y/x)$$

intersect all rays through the origin at an angle of $45°$. (*Hint:* Show that if (x_0, y_0) is any point on a level curve $f(x, y) = c$, $\tan \alpha$ is the slope of the curve at that point, and $\tan \theta = y_0/x_0$, then $\tan \alpha = \tan (\theta + \tfrac{1}{4}\pi)$.)

8.8   Explain Exs. 6.5 and 6.6 in the light of Th. 8.1.

8.9   For each of the functions of Ex. 6.3, sketch the level curve through the point $(1, 1)$ and verify Th. 8.1.

8.10   If $f(x, y) = \sin \log \cosh \tan e^{x^2 + y^2}$, find $f_x(0, 1)$. (*Hint:* do not calculate. Think!)

8.11   For each of the following pairs of functions $f(x, y)$, $g(x, y)$ show that each level curve of $f(x, y)$ intersects at right angles all level curves of $g(x, y)$. (More precisely, this property holds at all points of intersection where both level curves are regular curves.)

a. $f(x, y) = x^2 - y^2$, $g(x, y) = xy$
b. $f(x, y) = x^2 + y^2$, $g(x, y) = y^2/x^2$
c. $f(x, y) = e^x \cos y$, $g(x, y) = e^x \sin y$

8.12 a. Write an equation involving the partial derivatives of $f(x, y)$ and $g(x, y)$ that is equivalent to the condition that the level curves of f and g intersect at right angles wherever f and g have nonzero gradient.

b. Show that any pair of functions satisfying the Cauchy-Riemann equations $f_x = g_y$, $f_y = -g_x$, have the property that their level curves intersect at right angles.

c. Show that the converse of part b is false.

8.13 An important quantity in meteorology is *atmospheric pressure*. When the atmospheric pressure is considered as a function of position over a portion of the earth's surface, the level curves are called *isobars*. In the absence of other forces, the gradient of this function would indicate the magnitude and direction of winds. The air would tend to travel from higher to lower pressure areas, the speed would be proportional to the rate of change of pressure and the direction would be that of the maximum rate of change of pressure. Thus, by Th. 8.1, the direction of winds would be perpendicular to the isobars. In fact, other factors, such as the rotation of the earth, deflect the winds, but a map showing the isobars is still a basic component for predicting winds and weather in general. (See Fig. 8.7 where atmospheric pressure is given in inches of mercury (29.2 to 30.6) and in millibars (989 to 1036).)

Explain why the wind velocity tends to be small near the center of either a high pressure or a low pressure area. (The "doldrums" are the low-pressure areas near the equator, and the "horse latitudes" refer to the high-pressure belts over the oceans at around 30° latitude North and South of the equator.)

8.14 For each of the following functions $f(x, y)$ and points (x_0, y_0), check whether the condition $f_y(x_0, y_0) \neq 0$ is satisfied. If not, decide whether the conclusion of Th. 8.2 is valid, and if it is not explain what part of it fails.

    a. $f(x, y) = x^2 + y^2$, $(x_0, y_0) = (-5, 0)$
    b. $f(x, y) = x^2 + y^2$, $(x_0, y_0) = (-4, 3)$
    c. $f(x, y) = x + y^3$, $(x_0, y_0) = (0, 0)$
    d. $f(x, y) = x^2 + 2xy + y^2$, $(x_0, y_0) = (0, 0)$

8.15 Show that the equation $(x + y)^5 - xy = 1$ defines a regular curve near each of its points. (*Hint:* show that a simultaneous solution of $f_x = 0$ and $f_y = 0$ cannot lie on the curve.)

8.16 For each of the following functions $f(x, y)$, sketch and describe in words the level curve $f(x, y) = 0$. Check in particular any points on the curve where $\nabla f = 0$, and describe the curve near those points.

    a. $f(x, y) = x^2 - y^3$ *d.* $f(x, y) = (x^2 + y^2 - 1)^2$
    b. $f(x, y) = x^4 + y^4$ *e.* $f(x, y) = x^2 + y^2 + e^{xy}$
    c. $f(x, y) = (x^2 + y^2 - 1)(x^2 + y^2)$ **f.* $f(x, y) = x^3 - x^2 + xy^2 + y^2$

(*Hint:* in part f, either solve for y as a function of x, or else use the parametric form given in Ex. 2.13.)

8.17 The *inverse function theorem* for functions of one variable may be stated as follows. If $h(y)$ is differentiable and $h'(y_0) \neq 0$, then the equation $x = h(y)$

FIGURE 8.7 Mean January sea-level pressures and wind directions of the world [Thomas A. Blair and Robert C. Fite, *Weather Elements*, 4th Ed., © 1957. Reprinted by permission of Prentice-Hall, Inc., Englewood Cliffs, New Jersey. Base map Copyright Denoyer-Geppert Company, Chicago, used by permission.]

can be solved near $x_0 = h(y_0)$ in the form $y = g(x)$, where $g(x)$ is differentiable and $g'(x_0) = 1/h'(y_0)$. By considering the function $f(x, y) = h(y) - x$, show how this theorem is related to the implicit function theorem, Th. 8.2.

8.18 Kepler's laws of planetary motion (see Ex. 7.25) do not give the position of a planet explicitly as a function of time. However, if a planet moves along the ellipse $x = a \cos \varphi$, $y = b \sin \varphi$, where the sun is at the focus $(c, 0)$, and if t is the time from the moment the planet passes the point $(a, 0)$ (the *perihelion*) to the position corresponding to a given value of φ, then it follows from Kepler's second law that φ and t are related by *Kepler's equation*

$$\varphi - kt - \epsilon \sin \varphi = 0,$$

where k is a positive constant. (We are using the standard notation for an ellipse: $c = (a^2 - b^2)^{1/2}$, $\epsilon = c/a$. Thus, $0 < \epsilon < 1$.)

a. Show that for an arbitrary pair of values t_0, φ_0 satisfying Kepler's equation, there exists a solution of that equation in the form $\varphi = g(t)$ near (t_0, φ_0).

b. Show that if $g(t)$ is the function of part a, then

$$\frac{d\varphi}{dt} = g'(t) = \frac{k}{1 - \epsilon \cos \varphi}.$$

c. Deduce that $d\varphi/dt$ is maximum at perihelion $(a, 0)$, and minimum at aphelion $(-a, 0)$.

d. Find the velocity vector $\mathbf{v}(t) = \langle x'(t), y'(t) \rangle$ at an arbitrary point.

*e. Show that the speed $|\mathbf{v}(t)|$ decreases continually from perihelion to aphelion, and then increases from aphelion to perihelion. (*Hint:* express $|\mathbf{v}(t)|^2$ as a function of $\cos \varphi$. Call this function $f(u)$, where $u = \cos \varphi$. Show that $f'(u) > 0$ for all u, so that $f(u)$ increases when u increases and decreases when u decreases.)

8 Maximum and minimum along a curve

A railroad train travels from San Francisco to Los Angeles. Plotting its route on a contour map, how do we find the point of maximum altitude? We might guess that at such a point the direction of the tracks must be tangent to a contour line (Fig. 9.1). Indeed, where the tracks cross a contour line, one would expect the altitude to be either increasing or decreasing. With slight qualifications, this observation leads to very simple solutions of many maximum and minimum problems, and to a general method for treating problems of this type.[5] The method may be stated roughly as follows:

---

[5] The approach used in this section was inspired by the treatment in Chapter 8 of Polya's book [30]. The reader is referred there for further interesting comments and examples.

FIGURE 9.1   Highest point along track, and tangent contour line

if a function f has a maximum or minimum along a curve C at a point (x_0, y_0), *then the level curve of f through* (x_0, y_0) *is tangent to C.* In the above discussion, the function $f(x, y)$ is the altitude, and the curve C is the route of the train.

Theorem 9.1   Method of Level Curves   *Let* $f(x, y)$ *be continuously differentiable in D, and let* $C:x(t)$, $y(t)$, $a \le t \le b$, *be a regular curve in D. Suppose* $f(x, y)$ *has a maximum or minimum along C at the point* $(x_0, y_0) = (x(t_0), y(t_0))$, *where* $a < t_0 < b$. *Then either*

1. $\nabla f(x_0, y_0) = 0$

or

2.   *the level curve* $f(x, y) = f(x_0, y_0)$ *is tangent to C at* (x_0, y_0).

PROOF.   We consider the function $f_C(t)$. If it has a maximum or minimum at t_0, then $f_C'(t_0) = 0$. But by the chain rule,

$$(9.1) \qquad 0 = f_C'(t_0) = f_x(x_0, y_0)x'(t_0) + f_y(x_0, y_0)y'(t_0)$$
$$= \nabla f(x_0, y_0) \cdot \langle x'(t_0), y'(t_0) \rangle.$$

There are two alternatives. Either $\nabla f(x_0, y_0) = 0$ (case 1 in the statement of the theorem), or else $\nabla f(x_0, y_0) \ne 0$, in which case we know from the Corollary to Th. 8.2 that the level curve $f(x, y) = f(x_0, y_0)$ is a regular curve near (x_0, y_0). Furthermore, by Th. 8.1 the tangent to this level curve is perpendicular to $\nabla f(x_0, y_0)$. But Eq. (9.1) states that the tangent to C is perpendicular to $\nabla f(x_0, y_0)$. Thus, the curve C and the level curve of f have the same tangent line at (x_0, y_0), and case 2 holds in the statement of the theorem.   ◆

Remark   In practice, one often ignores case 1, and simply looks for points of tangency (as in our rough statement of the theorem). However, it should be noted that case 1 is a real possibility, and not just a technical complication. For example, should the train discussed in the first paragraph of this section choose to go right over the top of a mountain, then at its highest point the level curve would reduce to a point, and case 2 would be meaningless. But at a mountain top $\nabla f = 0$. (If our curve represented a trail rather than a train route, this would be a very likely possibility.) An even more frequent occurrence would be for the highest point to be the summit of a pass. It is intuitively clear that a contour line through such a point generally splits into a pair of crossing curves, so that again case 2 is meaningless (Fig. 9.2). But we have observed earlier (following Def. 7.1), that case 1 must necessarily hold at the summit of a pass.

Thus, to apply the method of Th. 9.1 one should first check if there are any points on the curve C where $\nabla f(x_0, y_0) = 0$, and then look for points of tangency.

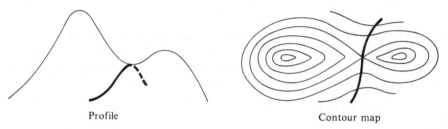

Profile   Contour map

FIGURE 9.2   Summit of a pass

Example 9.1

Find the point along a coastline nearest to a fixed point inland.

If we let (x_1, y_1) be the fixed point and C the coastline, then we have to minimize the function

$$f(x, y) = \sqrt{(x - x_1)^2 + (y - y_1)^2},$$

subject to the restriction that (x, y) lies on C. Since

$$\nabla f = \frac{\langle x - x_1, y - y_1 \rangle}{\sqrt{(x - x_1)^2 + (y - y_1)^2}},$$

which is zero only at the point (x_1, y_1), we have $\nabla f(x_0, y_0) \neq 0$, where (x_0, y_0) is the point on C nearest (x_1, y_1). Thus the level curve of f through (x_0, y_0) must be tangent to C at (x_0, y_0). But the level curves of f are circles about (x_1, y_1), and since the radius of a circle is always perpendicular to its tangent line, we may state the result as follows: the nearest point to (x_1, y_1) along C, must be a point (x_0, y_0) such that the line from (x_1, y_1) to (x_0, y_0) is perpendicular to C at (x_0, y_0) (Fig. 9.3).

A "practical" way to find this point would be to place a light at (x_1, y_1) and then sail along the coast holding a mirror parallel to the direction of motion. At the nearest point, the light would be reflected directly back to (x_1, y_1).

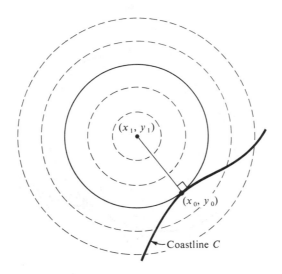

FIGURE 9.3   Shortest distance from a point to a curve

Example 9.2

The milkmaid problem   A milkmaid has to go from her house to a river, fill a pail with water, and then take it to the barn. For what point P along the river is her total path the shortest?

Denote by H and B the points at which the house and barn are located. For any point P, the function we are minimizing is the sum of the distances from H to P and from P to B (Fig. 9.4). The level curves of this function are ellipses with foci at H and B. At the point P along the river where this function is a minimum, the corresponding ellipse must be tangent to the river.

Consider first the case where the river is a straight line l. Then this line is tangent to the ellipse at the point P. In this case there is an elementary solution to the problem. If we reflect B in the straight line we get a point B', and the total distance from H to P to B is the same as the distance from H to P to B' (Fig. 9.5). To minimize this distance, we clearly must choose P on the straight line from H to B'. Then the angle between PH and l must equal the angle between PB' and l, which equals the angle between PB and l. The equality of these angles is a basic fact of reflection of light, and follows from the assumption that reflected light follows the shortest path. The above reasoning also shows that if P is the point where a tangent line touches an ellipse, then the lines from P to the foci make equal angles with l.

Finally, if we consider the general case, we see that the river must have a direction at P tangent to the ellipse, and hence that the angles made by PH and PB with the river must be equal. We may therefore formulate an equally practical solution to this problem as to the last. We place a bright light at the house, sail down the river with a mirror, and at the desired point the reflected light will hit the barn.

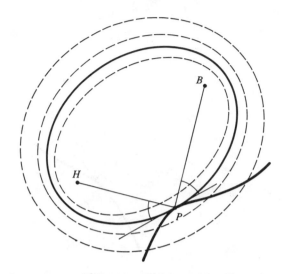

FIGURE 9.4 Shortest distance between two points, with stop-off on a curve

Example 9.3

If the sum of two numbers is a given positive number s, what is the maximum of their product?

Let the numbers be x and y. We are given that $x + y = s$, for some fixed number s. The function we wish to maximize is $f(x, y) = xy$. We must therefore find the point on the line $x + y = s$ that maximizes xy. We note that the function xy is negative where the line $x + y = s$ is in the second and fourth quadrants,

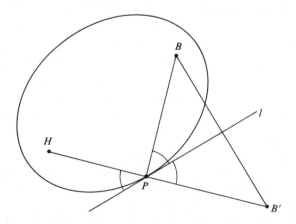

FIGURE 9.5 Reflection method for finding shortest path

zero where it crosses the x and y axes, and positive in the first quadrant. It therefore has a maximum at some point (x_0, y_0) in the first quadrant, and the level curve $xy = c$ through this point must be tangent to $x + y = s$. This means the slope of $xy = c$ at (x_0, y_0) must be -1. But if

$$y = \frac{c}{x},$$

$$\frac{dy}{dx} = -\frac{c}{x^2} = -1 \Leftrightarrow x^2 = c \Leftrightarrow x = \sqrt{c}$$

$$\Leftrightarrow y = c/x = \sqrt{c}.$$

Thus $x_0 = y_0$, and since $x_0 + y_0 = s$, $x_0 = y_0 = s/2$, and $x_0 y_0 = s^2/4$. Thus for any point (x, y) on $x + y = s$ (Fig. 9.6),

$$xy \leq x_0 y_0 = \left(\frac{s}{2}\right)^2 = \left(\frac{x + y}{2}\right)^2.$$

In particular, if x and y are *any* two positive numbers, then since they lie on some line $x + y = s$, we have the inequality

(9.2) $$\sqrt{xy} \leq \frac{x + y}{2},$$ equality only if $y = x$.

This is called the inequality of the *geometric* mean and the *arithmetic* mean, these being the designation of the left- and right-hand sides of (9.2), respectively.

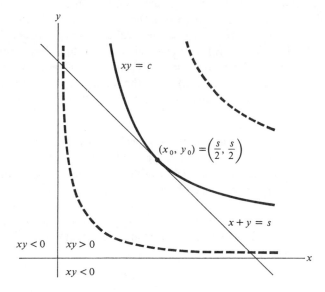

FIGURE 9.6 Minimizing xy for $x + y = s$

We now wish to consider a special case of the type of problem we have been considering. We assume that the curve C is given implicitly by an equation $g(x, y) = k$. Example 9.3 was of this nature, where the equation $g(x, y) = k$ was $x + y = s$. The problem can be formulated as follows:

"find the maximum (or minimum) of a function $f(x, y)$ subject to the condition $g(x, y) = k$."

Such a problem is called a maximum problem with a *side condition* or with a *constraint*. The solution consists in observing that the side condition $g(x, y) = k$ is merely the implicit form of a curve C, and the problem therefore reduces to our previous one. The result is then usually stated as follows.

Theorem 9.2 Lagrange Multiplier Method *If a function $f(x, y)$ has a maximum or minimum at (x_0, y_0) subject to the condition $g(x, y) = k$, then either*

    1. $\nabla g(x_0, y_0) = 0$

or

    2. $\nabla f(x_0, y_0) = \lambda \nabla g(x_0, y_0),$ *for some λ.*

PROOF. It is sufficient to show that if case 1 does not hold, then case 2 must hold. Suppose, therefore, $\nabla g(x_0, y_0) \neq 0$. Then the equation $g(x, y) = g(x_0, y_0) = k$ defines a regular curve C near (x_0, y_0). Since $f(x, y)$ has a maximum or minimum along C at (x_0, y_0), either $\nabla f(x_0, y_0) = 0$, in which case conclusion 2 holds with $\lambda = 0$, or else the level curve of f through (x_0, y_0) is tangent to C. But $\nabla f(x_0, y_0)$ is perpendicular to this level curve, and $\nabla g(x_0, y_0)$ is perpendicular to the curve C (which is a level curve of $g(x, y)$) and hence $\nabla f(x_0, y_0)$ and $\nabla g(x_0, y_0)$ have the same (or opposite) direction. But that is precisely what conclusion 2 asserts (Fig. 9.7).   ◆

Remark The natural way to attack a problem of this type would be to solve the equation $g(x, y) = k$ for x or y, and substitute in the function f, thus obtaining a function of a single variable. For instance, in Example 9.3, we had $x + y = s$ or $y = s - x$, and $f(x, y) = xy = sx - x^2$. Differentiation with respect to x yields $x = s/2$. However, as we have observed earlier, there are very few cases where an equation $g(x, y) = k$ can be solved explicitly. An alternative approach is to differentiate the equation $g(x, y) = k$ implicitly, giving

$$g_x + g_y \frac{dy}{dx} = 0.$$

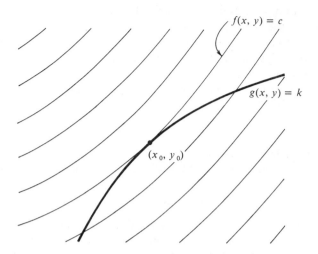

FIGURE 9.7 Geometric interpretation of Lagrange multiplier method

Then considering y to be a function of x by the equation $g(x, y) = k$, we have that $f(x, y(x))$ has a maximum or minimum, and hence

$$f_x + f_y \frac{dy}{dx} = 0.$$

Eliminating dy/dx, gives

(9.3)
$$\frac{g_x}{g_y} = \frac{f_x}{f_y},$$

which can sometimes be solved simultaneously with $g(x, y) = k$ to yield the desired point. It is a question of solving two equations in two unknowns. If we rewrite Eq. (9.3) in the form

(9.4)
$$\frac{f_x}{g_x} = \frac{f_y}{g_y},$$

and if we denote the common value of both sides of this equation by λ, then we have again case 2 in Th. 9.2, which may be written as

(9.5) $f_x(x_0, y_0) = \lambda g_x(x_0, y_0),$ $f_y(x_0, y_0) = \lambda g_y(x_0, y_0).$

These two equations, combined with

(9.6) $g(x_0, y_0) = k$

yield three equations in the three unknowns x_0, y_0, λ. Although it may seem to complicate matters to introduce an extra unknown and an extra equation, the fact is that this device often proves useful. (The application made in the following section is a good case in point.) The new quantity λ, incidentally, is what is referred to as the *Lagrange multiplier*. One interpretation of this method is to consider the function $h(x, y) = f(x, y) - \lambda g(x, y)$, formed by changing f by a fixed multiple λ of the constraint function g. Then Eqs. (9.5) simply say that $\nabla h(x_0, y_0) = 0$, and the problem is to find a multiple λ such that this equation holds simultaneously with Eq. (9.6).

Example 9.4

Find the minimum perimeter for rectangles having a fixed area A.

If the rectangle has length x and width y, then the perimeter is $2x + 2y$ and the area is xy. If we let

$$f(x, y) = 2(x + y), \qquad g(x, y) = xy,$$

the problem is to minimize $f(x, y)$ under the side condition $g(x, y) = A$.

Method 1. Level curves The equation $xy = A$ defines a hyperbola. Since the quantities x and y represent the sides of a rectangle, they cannot be negative. Thus we choose the curve C to be the branch of $xy = A$ lying in the first quadrant, and we wish to find the minimum of $f(x, y)$ along C. Since $\nabla f = \langle 2, 2 \rangle$, which is never zero, at the point (x_0, y_0) of C where the minimum occurs, the curve C must be tangent to the level curve of $f(x, y)$. But the level curves of $f(x, y)$ are the straight lines $x + y = c$. It is clear geometrically (see Fig. 9.8), and it is easy to verify that tangency can only occur if $x_0 = y_0$. Since $x_0 y_0 = A$, we find $x_0 = y_0 = \sqrt{A}$, and $f(x_0, y_0) = 4\sqrt{A}$.

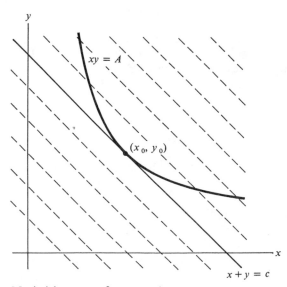

FIGURE 9.8 Maximizing $x + y$ for $xy = A$

Method 2.   Elimination of one variable   The condition $xy = A$ yields $y = A/x$. Substituting this in $f(x, y)$, we find that the function to be minimized becomes $2(x + A/x)$. Setting the derivative of this function equal to zero yields $2(1 - A/x^2) = 0$, or $x^2 = A$, $x = \pm\sqrt{A}$. Since x is positive, $x = \sqrt{A}$, and since $xy = A$, $y = A/x = A/\sqrt{A} = \sqrt{A}$.

Method 3.   Implicit differentiation   Without solving the equation $xy = A$ for y in terms of x, we can differentiate implicitly and obtain

$$y + x\frac{dy}{dx} = 0 \qquad \text{or} \qquad \frac{dy}{dx} = -\frac{y}{x}.$$

We consider the function

$$h(x) = f(x, y(x)) = 2(x + y(x)),$$

where $y(x)$ is determined from the relation $xy = A$. Then

$$h'(x) = 2\left(1 + \frac{dy}{dx}\right) = 2\left(1 - \frac{y}{x}\right).$$

Setting $h'(x) = 0$ yields $y = x$, which gives the same solution to the problem as the previous two methods.

Method 4.   Lagrange multipliers   We have

$$\nabla f = \langle 2, 2 \rangle, \qquad \nabla g = \langle y, x \rangle.$$

Since ∇g is zero only at the origin, which does not lie on the curve $xy = A$ for $A > 0$, we conclude that at the extreme point (x_0, y_0) on the curve $xy = A$ we must have $\nabla f(x_0, y_0) = \lambda \nabla g(x_0, y_0)$ for some λ. This yields three equations (corresponding to Eqs. (9.5) and (9.6)):

$$2 = \lambda y_0, \qquad 2 = \lambda x_0, \qquad x_0 y_0 = A.$$

From the first two equations we have $\lambda = 2/x_0 = 2/y_0$, whence $x_0 = y_0$, and from the third, $x_0 = y_0 = A^{1/2}$.

Exercises

9.1   Solve each of the following problems first by using Lagrange multipliers, and then by eliminating y and treating the problem as an extremum problem in one variable.

    a. Find the maximum and minimum of $x + 2y$ on $x^2 + y^2 = 1$.

    b. Find the maximum and minimum of $x^4 + y^4 - x^2 - y^2 + 1$ on $x^2 + y^2 = 4$.

    c. Find the minimum of $x + y$ on the branch of the curve $1/x + 1/y = 1$ lying in the first quadrant.

    d. Find the maximum of $1/x + 1/y$ on the part of the curve $1/x^2 + 1/y^2 = 2$ lying in the first quadrant.

    e. Find the minimum of $x^2 + y^2$ on the curve $x = y^2 + 1$. Explain why the second method does not work. (*Hint:* interpret the problem geometrically and indicate the answer by means of a sketch.)

Note: the Lagrange multiplier method is based on the *assumption* that (x_0, y_0) represents a maximum or minimum. It then gives a means of finding (x_0, y_0). However, if a point is obtained by the Lagrange multiplier method, it is sometimes difficult to decide whether that point represents a maximum or a minimum of the given function (for example, in Ex. 9.1d). In fact it may represent neither, but rather a kind of "inflection point" (see Ex. 9.2). Furthermore it may represent a local maximum or minimum but not an absolute maximum or minimum (see Ex. 9.3). In practice, some other reasoning is usually used to show that a maximum or minimum exists, and then the Lagrange multiplier method is used to find it.

9.2 Use Lagrange multipliers to find the points on the curve $x + x^5 - y = 2$ where the function $f(x, y) = x - y$ may have a maximum or minimum. Then sketch the curve and sketch the level curves of $f(x, y)$, and show that the point or points you have found do not represent either a local maximum or a local minimum of f.

9.3 Use Lagrange multipliers to find the points on the curve $x^3 - 3x - y = 0$ where the function e^y may have a maximum or minimum. Show by a sketch that these points represent a local maximum or minimum but not an absolute maximum or minimum.

9.4 Use Lagrange multipliers to find the angle of a circular sector, which yields the minimum perimeter for a given area. (The angle and the radius are both allowed to vary.)

9.5 Find the distance from the point (x_1, y_1) to the line $ax + by + c = 0$, using Lagrange multipliers.

9.6 Solve parts a and b below by the method of Lagrange multipliers.

   a. Find the point on $x^2 + y^2 = 25$ where the function $3x + 4y$ is maximum.
   b. Find the point on $3x + 4y = 25$ where the function $x^2 + y^2$ is minimum.
   c. Give a geometric interpretation of part b.
   d. Give a geometric interpretation of part a. (*Hint:* note that the function to be maximized may be interpreted as a dot product of vectors.)
   e. Explain why in both parts a and b, the point sought must lie on the line $y = \frac{4}{3}x$.

Note: It happens quite frequently that the problem of maximizing a function $f(x, y)$ subject to the condition that $g(x, y)$ be constant, is equivalent to minimizing $g(x, y)$ with $f(x, y)$ held constant. This is illustrated by Exs. 9.5a, b, and also by Examples 9.3 and 9.4 (see also Exs. 10.22–10.24). The method of level curves provides a good explanation of this phenomenon, since it states that in both cases, the level curves of f and g through the given point must be tangent.

9.7 Illustrate the solutions to Exs. 9.6a, b by the method of level curves, using sketches similar to Figs. 9.6 and 9.8.

9.8   *a.* Let $f(x, y)$ be the distance from the point (x, y) to a fixed straight line. What are the level curves of f?

   b. Use the method of level curves to find the point on the ellipse $x^2/4 + y^2 = 1$ nearest to the line $y = x + 5$.

9.9   *a.* Let $f(x, y)$ be the distance from the point (x, y) to a fixed circle. What are the level curves of f?

   b. Let C_1 and C_2 be nonconcentric circles. Show that the point on C_1 farthest from C_2 must lie on the line through the centers of C_1 and C_2.

9.10   Let $f(x, y) = 4 - (x + y)^2$ and let C be the circle $x^2 + y^2 = 2$.

   a. What are the level curves of $f(x, y)$?

   b. At what points of C are the level curves of $f(x, y)$ tangent to C?

   c. What is the maximum value of $f(x, y)$ for all x and y, and at which points is it attained?

   d. What is the maximum value of $f(x, y)$ on C, and at which points is it attained?

   e. Are the level curves of (x, y) tangent to C at the points found in part d?

   f. Explain how the method of level curves can be used to find the maximum and minimum of $f(x, y)$ on C.

   g. Sketch the surface $z = f(x, y)$ and sketch the curve on that surface lying over the circle C, indicating the highest and lowest points on that curve, and the level curves at those points.

*9.11   Let L be the line segment joining the points $(0, 1)$ and $(0, 3)$. Let $f(x, y)$ be the angle subtended by L at the point (x, y).

   a. Describe the level curves of $f(x, y)$. (You may use results from plane geometry.)

   b. Use the method of level curves to find the point on the positive x axis where $f(x, y)$ is maximum.

9.12   Let $f(x, y) = x^5 + x + 1$, $g(x, y) = x^3 + y^2$.

   a. Use the method of Lagrange multipliers to find the maximum of $f(x, y)$ subject to the condition $g(x, y) = 0$.

   b. What are the level curves of $f(x, y)$?

   c. Sketch the curve $g(x, y) = 0$, and indicate the point where $f(x, y)$ is maximum.

   d. Explain why the method of level curves fails here and why Eqs. (9.5) and (9.6) do not have any common solutions in this case.

10 Quadratic forms

The study of quadratic forms affords a striking illustration of the usefulness of the method of Lagrange multipliers. Since quadratic forms play a basic role in many parts of mathematics, including the study of maxima and minima for functions of several variables, it is well worth deriving some of their basic properties.

Definition 10.1 A *quadratic form* is a homogeneous quadratic polynomial.

Thus, a quadratic form $q(x, y)$ in two variables has the form

$$(10.1) \qquad q(x, y) = ax^2 + 2bxy + cy^2,$$

where a, b, and c are constants. As we remarked in the case of general homogeneous functions, it is sufficient to know the values of the function on the unit circle $x^2 + y^2 = 1$ in order to know it everywhere. Specifically, we may write any point (x, y) as $(r \cos \alpha, r \sin \alpha)$, and we have from Eq. (10.1) that

$$(10.2) \quad q(x, y) = q(r \cos \alpha, r \sin \alpha) = r^2 q(\cos \alpha, \sin \alpha), \quad r^2 = x^2 + y^2$$

We therefore pose the following problem:

"Find the maximum and minimum values of $q(x, y)$ on the unit circle $x^2 + y^2 = 1$."

We note first that every point (x, y) on the unit circle may be written as $x = \cos t$, $y = \sin t$, and therefore if $x^2 + y^2 = 1$, we have

$$q(x, y) = q(\cos t, \sin t) = a \cos^2 t + 2b \cos t \sin t + c \sin^2 t,$$
$$0 \le t \le 2\pi,$$

which is a continuous function of t for $0 \le t \le 2\pi$, and hence assumes a maximum and minimum.[6] We may therefore apply the method of Lagrange multipliers. We set

$$g(x, y) = x^2 + y^2$$

and state the problem as follows:

"Find the maximum and minimum of $q(x, y)$ if $g(x, y) = 1$."

We first check the gradient of g,

$$(10.3) \qquad \nabla g = \langle 2x, 2y \rangle = 2 \langle x, y \rangle.$$

Thus $\nabla g = \mathbf{0}$ only at the origin, and in particular $\nabla g \ne \mathbf{0}$ on the level curve $g(x, y) = 1$. We conclude that if (x_0, y_0) is a point on $x^2 + y^2 = 1$ where $q(x, y)$ has a maximum or minimum, then there exists a value λ for which

$$(10.4) \qquad \nabla q(x_0, y_0) = \lambda \nabla g(x_0, y_0).$$

---

[6] This fact may appear to be intuitively obvious. For a complete proof see [3], Vol. I, Section 8.8.

But

(10.5) $$\nabla q = 2\langle ax + by, bx + cy\rangle,$$

and comparing this with Eq. (10.3), we may write Eq. (10.4) in the form

(10.4a)
$$ax_0 + by_0 = \lambda x_0,$$
$$bx_0 + cy_0 = \lambda y_0.$$

These equations must be solved simultaneously with

(10.6) $$x_0^2 + y_0^2 = 1.$$

There is a useful trick which may be used whenever the Lagrange multiplier method is applied to homogeneous functions (see Ex. (10.26)). We multiply the first equation of Eqs. (10.4a) by x_0, the second by y_0, and add; this yields

$$ax_0^2 + 2bx_0 y_0 + cy_0^2 = \lambda(x_0^2 + y_0^2),$$

or, by virtue of Eq. (10.6),

(10.7) $$q(x_0, y_0) = \lambda.$$

This result may be described as follows: *The value of the Lagrange multiplier λ is in this case precisely the value of the function $q(x, y)$ that we are seeking.*

We may now easily determine the value of λ from Eqs. (10.4a). To do so, we rewrite these equations by transposing the right-hand sides:

(10.4b)
$$(a - \lambda)x_0 + \qquad by_0 = 0,$$
$$bx_0 + (c - \lambda)y_0 = 0.$$

These are two linear homogeneous equations in the unknowns x_0, y_0 which by Eq. (10.6) have a nontrivial solution $(x_0, y_0) \neq (0, 0)$. This can only happen if the determinant of the system is zero; that is,

$$\det \begin{pmatrix} a - \lambda & b \\ b & c - \lambda \end{pmatrix} = (a - \lambda)(c - \lambda) - b^2 = 0,$$

or

(10.8) $$\lambda^2 - (a + c)\lambda + ac - b^2 = 0.$$

But the value of λ at the maximum point as well as the value at the minimum point must satisfy this equation. We have therefore proved the following theorem.

Theorem 10.1   *The roots of Eq. (10.8) represent the maximum and minimum values of $ax^2 + 2bxy + cy^2$ on $x^2 + y^2 = 1$.*

Example 10.1

$$q(x, y) = x^2 + 2xy + y^2.$$

Here

$$a = b = c = 1, \qquad a + c = 2, \qquad ac - b^2 = 0.$$

Equation (10.8) becomes

$$\lambda^2 - 2\lambda = 0 \qquad \text{or} \qquad \lambda(\lambda - 2) = 0.$$

Hence

$$\max q(x, y) = 2 \qquad \text{and} \qquad \min q(x, y) = 0$$

on $x^2 + y^2 = 1$.

Example 10.2

$$q(x, y) = x^2 + 4xy + y^2.$$

In this case

$$a = c = 1, \qquad b = 2,$$

and Eq. (10.8) becomes

$$\lambda^2 - 2\lambda - 3 = 0 \qquad \text{or} \qquad (\lambda - 3)(\lambda + 1) = 0.$$

Thus

$$\max q(x, y) = 3, \qquad \min q(x, y) = -1$$

on $x^2 + y^2 = 1$.

Example 10.3

$$q(x, y) = Ax^2 + Cy^2, \qquad\qquad\qquad A \geq C.$$

Here

$$a = A, \qquad b = 0, \qquad c = C,$$

and Eq. (10.8) becomes

$$\lambda^2 - (A + C)\lambda + AC = 0 \qquad \text{or} \qquad (\lambda - A)(\lambda - C) = 0.$$

Thus

$$\max q(x, y) = A \qquad \text{and} \qquad \min q(x, y) = C$$

on $x^2 + y^2 = 1$.

There are several remarkable features that distinguish the solution provided by Th. 10.1 from that of a typical maximum-minimum problem in calculus. We note the following:

1. The standard procedure for finding the maximum of a function requires us first to find the point at which the maximum occurs, and then to

evaluate the function at that point. Using Th. 10.1, we obtain the maximum value directly. If (as is usually *not* the case) we wish to know the point at which the maximum is assumed, we may return to Eqs. (10.4b), substitute in the value of λ, and solve. Thus, in Example 10.1, the maximum value is $\lambda = 2$, and since $a = b = c = 1$, Eqs. (10.4b) become $-x_0 + y_0 = 0$, $x_0 - y_0 = 0$; these equations are equivalent: $y_0 = x_0$, and since $x_0^2 + y_0^2 = 1$, we have $2x_0^2 = 1$, so that $x_0 = \pm 1/\sqrt{2}$. The maximum of $x^2 + 2xy + y^2$ occurs therefore at $(1/\sqrt{2}, 1/\sqrt{2})$ and $(-1/\sqrt{2}, -1/\sqrt{2})$.

2.   Calculus is used to reduce this maximum-minimum problem to the solution of an algebraic equation. For each case the solution is then obtained by simple algebra.

3.   In a large class of problems it is not even necessary to find the precise maximum and minimum values, but merely to know if they are positive or negative. Again a simple algebraic procedure provides the answer.

The following observations illustrate point 3 above.

Let λ_1 and λ_2 be the maximum and minimum, respectively, of $q(x, y)$ on $x^2 + y^2 = 1$. Thus,

$$(10.9) \qquad\qquad \lambda_2 \le q(x, y) \le \lambda_1$$

for $x^2 + y^2 = 1$. Since λ_1 and λ_2 are the roots of Eq. (10.8), we have

$$(10.10) \qquad\qquad \lambda_1\lambda_2 = ac - b^2, \qquad \lambda_1 + \lambda_2 = a + c.$$

We note next that the following statements are equivalent.

Conditions I
a. $q(x, y) > 0$,   for all $(x, y) \ne (0, 0)$
b. $q(x, y) > 0$,   for $x^2 + y^2 = 1$
c. $\lambda_1 > 0$, $\lambda_2 > 0$
d. $\lambda_1\lambda_2 > 0$, $\lambda_1 + \lambda_2 > 0$
e. $ac - b^2 > 0$, $a + c > 0$
f. $ac - b^2 > 0$, $a > 0$

That statements a and b are equivalent follows from Eq. (10.2), since $r^2 > 0$ unless $(x, y) = (0, 0)$. That statements b and c are equivalent follows from Eq. (10.9). The equivalence of statements c and d is trivial, and so is that of statements d and e, using Eq. (10.10). Finally, if $ac - b^2 > 0$, then $ac > b^2 \ge 0$, so that a and c have the same sign. Thus,

$$a > 0 \Leftrightarrow c > 0 \Leftrightarrow a + c > 0.$$

Definition 10.2   A quadratic form is called *positive definite* if it satisfies condition I.a.

The equivalence of the various statements listed under Conditions I allow us to check at a glance whether a given quadratic form is positive definite or not. Generally speaking, condition I.f is the quickest to verify.

Example 10.4

$$q(x, y) = 2x^2 + 4xy + y^2.$$

Here

$$a = 2, \quad b = 2, \quad c = 1;$$
$$ac - b^2 = 2 - 4 = -2.$$

Hence, $q(x, y)$ is *not* positive definite.

Example 10.5

$$q(x, y) = x^2 - 3xy + 4y^2.$$

Here

$$a = 1, \quad b = -\tfrac{3}{2}, \quad c = 4;$$
$$ac - b^2 = 4 - \tfrac{9}{4} > 0.$$

Also $a > 0$. Hence, $q(x, y)$ is positive definite.

Example 10.6

$$q(x, y) = 2x^2 + 2xy - y^2.$$

Here

$$a = 2, \quad b = 1, \quad c = -1;$$
$$ac - b^2 = -3.$$

Thus, $q(x, y)$ is not positive definite.

In precisely the same way as outlined above, one may prove that the following sets of conditions are equivalent.

Conditions II
a. $q(x, y) < 0$,   for all $(x, y) \neq (0, 0)$
b. $q(x, y) < 0$,   for $x^2 + y^2 = 1$
c. $\lambda_1 < 0$, $\lambda_2 < 0$
d. $\lambda_1 \lambda_2 > 0$, $\lambda_1 + \lambda_2 < 0$
e. $ac - b^2 > 0$, $a + c < 0$
f. $ac - b^2 > 0$, $a < 0$

Definition 10.3 A quadratic form satisfying condition II.a is called *negative definite*.

Conditions III

a. $q(x, y) \geq 0,$ for all (x, y)
    $q(x_0, y_0) = 0,$ for some $(x_0, y_0) \neq (0, 0)$
b. $q(x, y) \geq 0,$ for $x^2 + y^2 = 1$
    $q(x_0, y_0) = 0,$ for some $(x_0, y_0),$ $x_0^2 + y_0^2 = 1$
c. $\lambda_1 \geq 0,$ $\lambda_2 = 0$
d. $\lambda_1 \lambda_2 = 0,$ $\lambda_1 + \lambda_2 \geq 0$
e. $ac - b^2 = 0,$ $a + c \geq 0$

Conditions IV

a. $q(x, y) \leq 0,$ for all (x, y)
    $q(x_0, y_0) = 0,$ for some $(x_0, y_0) \neq (0, 0)$
b. $q(x, y) \leq 0,$ for $x^2 + y^2 = 1$
    $q(x_0, y_0) = 0,$ for some $(x_0, y_0),$ $x_0^2 + y_0^2 = 1$
c. $\lambda_1 = 0,$ $\lambda_2 \leq 0$
d. $\lambda_1 \lambda_2 = 0,$ $\lambda_1 + \lambda_2 \leq 0$
e. $ac - b^2 = 0,$ $a + c \leq 0$

Definition 10.4   A quadratic form is called *positive semidefinite* if it satisfies condition III.a and *negative semidefinite* if it satisfies condition IV.a.

Finally, the following conditions are equivalent.

Conditions V
a. $q(x, y)$ takes on both positive and negative values
b. $q(x, y)$ takes on both positive and negative values on $x^2 + y^2 = 1$
c. $\lambda_1 > 0,$ $\lambda_2 < 0$
d. $\lambda_1 \lambda_2 < 0$
e. $ac - b^2 < 0$

We may summarize the situation as follows.

Case 1. $ac - b^2 < 0$
$q(x, y)$ changes sign

Case 2. $ac - b^2 > 0$
$q(x, y)$ does not change sign
$q(x, y)$ is always positive if $a > 0$, and always negative if $a < 0$
$q(x, y)$ is a *definite* form

Case 3. $ac - b^2 = 0$
$q(x, y)$ does not change sign
$q(x, y)$ does take on the value zero
$q(x, y)$ is a *semidefinite* form

Rotation of Coordinates

The above properties of quadratic forms are all that we need in the remainder of this chapter. However, we can obtain considerably more information, as well as new insight into the facts already noted, if we ask the simple question

"what happens to a quadratic form under rotation of coordinates?"

Let us consider, then, new coordinates (X, Y) obtained by a rotation through an angle θ about the origin (Fig. 10.1). The relation between the old and the new coordinates of a point is given by

(10.11)
$$x = X \cos \theta - Y \sin \theta,$$
$$y = X \sin \theta + Y \cos \theta.$$

If we substitute these expressions for x and y, we obtain

(10.12) $$q(x, y) = Q(X, Y) = AX^2 + 2BXY + CY^2,$$

where A, B, C are new constants that depend on a, b, c, and θ. We are not interested in the precise expressions relating A, B, C to a, b, c, but only in the fact that what we obtain is again a quadratic form in the new coordinates. Further, it follows directly from Eq. (10.11) that

(10.13) $$x^2 + y^2 = X^2 + Y^2,$$

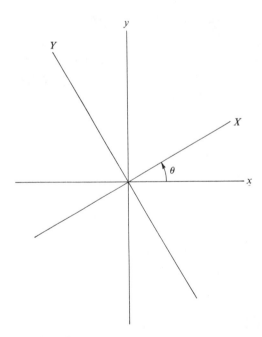

FIGURE 10.1 Rotation of axes

so that

$$\lambda_1 = \max_{x^2+y^2=1} q(x, y) = \max_{X^2+Y^2=1} Q(X, Y),$$

(10.14)

$$\lambda_2 = \min_{x^2+y^2=1} q(x, y) = \min_{X^2+Y^2=1} Q(X, Y),$$

expressing the fact that $Q(X, Y)$ and $q(x, y)$ represent exactly the same function, but in different coordinates, so that their maximum and minimum on the unit circle ($x^2 + y^2 = 1$ or $X^2 + Y^2 = 1$) are the same. We may therefore apply all our previous results to the form $Q(X, Y)$, and in particular, from Eq. (10.10), we have

(10.15) $ac - b^2 = \lambda_1\lambda_2 = AC - B^2,$ $a + c = \lambda_1 + \lambda_2 = A + C.$

Equation (10.15) is usually expressed by saying that $ac - b^2$ and $a + c$ are *invariants* of a quadratic form under rotation of coordinates.

We now make a special rotation that simplifies the expression for $Q(X, Y)$. We suppose that $q(x, y)$ has its maximum on $x^2 + y^2 = 1$ at (x_0, y_0) and we choose the positive X axis to pass through this point. This means that in the new coordinates, the point $x = x_0$, $y = y_0$ corresponds to $X = 1$, $Y = 0$. Thus,

$$\lambda_1 = q(x_0, y_0) = Q(1, 0) = A$$

by Eq. (10.12). But substituting $\lambda_1 = A$ in the second of Eqs. (10.15), $\lambda_1 + \lambda_2 = A + C$, yields $\lambda_2 = C$. Then $\lambda_1\lambda_2 = AC$, and from the first of Eqs. (10.15), $\lambda_1\lambda_2 = AC - B^2$, we conclude that $B^2 = 0$, or $B = 0$. We may summarize as follows.

Theorem 10.2   Reduction to Diagonal Form   *Every quadratic form $q(x, y)$ may be transformed into the normal form*

(10.16) $Q(X, Y) = \lambda_1 X^2 + \lambda_2 Y^2,$ $\lambda_1 \geq \lambda_2,$

by a rotation of coordinates. The coefficients λ_1, λ_2 represent the maximum and minimum values of the quadratic form on the unit circle.

Theorem 10.2 has many important consequences. We list the following.

Corollary 1   *A quadratic form $q(x, y)$ is constant on $x^2 + y^2 = 1 \Leftrightarrow q(x, y) = c(x^2 + y^2)$; that is, $a = c$, $b = 0$.*

PROOF.   Clearly $c(x^2 + y^2)$ is constant for $x^2 + y^2 = 1$. Conversely, suppose $q(x, y) \equiv c$ on $x^2 + y^2 = 1$. Then $\lambda_1 = \lambda_2 = c$, and after rotating coordinates to obtain Eq. (10.16), we have

$$q(x, y) = Q(X, Y) = c(X^2 + Y^2) = c(x^2 + y^2),$$

using Eq. (10.13).   ◆

Corollary 2 *If $q(x, y)$ is not constant on $x^2 + y^2 = 1$, then the points where $q(x, y)$ assumes its maximum and minimum lie on a pair of perpendicular lines through the origin.*

PROOF. If $q(x, y)$ is not constant, then $\lambda_1 > \lambda_2$, and

$$q(x, y) = \lambda_1 \Leftrightarrow Q(X, Y) = \lambda_1 \Leftrightarrow X = \pm 1, \, Y = 0;$$

$$q(x, y) = \lambda_2 \Leftrightarrow Q(X, Y) = \lambda_2 \Leftrightarrow X = 0, \, Y = \pm 1.$$

Thus the lines through the maximum and minimum points are the X and Y axes. $\blacklozenge$

Corollary 3 *If $q(x, y)$ is positive definite, then the equation $q(x, y) = 1$ defines an ellipse (or circle). If $ac - b^2 < 0$, then $q(x, y) = 1$ defines a hyperbola.*

PROOF. If $q(x, y)$ is positive definite, then $\lambda_1 \geq \lambda_2 > 0$, and the equation $\lambda_1 X^2 + \lambda_2 Y^2 = 1$ defines an ellipse if $\lambda_1 > \lambda_2$, a circle if $\lambda_1 = \lambda_2$. If $ac - b^2 < 0$, then $\lambda_1 \lambda_2 < 0$, and we have a hyperbola. $\blacklozenge$

We examine somewhat more closely the positive definite case. If

$$q(x, y) = ax^2 + 2bxy + cy^2$$

is positive definite, then by rotating coordinates, we have

$$(10.17) \qquad q(x, y) = 1 \Leftrightarrow \lambda_1 X^2 + \lambda_2 Y^2 = 1 \Leftrightarrow \frac{X^2}{m^2} + \frac{Y^2}{n^2} = 1,$$

where

$$(10.18) \qquad m = \frac{1}{\sqrt{\lambda_1}}, \qquad n = \frac{1}{\sqrt{\lambda_2}},$$

are the semiminor and semimajor axes, respectively, of the ellipse (Fig. 10.2). Since

$$(10.19) \quad ac - b^2 = \lambda_1 \lambda_2 = \frac{1}{m^2 n^2}, \qquad a + c = \lambda_1 + \lambda_2 = \frac{1}{m^2} + \frac{1}{n^2},$$

we see the geometric significance of Eq. (10.15). These quantities are invariant because they may be expressed in terms of the fundamental dimensions of the ellipse, which have nothing to do with coordinates. In particular, the area inside the ellipse defined by Eq. (10.17) is equal to πmn. We have therefore the following consequence of Eq. (10.19).

Corollary 4 *If the equation $ax^2 + 2bxy + cy^2 = 1$ represents an ellipse, then the area inside the ellipse is equal to $\pi/(ac - b^2)^{1/2}$.*

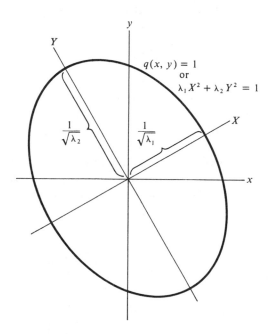

FIGURE 10.2   Geometric significance of the quantities λ_1, λ_2 associated with a positive definite quadratic form

We conclude this section with several remarks directed toward those who have some familiarity with matrix theory.

With the quadratic form $ax^2 + 2bxy + cy^2$, one associates the symmetric matrix

(10.20) $$M = \begin{pmatrix} a & b \\ b & c \end{pmatrix}.$$

Its determinant is $ac - b^2$, and its *trace* is $a + c$. The roots λ_1, λ_2 of Eq. (10.8) are called the *characteristic values* or *eigenvalues* of the matrix M. If we substitute these values in Eq. (10.4a) and solve for x_0, y_0, the resulting vectors $\langle x_0, y_0 \rangle$ are called the *characteristic vectors* or *eigenvectors* of M. We may summarize the results of this section as follows.

To each symmetric matrix M one may associate a quadratic form $q(x, y)$. (Namely, to the matrix (10.20) one associates the form $q(x, y) = ax^2 + 2bxy + cy^2$.) The matrix M has real characteristic values, which are the maximum and minimum of $q(x, y)$ on $x^2 + y^2 = 1$. The characteristic vectors of M are orthogonal. Under a rotation of coordinates, the same quadratic form is associated with a new symmetric matrix $\tilde{M}$, which has the same determinant and trace as M. One may always choose coordinates so that the matrix $\tilde{M}$ is in diagonal form:

$$\tilde{M} = \begin{pmatrix} \lambda_1 & 0 \\ 0 & \lambda_2 \end{pmatrix}.$$

To do this the new coordinate axes should be chosen in the direction of the characteristic vectors of M. The elements λ_1, λ_2 on the diagonal of $\tilde{M}$ are precisely the characteristic values of M.

All of these results for 2×2 matrices extend to symmetric matrices of any size. They are proved for the general case in books on linear algebra. The geometric interpretation of these facts in the two-dimensional case can be of great value in understanding the general situation in higher dimensions.

Exercises

10.1 Using Th. 10.1, find the maximum and minimum on $x^2 + y^2 = 1$ of each of the following quadratic forms.

a. $2x^2 + 2xy + 2y^2$ *g.* $-x^2 + 2xy + y^2$
b. $2x^2 - 2xy + 2y^2$ *h.* $-4x^2 - 12xy - 9y^2$
c. $x^2 - 4xy + y^2$ *i.* $-2x^2 - 2xy + y^2$
d. $x^2 - 4xy + 4y^2$ *j.* $x^2 - xy + y^2$
e. $-3x^2 + 4xy - 3y^2$ *k.* $x^2 - 3xy + 4y^2$
f. $3x^2 - 8xy + 3y^2$ *l.* $5x^2 - 3y^2$

10.2 Verify Eq. (10.10) directly for each part of Ex. 10.1.

10.3 Using Eqs. (10.4a) (or (10.4b)), find the points at which the maximum and minimum occur in Ex. 10.1a, b, c, d.

10.4 For each part of Ex. 10.3 show that:

a. The maximum occurs at a pair of points lying on a straight line through the origin.
b. The minimum occurs at a pair of points lying on a straight line through the origin.
c. These two lines are perpendicular.

10.5 Classify each of the following quadratic forms according to the five categories listed in the text: I—positive definite; II—negative definite; III—positive semidefinite; IV—negative semidefinite; V—takes on both positive and negative values.

a. $x^2 - 2xy + y^2$ *d.* $-x^2 + 3xy - 3y^2$
b. $x^2 - 2xy - y^2$ *e.* $-x^2 + 3xy - 2y^2$
c. $2x^2 - 2xy + y^2$ *f.* $-x^2 + 4xy - 4y^2$

10.6   Let λ_1 be the maximum and λ_2 the minimum of the quadratic form $q(x, y) = ax^2 + 2bxy + cy^2$ on $x^2 + y^2 = 1$. Find the explicit expressions for λ_1 and λ_2 in terms of a, b, c. (*Hint:* use Eq. (10.8).) Use these expressions to verify your answers to Ex. 10.1a, b, c, d.

10.7   Using Ex. 10.6, show that $\lambda_1 = \lambda_2$ (that is, $q(x, y)$ is constant on $x^2 + y^2 = 1$) if and only if $a = c$, $b = 0$ (that is, $q(x, y) = a(x^2 + y^2)$).

10.8   By virtue of the first of Eqs. (10.4b), the point on $x^2 + y^2 = 1$ where $q(x, y)$ attains its maximum λ_1 must lie on the straight line $(a - \lambda_1)x + by = 0$, while the point where it attains its minimum λ_2 must lie on the straight line $(a - \lambda_2)x + by = 0$.

    a. Write down the condition for these two lines to be perpendicular.
    b. Verify that this condition is satisfied. (*Hint:* use Eq. (10.10).)

10.9   Let $q(x, y) = (\alpha x + \beta y)^2$.

    a. Find the coefficients a, b, c in terms of α, β.
    b. Show that $q(x, y)$ is positive semidefinite.

10.10   Let $q(x, y) = ax^2 + 2bxy + cy^2$ be positive semidefinite.

    a. Show that the maximum of $q(x, y)$ on $x^2 + y^2 = 1$ is $\lambda_1 = a + c$.
    b. Show that $a \geq 0$ and $c \geq 0$. (Note that $a = q(1, 0)$, $c = q(0, 1)$.)
    c. Show that $q(x, y) = (\alpha x \pm \beta y)^2$, where $\alpha = a^{1/2}$, $\beta = b^{1/2}$.
    d. What are the level lines of $q(x, y)$?
    e. Sketch the surface $z = q(x, y)$.

10.11   Let $q(x, y)$ be negative semidefinite. What would be the analogous statements for Ex. 10.10a, b, c? Answer Ex. 10.10d, e in this case.

10.12   Prove the equivalence of the statements listed under Conditions V in the text, for quadratic forms that take on both positive and negative values.

10.13   Important use was made in this section of the following fact: if a pair of homogeneous linear equations

$$\alpha x + \beta y = 0,$$
$$\gamma x + \delta y = 0$$

has a simultaneous solution (x_0, y_0) different from $(0, 0)$, then the determinant $\alpha\delta - \beta\gamma$ must be zero. Explain carefully why this is true.

10.14   If $q(x, y) = ax^2 + 2bxy + cy^2$, and if $Q(X, Y) = AX^2 + 2BXY + CY^2$ is the corresponding form obtained by a rotation of coordinates, Eq. (10.11), find the explicit expressions for A, B, and C in terms of a, b, c, and θ.

10.15   Using the expressions in Ex. 10.14, show directly that $A + C = a + c$.

*10.16   Using the expressions in Ex. 10.14, show directly that $AC - B^2 = ac - b^2$.

10.17 Use the notation and results of Ex. 10.14.

a. Show that $B = \frac{1}{2}(c - a) \sin 2\theta + b \cos 2\theta$.

b. Show that the angle θ through which coordinates must be rotated in order to reduce $q(x, y)$ to the normal form of Eq. (10.16) satisfies

$$\tan 2\theta = 2b/(a - c).$$

c. Show that if $ax^2 + 2bxy + cy^2 = 1$ defines an ellipse, then the angle between its major axis and the x axis is the angle θ of part b.

d. Show that for the ellipse of part c, if $a = c$, then the major and minor axes lie along the diagonals between the x and y axes.

10.18 Note that in order to write down the normal form Eq. (10.16) of a quadratic form, obtained by a rotation of coordinates, it is *not* necessary to carry out the rotation of coordinates, or even to know the angle through which they are rotated. The final result can be written down directly as soon as the quantities λ_1 and λ_2 are known. Use this to write down the normal form of each of the following quadratic forms.

a. $x^2 + 4xy + y^2$ *d.* $-3x^2 + 4xy - 3y^2$
b. $x^2 + 4xy + 4y^2$ *e.* xy
c. $x^2 + xy + y^2$

10.19 Describe the curve defined by each of the following equations, and in case it is an ellipse give the area enclosed by it.

a. $x^2 + 9y^2 = 1$ *d.* $4x^2 + 4xy + 5y^2 = 1$
b. $2x^2 = 1$ *e.* $2x^2 - 4xy + 2y^2 = 1$
c. $2x^2 - 5xy + 3y^2 = 1$ *f.* $-3x^2 + 7xy - 5y^2 = 1$

10.20 Describe all the level curves of $q(x, y) = ax^2 + 2bxy + cy^2$, making sure to consider $q(x, y) = k$ for k positive, negative, and zero, where:

a. $q(x, y)$ is positive definite
b. $q(x, y)$ is negative definite
c. $q(x, y)$ is positive semidefinite
d. $q(x, y)$ is negative semidefinite
e. $ac - b^2 < 0$

(*Hint:* it may be easiest to consider a rotation of coordinates that transforms $q(x, y)$ into normal form.)

10.21 Sketch the surface $z = q(x, y)$ for each of the parts of Ex. 10.19.

10.22 Using the methods developed in this section, find the maximum and minimum of $x^2 + y^2$ under the condition $2x^2 + 2xy + 2y^2 = 1$.

10.23 *a.* What is the relation between the answers to Ex. 10.22 and to Ex. 10.1a?

b. What is the geometric interpretation of Ex. 10.22?

*10.24 Let $q(x, y)$ be a positive definite quadratic form, and let λ_1, λ_2 be the maximum and minimum of $q(x, y)$ on $x^2 + y^2 = 1$. Show that

a. $q(x, y)/(x^2 + y^2)$ is constant on every ray through the origin.

b. $\lambda_1 = \max \dfrac{q(x, y)}{x^2 + y^2}$,   for all $(x, y) \neq (0, 0)$

c. $\dfrac{1}{\lambda_1} = \min \dfrac{x^2 + y^2}{q(x, y)}$,   for all $(x, y) \neq (0, 0)$

d. $\dfrac{1}{\lambda_1} = \min (x^2 + y^2)$,   for $q(x, y) = 1$

e. $\dfrac{1}{\sqrt{\lambda_1}}$ is the length of the semiminor axis of the ellipse $q(x, y) = 1$.

10.25 **a.** Show that if $q(x, y)$ is a positive definite quadratic form, then the gradient of $q(x, y)$ vanishes only at the origin.

b. Let $ex^2 + 2fxy + gy^2$ be a positive definite quadratic form. Using the methods of this section, show that the maximum and minimum of $ax^2 + 2bxy + cy^2$ under the condition $ex^2 + 2fxy + gy^2 = 1$ are the roots of the equation

$$(eg - f^2)\lambda^2 - (ag + ce - 2bf)\lambda + (ac - b^2) = 0.$$

10.26 Let $f(x, y)$ and $g(x, y)$ be homogeneous functions of degree $k \neq 0$. Suppose that $f(x, y)$ has a maximum or minimum under the condition $g(x, y) = 1$ at a point (x_0, y_0) where $\nabla g(x_0, y_0) \neq 0$. Then by the Lagrange multiplier method there exists a λ such that

$$\nabla f(x_0, y_0) = \lambda \nabla g(x_0, y_0), \qquad\qquad g(x_0, y_0) = 1.$$

Show that $\lambda = f(x_0, y_0)$. (*Hint:* use Euler's theorem for homogeneous functions.)

10.27 Find the minimum of $\cos^2 t - 2 \cos t \sin t + 3 \sin^2 t$ for $0 \leq t \leq 2\pi$.

10.28 Show that every quadratic polynomial $ax^2 + 2bxy + cy^2 + 2dx + 2ey + f$ can be reduced to the form $AX^2 + BY^2 + C$ by a rotation and translation of coordinates in the x, y plane. (*Hint:* first remove the xy term by a rotation, and then complete the squares.)

10.29 The following approach to quadratic forms yields their principal properties without the use of calculus.

a. Show that every quadratic form

$$q(x, y) = ax^2 + 2bxy + cy^2$$

can be expressed in either of the two forms

$$q(x, y) = r^2(A \cos 2\theta + B \sin 2\theta + C)$$

or

$$q(x, y) = r^2(D \cos (2\theta - \alpha) + C)$$

using polar coordinates $x = r \cos \theta$, $y = r \sin \theta$. Find the constants A, B, C, D explicitly in terms of a, b, c.

b. Using part a, find the maximum and minimum of $q(x, y)$ on $x^2 + y^2 = 1$.

c. Show that if $q(x, y)$ takes on its maximum on $x^2 + y^2 = 1$ for $\theta = \theta_0$, then it takes on its minimum for $\theta = \theta_0 + \frac{1}{2}\pi$.

11   Higher order derivatives

If $f(x, y)$ is continuously differentiable, then f_x and f_y are by definition continuous functions of x and y. We may ask whether they in turn are differentiable. If so, we may form $(f_x)_x$, $(f_x)_y$, etc. We adopt the following notation:

$$f_{xx} = (f_x)_x = \frac{\partial^2 f}{\partial x^2} = \frac{\partial}{\partial x}\left(\frac{\partial f}{\partial x}\right)$$

$$f_{xy} = (f_x)_y = \frac{\partial^2 f}{\partial y \partial x} = \frac{\partial}{\partial y}\left(\frac{\partial f}{\partial x}\right)$$

$$f_{yx} = (f_y)_x = \frac{\partial^2 f}{\partial x \partial y} = \frac{\partial}{\partial x}\left(\frac{\partial f}{\partial y}\right)$$

$$f_{yy} = (f_y)_y = \frac{\partial^2 f}{\partial y^2} = \frac{\partial}{\partial y}\left(\frac{\partial f}{\partial y}\right).$$

These are called the *second-order partial derivatives of $f(x, y)$*.

Example 11.1

$$f(x, y) = x^4 + 3xy^3$$

$$f_x = 4x^3 + 3y^3, \qquad f_y = 9xy^2$$

$$f_{xx} = 12x^2, \qquad f_{xy} = 9y^2, \qquad f_{yx} = 9y^2, \qquad f_{yy} = 18xy$$

Example 11.2

$$f(x, y) = e^{x \sin y}$$

$$f_x = e^{x \sin y} \sin y, \qquad f_y = e^{x \sin y} x \cos y$$

$$f_{xx} = e^{x \sin y} \sin^2 y,$$

$$f_{xy} = e^{x \sin y}(\cos y + x \sin y \cos y),$$

$$f_{yx} = e^{x \sin y}(\cos y + x \sin y \cos y),$$

$$f_{yy} = e^{x \sin y}(-x \sin y + x^2 \cos^2 y).$$

The equality of the so-called "mixed partial derivatives" f_{xy} and f_{yx} in the first example might pass for a coincidence, but it seems improbable in the second example unless there is a general rule. Indeed there is, and it is the following.

Theorem 11.1   *If f_{xy} and f_{yx} exist and are continuous, then they are equal.*

PROOF.   Choose any point (x_0, y_0), and assume that the hypotheses of the theorem are satisfied inside some circle about (x_0, y_0). Let (x_1, y_1) be any other point in this circle, and form the function (see Fig. 11.1)

(11.1) $g(y) = f(x_1, y) - f(x_0, y)$.

By the mean-value theorem there is a y_2 between y_0 and y_1 such that

(11.2) $g(y_1) - g(y_0) = g'(y_2)(y_1 - y_0)$.

But by Eq. (11.1),

(11.3) $g'(y_2) = f_y(x_1, y_2) - f_y(x_0, y_2) = (x_1 - x_0)f_{yx}(x_2, y_2)$,

where the second equality is obtained by applying the mean-value theorem to the function $f_y(x, y_2)$. Combining Eqs. (11.1), (11.2), and (11.3) yields

(11.4) $f(x_1, y_1) - f(x_0, y_1) - f(x_1, y_0) + f(x_0, y_0)$
$$= (x_1 - x_0)(y_1 - y_0)f_{yx}(x_2, y_2)$$

where (x_2, y_2) is some point in the rectangle indicated in Fig. 11.1.

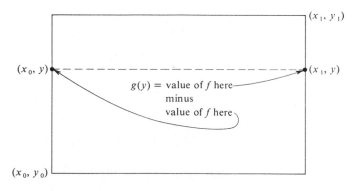

FIGURE 11.1   Notation for proof that $f_{xy} = f_{yx}$

The same reasoning applied to the function
$$h(x) = f(x, y_1) - f(x, y_0)$$
yields
$$h(x_1) - h(x_0) = h'(x_3)(x_1 - x_0)$$
and
$$h'(x_3) = f_x(x_3, y_1) - f_x(x_3, y_0) = (y_1 - y_0)f_{xy}(x_3, y_3),$$
whence

(11.5) $f(x_1, y_1) - f(x_1, y_0) - f(x_0, y_1) + f(x_0, y_0)$
$$= (x_1 - x_0)(y_1 - y_0)f_{xy}(x_3, y_3),$$

where (x_3, y_3) is again a point in the same rectangle.

Comparing Eqs. (11.4) and (11.5), we see that

(11.6) $$f_{yx}(x_2, y_2) = f_{xy}(x_3, y_3).$$

Finally, as $(x_1, y_1) \to (x_0, y_0)$ we have $(x_2, y_2) \to (x_0, y_0)$ and $(x_3, y_3) \to (x_0, y_0)$, so that the continuity of f_{xy} and f_{yx} applied to Eq. (11.6) yields $f_{yx}(x_0, y_0) = f_{xy}(x_0, y_0)$, which proves the theorem. ◆

Remark Lest it be thought that equality of f_{xy} and f_{yx} is too obvious to need a proof, we give the following example.

Let

$$f(x, y) = \begin{cases} 0 & \text{if } (x, y) = (0, 0) \\ xyg(x, y) & \text{if } (x, y) \neq (0, 0), \end{cases}$$

where $g(x, y)$ is defined and differentiable everywhere except at the origin. We shall see how to choose $g(x, y)$ so that $f_{xy} \neq f_{yx}$. First we note that $f \equiv 0$ on the x and y axes so that $f_x(0, 0) = f_y(0, 0) = 0$. Next, if $(x, y) \neq (0, 0)$, we have

$$f_x(x, y) = yg(x, y) + xyg_x(x, y),$$

$$f_y(x, y) = xg(x, y) + xyg_y(x, y).$$

In particular,

$$f_x(0, y) = yg(0, y), \qquad f_y(x, 0) = xg(x, 0);$$

so that

$$f_{xy}(0, 0) = \lim_{y \to 0} \frac{f_x(0, y) - f_x(0, 0)}{y} = \lim_{y \to 0} g(0, y),$$

$$f_{yx}(0, 0) = \lim_{x \to 0} \frac{f_y(x, 0) - f_y(0, 0)}{x} = \lim_{x \to 0} g(x, 0),$$

and the question reduces to the equality of these two limits. Now if $g(x, y)$ is continuous at the origin, then both limits are the same and equal $g(0, 0)$. However, a simple choice, such as

$$g(x, y) = \frac{x^2 - y^2}{x^2 + y^2},$$

yields

$$g(x, 0) \equiv 1, \qquad g(0, y) \equiv -1,$$

and

$$f_{xy}(0, 0) = -1, \qquad f_{yx}(0, 0) = 1.$$

In the example that we have just given, the partial derivatives f_{xy} and f_{yx} exist everywhere, but they are not continuous at the origin, and hence Theorem 11.1 does not apply. The case where some derivative exists but is not continuous may be considered exceptional. In most problems, either derivatives of all orders exist and are continuous, as in the case of polynomials, trigonometric functions, and exponential functions, or else some derivative fails to exist, as in the case of fractional exponents. It is convenient to introduce the notation

$$f \in \mathscr{C}^1,$$

meaning f is continuously differentiable, and

$$f \in \mathscr{C}^2,$$

meaning all second-order derivatives of f exist and are continuous. Thus, if $f \in \mathscr{C}^2$, Th. 11.1 applies, and $f_{xy} = f_{yx}$, so that f has essentially three distinct second-order derivatives f_{xx}, f_{xy}, f_{yy}.

In a similar manner we can form higher order derivatives, where an *nth-order partial derivative* is obtained by n successive differentiations, each with respect to either x or y. We introduce the notation

$$\mathscr{C}^n \quad \text{equals} \quad \begin{bmatrix} \text{the class of functions } f(x, y), \text{ all of whose } n\text{th-order} \\ \text{partial derivatives exist and are continuous} \end{bmatrix},$$

$f \in \mathscr{C}^n$ means [f is a member of the class $\mathscr{C}^n$].

Now if $f \in \mathscr{C}^n$, then all the $(n - 1)$-order partial derivatives of f are continuously differentiable, hence continuous, and therefore $f \in \mathscr{C}^{n-1}$. Thus, the classes $\mathscr{C}^n$ are each included in the previous ones:

$$\mathscr{C}^n \subset \mathscr{C}^{n-1} \subset \cdots \subset \mathscr{C}^2 \subset \mathscr{C}^1 \subset \mathscr{C}^0,$$

where the notation $\mathscr{C}^0$ denotes the class of continuous functions. Furthermore, we have

$$f \in \mathscr{C}^n \Leftrightarrow f_x, f_y \in \mathscr{C}^{n-1} \Leftrightarrow f_{xx}, f_{xy}, f_{yy} \in \mathscr{C}^{n-2} \Leftrightarrow \cdots$$

Applying Th. 11.1, which allows interchanging the order of differentiation at each step, we come to the following conclusion.

If $f \in \mathscr{C}^n$, then the nth-order partial derivatives do not depend on the order of differentiation, but only on the total number of times f is differentiated with respect to x and y. If f is differentiated k times with respect to x and l times with respect to y, we obtain an *n*th-order derivative if $k + l = n$. The notation for this is

$$\frac{\partial^n f}{\partial x^k \, \partial y^l}, \qquad\qquad k + l = n.$$

Thus the third derivatives of $f(x, y)$ are

$$\frac{\partial^3 f}{\partial x^3}, \qquad \frac{\partial^3 f}{\partial x^2 \, \partial y}, \qquad \frac{\partial^3 f}{\partial x \, \partial y^2}, \qquad \frac{\partial^3 f}{\partial y^3}.$$

In general, a function $f(x, y) \in \mathscr{C}^n$ has $n + 1$ distinct nth-order derivatives

$$\frac{\partial^n f}{\partial x^k \, \partial y^{n-k}},$$

$k = 0, 1, 2, \ldots, n$.

Example 11.3

$$f = xe^y,$$

$$f_x = e^y, \qquad\qquad f_y = xe^y,$$

$$f_{xx} = 0, \qquad\qquad f_{xy} = e^y, \qquad\qquad f_{yy} = xe^y,$$

$$f_{xxx} = 0, \qquad f_{xxy} = 0, \qquad f_{xyy} = e^y, \qquad f_{yyy} = xe^y$$

$$\vdots$$

$$\frac{\partial^n f}{\partial x^n} = 0, \quad \frac{\partial^n f}{\partial x^{n-1} \, \partial y} = 0, \quad \ldots, \quad \frac{\partial^n f}{\partial x^2 \, \partial y^{n-2}} = 0, \quad \frac{\partial^n f}{\partial x \, \partial y^{n-1}} = e^y, \quad \frac{\partial^n f}{\partial y^n} = xe^y.$$

Thus, for this particular function, partial differentiation any number of times with respect to the variable y leaves the function unchanged; all derivatives involving one differentiation with respect to x and the rest with respect to y are equal to e^y; and all derivatives involving two or more differentiations with respect to x are equal to zero.

Higher Order Directional Derivatives

The meaning of the second-order partial derivatives $f_{xx}(x_0, y_0)$, and $f_{yy}(x_0, y_0)$ is quite clear, since they are the ordinary second derivatives of the functions $f(x, y_0)$ and $f(x_0, y)$, respectively. The necessity for considering the quantity f_{xy} becomes clear if we try to compute the ordinary second-derivative of f with respect to arc length along an arbitrary straight line through (x_0, y_0). If we let

$$C: x = x_0 + s \cos \alpha, y = y_0 + s \sin \alpha \qquad \text{for a fixed } \alpha$$

denote such a line, and use again the notation

$$f_C(s) = f(x_0 + s \cos \alpha, y_0 + s \sin \alpha),$$

then

$$f'_C(s) = \nabla_\alpha f.$$

We now introduce the notation

$$\nabla_\alpha^2 f = f''_C(s), \ldots, \nabla_\alpha^n f = f_C^{(n)}(s),$$

wherever the derivatives on the right exist. Thus, *the nth-order directional derivative of f in the direction α is the ordinary nth-order derivative of f with respect to arc length along the straight line in the direction α.*

Suppose now that $f \in \mathscr{C}^2$. Then

$$f_C'(s) = \nabla_\alpha f = f_x \cos \alpha + f_y \sin \alpha,$$

and since the right-hand side is continuously differentiable, we have

$$f_C''(s) = \nabla_\alpha(\nabla_\alpha f) = \nabla_\alpha(f_x \cos \alpha + f_y \sin \alpha)$$
$$= (f_x \cos \alpha + f_y \sin \alpha)_x \cos \alpha + (f_x \cos \alpha + f_y \sin \alpha)_y \sin \alpha,$$

or

(11.7) $$\nabla_\alpha^2 f = f_{xx} \cos^2 \alpha + 2f_{xy} \cos \alpha \sin \alpha + f_{yy} \sin^2 \alpha.$$

The importance of this formula is that it allows us to describe very accurately how the surface $z = f(x, y)$ behaves near any given point (x_0, y_0). We need only observe that if the left-hand side, $\nabla_\alpha^2 f(x_0, y_0)$, is positive, then $f_C'' > 0$, so that the curve on the surface lying over the straight line C is concave upwards (Fig. 11.2). Similarly, $\nabla_\alpha^2 f < 0$ means that this curve is concave downwards. Suppose now that we fix (x_0, y_0) and set

(11.8) $$a = f_{xx}(x_0, y_0), \qquad b = f_{xy}(x_0, y_0), \qquad c = f_{yy}(x_0, y_0);$$
$$X = \cos \alpha, \qquad Y = \sin \alpha.$$

Then a, b, c are constants, and as α varies from 0 to 2π, (X, Y) goes once around the circle $X^2 + Y^2 = 1$. Equation (11.7) becomes

(11.9) $$\nabla_\alpha^2 f(x_0, y_0) = aX^2 + 2bXY + cY^2, \qquad X^2 + Y^2 = 1,$$

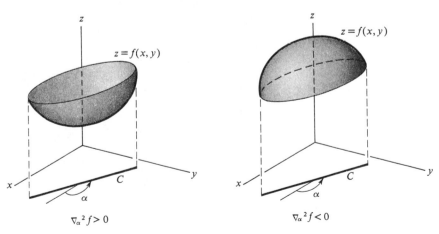

$$\nabla_\alpha^2 f > 0 \qquad\qquad\qquad \nabla_\alpha^2 f < 0$$

FIGURE 11.2 Geometric interpretation of the sign of the second directional derivative

and we see that the totality of values of $\nabla_\alpha^2 f(x_0, y_0)$ is precisely the set of values of a quadratic form on the unit circle. We may apply all the results of the previous section to describe the behavior of the surface $z = f(x, y)$. For example, if the quadratic form is positive definite, then over each straight line the surface will be convex upwards, while if it is negative definite, the surface will be convex downwards (Fig. 11.3). If the quadratic form takes on both positive and negative values, then the surface will be convex upwards over certain lines through (x_0, y_0) and convex downwards over others. Such a point is called a *saddle point* (Fig. 11.4).

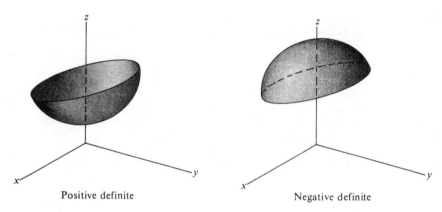

Positive definite   Negative definite

FIGURE 11.3 Picture of surfaces whose second directional derivatives are positive definite or negative definite

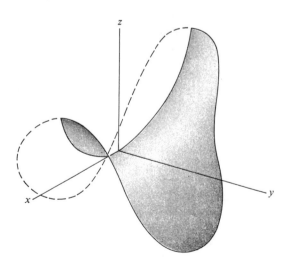

FIGURE 11.4 Saddle surface

Example 11.4

$$f(x, y) = x^2 - y^2;$$

$$f_x = 2x, \qquad f_y = -2y;$$

$$f_{xx} \equiv 2, \qquad f_{xy} \equiv 0, \qquad f_{yy} \equiv -2.$$

This surface is concave up in the x direction and concave down in the y direction. Every point is a saddle point.

We shall discuss these cases in more detail in the next section, and also their use in maximum-minimum problems. We give now just one illustration of this line of reasoning.

Theorem 11.2   *Suppose $f(x, y) \in \mathscr{C}^2$ and*

(11.10) $$f_{xx} f_{yy} - f_{xy}^2 < 0, \qquad\qquad at \quad (x_0, y_0).$$

Then $f(x, y)$ can have neither a local maximum nor a local minimum at (x_0, y_0).

Remark   This is a result that has no analog for functions of a single variable, since there is no condition analogous to (11.10).

PROOF.   Condition (11.10) is precisely the condition $ac - b^2 < 0$ that guarantees that the quadratic form (11.9) takes on both positive and negative values. In other words, there are a pair of lines C_1, C_2 through (x_0, y_0) such that for the corresponding directions α_1, α_2 we have $\nabla_{\alpha_1}^2 f > 0$, $\nabla_{\alpha_2}^2 f < 0$. Thus, $f_{C_1}'' > 0$ and $f_{C_2}'' < 0$. But from the theory of functions of a single variable, this means that f_{C_1} cannot have a local maximum and f_{C_2} cannot have a local minimum. But if $f(x, y)$ had a local maximum or minimum at (x_0, y_0), then the same would be true of both f_{C_1} and f_{C_2}, which are merely restrictions of f to straight lines. Hence f can have neither, and the theorem is proved.   ◆

Remark   Condition (11.10) means geometrically that (x_0, y_0) is a saddle point, so that the theorem is intuitively obvious.

Example 11.5

$$f(x, y) = x^2 - y^2.$$

Here

$$f_{xx} f_{yy} - f_{xy}^2 = -4 < 0.$$

This function cannot have a local maximum or minimum anywhere.

Example 11.6

$f(x, y)$ is a *harmonic* function. This means that $f(x, y)$ satisfies the equation

$$(11.11) \qquad\qquad f_{xx} + f_{yy} = 0.$$

This equation is met in many areas of mathematics and its applications, and the study of its solutions has led to a theory, known as *potential theory*. We note that from Eq. (11.11) we have $f_{xx} = -f_{yy}$, and hence

$$f_{xx}f_{yy} - f_{xy}^2 = -f_{yy}^2 - f_{xy}^2 \leq 0$$

everywhere, with equality possible only if

$$f_{xx} = -f_{yy} = 0 \qquad \text{and} \qquad f_{xy} = 0.$$

We have the following basic property: *a harmonic function cannot have a local maximum or minimum unless it is constant*. In fact, by what we have just observed, either

$$f_{xx}f_{yy} - f_{xy}^2 < 0,$$

in which case f cannot have a local maximum or minimum by Th. 11.2 or else

$$f_{xx}f_{yy} - f_{xy}^2 = 0,$$

in which case all the second-partial derivatives are zero. We can also show in this case that f cannot have a local maximum or minimum unless all higher derivatives vanish too, in which case f is constant.[7] Examples of harmonic functions include all constants, all linear functions, and the quadratic function given in Example 11.5.

We conclude this section with the observation that Eq. (11.7) may be extended by induction to the general formula

$$(11.12) \quad \nabla_\alpha^n f = \frac{\partial^n f}{\partial x^n} \cos^n \alpha + n \frac{\partial^n f}{\partial x^{n-1} \partial y} \cos^{n-1} \alpha \sin \alpha + \cdots + \frac{\partial^n f}{\partial y^n} \sin^n \alpha$$

$$= \sum_{k=0}^{n} \binom{n}{k} \frac{\partial^n f}{\partial x^{n-k} \partial y^k} \cos^{n-k} \alpha \sin^k \alpha,$$

where $\binom{n}{k}$ is the *binomial coefficient*,

$$\binom{n}{k} = \frac{n(n-1)\cdots(n-k+1)}{1 \cdot 2 \cdots k} = \frac{n!}{k!(n-k)!}.$$

---

[7] See Exercise 26.16. For other proofs, see [29], Section 28, and [31], Chapter 2, Theorem 2. Compare also Corollary 2 of Theorem 12.3. Both that Corollary and the property mentioned here are known as *the maximum principle* for harmonic functions.

Exercises

11.1   Find all second-order partial derivatives of the following functions.

a. $f(x, y) = x^6 - 5x^4y^2 + 3xy^5$ *d.* $f(x, y) = \tan(x + y)$
b. $f(x, y) = e^x + \log y$ *e.* $f(x, y) = (x - y)^5$
c. $f(x, y) = \sin(y/x)$ *f.* $f(x, y) = \sqrt{x + y} + \log(x - y)$,

11.2   Show that each of the following functions $f(x, y)$ is harmonic, that is, $f_{xx} + f_{yy} \equiv 0$.

a. $f(x, y) = xy$ *c.* $f(x, y) = x/(x^2 + y^2)$
b. $f(x, y) = x^3 - 3xy^2$ *d.* $f(x, y) = \sin x \cosh y$

11.3   Suppose that the functions $f(x, y)$, $g(x, y)$ satisfy the Cauchy-Riemann equations

$$f_x = g_y, \qquad f_y = -g_x.$$

Show that if $f(x, y) \in \mathscr{C}^2$ and $g(x, y) \in \mathscr{C}^2$, then $f(x, y)$ and $g(x, y)$ are both harmonic.

11.4   Verify the statement of Ex. 11.3 for the following pairs of functions.

a. $f(x, y) = x^2 - y^2$, $g(x, y) = 2xy$
b. $f(x, y) = e^x \cos y$, $g(x, y) = e^x \sin y$
c. $f(x, y) = \log \sqrt{x^2 + y^2}$, $g(x, y) = \arctan(y/x)$.

11.5   *a.* Show that a quadratic form is a harmonic function if and only if it can be expressed as $a(x^2 - y^2) + 2bxy$, where a and b are arbitrary constants.
   b. Show that if $f(x, y) \in \mathscr{C}^3$ and $f(x, y)$ is harmonic, then

$$g(x, y) = f_x(x, y) \qquad \text{and} \qquad h(x, y) = f_y(x, y)$$

are harmonic also.

Note: in Exs. 11.6–11.9 assume that all functions are at least twice continuously differentiable.

11.6   Let $f(x, y) = g(x) + h(y)$.

   a. Show that $f_{xy} \equiv 0$.
   b. Show that if $g''(x) > 0$ and $h''(y) < 0$, then the surface $z = f(x, y)$ cannot have a local maximum or minimum at any point.
   c. Verify parts a and b for the function of Ex. 11.1b.

11.7   *a.* Show that the functions in Exs. 11.1d, e, f, satisfy $f_{xx} - f_{yy} \equiv 0$.
   b. Show that any function of the form

$$f(x, y) = g(x + y) + h(x - y)$$

satisfies the equation $f_{xx} - f_{yy} \equiv 0$.

11.8 Suppose that $f(x, y) = g(x^2 + y^2)$.

a. Find f_{xx}, f_{xy}, and f_{yy} in terms of the function g and its derivatives.
*b. Show that if $f(x, y)$ is a harmonic function that depends only on the distance r of the point (x, y) to the origin, then

$$f(x, y) = c \log r + d,$$

where c and d are constants. (*Hint:* write $f(x, y)$ in the form $g(x^2 + y^2)$. Using part a, show that, with $t = x^2 + y^2$, $g(t)$ satisfies the equation

$$\frac{d}{dt}(tg') = tg''(t) + g'(t) \equiv 0.)$$

11.9 Let $f(x, y)$ be homogeneous of degree k.

a. Show that $f(x, y)$ satisfies the equations

$$xf_{xx} + yf_{xy} = (k - 1)f_x, \qquad xf_{xy} + yf_{yy} = (k - 1)f_y.$$

b. Show that $f(x, y)$ satisfies

$$x^2f_{xx} + 2xyf_{xy} + y^2f_{yy} = k(k - 1)f.$$

11.10 Verify the equation in Ex. 11.9b for the functions of Exs. 11.1a, c, e.

11.11 Let $f(x, y) = e^x \sin y$.

a. Find all third-order derivatives of $f(x, y)$.
b. Find

$$\frac{\partial^4 f}{\partial y^4}, \qquad \frac{\partial^6 f}{\partial x^2\, \partial y^4}, \qquad \frac{\partial^9 f}{\partial x\, \partial y^8}.$$

c. Find

$$\frac{\partial^6 f}{\partial x^4\, \partial y^2}, \qquad \frac{\partial^6 f}{\partial y^6}, \qquad \frac{\partial^{10} f}{\partial y^{10}}, \qquad \frac{\partial^{50} f}{\partial y^{50}}.$$

d. Given any two integers k and l, with $k + l = n$, describe how $\partial^n f/\partial x^k\, \partial y^l$ depends on k and l.

11.12 Let $f(x, y) = e^{xy}$.

a. Find all fourth-order derivatives of $f(x, y)$.
b. What is $\partial^k f/\partial x^k$?
c. What is $\partial^{k+1} f/\partial x^k\, \partial y$?
d. What is $\partial^{k+1} f/\partial x\, \partial y^k$?
*e. What is the value at the origin of $\partial^{k+l} f/\partial x^k\, \partial y^l$, for arbitrary k and l?

11.13 Let $f(x, y) = 3x^2 - 8xy + 2y^2 + 12x - 4y + 17$.

a. Find $\nabla_\alpha^2 f$ for arbitrary α.
b. Find an α such that $\nabla_\alpha^2 f > 0$.
c. Find an α such that $\nabla_\alpha^2 f < 0$.
d. Can $f(x, y)$ have a local maximum or minimum at any point?

11.14   Let $f(x, y) = ax^2 + 2bxy + cy^2 + dx + ey + f$, where a and c have opposite signs.

    a. By examining the function on any pair of lines $x = x_0$ and $y = y_0$, show that it is concave upwards in one case and downwards in the other, so that $f(x, y)$ cannot have a local maximum or minimum at (x_0, y_0).

    b. Use Th. 11.2 to derive the same conclusion.

11.15   Let $f(x, y) = \log [1/(3x + 7xy + 2y + 1)]$.

    a. Find $\nabla^2_{\pi/4} f(0, 0)$.

    b. For which values of α is $\nabla^2_\alpha f(0, 0) > 0$?

11.16   Let $f(x, y) = \exp \{\frac{1}{2}[7x^2 + 10xy + 7y^2]\}$. Find the maximum and minimum of $\nabla^2_\alpha f(0, 0)$ for $0 \le \alpha \le 2\pi$.

11.17   Given $f_C(t) = f(x(t), y(t))$, find $f_C''(t)$.

11.18   Explain why the terms involving f_x and f_y in the answer to Ex. 11.17 do not appear in Eq. (11.7).

11.19   Let $r = (x^2 + y^2)^{1/2}$.

    a. Find r_{xx}, r_{xy}, r_{yy}.

    b. If x and y are functions of t, find d^2r/dt^2, first by using the answer to Ex. 11.17, and then by direct differentiation.

    c. If $\nabla^2_\alpha r(x_0, y_0)$ is considered as a quadratic form in the variables $X = \cos \alpha$, $Y = \sin \alpha$, classify it according to the categories of the previous section, that is, positive definite, positive semidefinite, etc.

    d. Sketch the surface $z = (x^2 + y^2)^{1/2}$, and interpret geometrically your answer to part c.

11.20   Let $z = (1 - x^2 - y^2)^{1/2}$, for $x^2 + y^2 < 1$.

    a. Find z_{xx}, z_{xy}, z_{yy}.

    b. Answer Ex. 11.19c for this function.

    c. Answer Ex. 11.19d for this function.

*11.21   *a.* Suppose the equation $f(x, y) = c$ satisfies the assumptions of the implicit function theorem and defines a curve $y = g(x)$. Show that if $f(x, y) \in \mathscr{C}^2$, then $g(x)$ is twice continuously differentiable and

$$g''(x) = \frac{-f_y^2 f_{xx} + 2f_x f_y f_{xy} - f_x^2 f_{yy}}{f_y^3}.$$

    b. Using the formula

$$R = \frac{[1 + (y')^2]^{3/2}}{|y''|}$$

for the radius of curvature of a curve $y = g(x)$, where $y'' \ne 0$, find a formula for the radius of curvature of a curve in the implicit form $f(x, y) = c$.

    c. Apply the formula of part b to find the radius of curvature at an arbitrary point of the circle $x^2 + y^2 = r^2$.

11.22 **a.** Find the expression for $\nabla_\alpha^3 f$ by taking the directional derivative ∇_α of the right-side of Eq. (11.7).

b. Find $\nabla_\alpha^3 f$ by substituting $n = 3$ in Eq. (11.12), and compare the result with your answer to part a.

11.23 Carry out Ex. 11.22a, b for $\nabla_\alpha^4 f$.

*11.24 Prove Eq. (11.12) by induction.

*11.25 Show that the function $f(x, y) = x^3 + x^2 y - 2xy^2 + \frac{1}{3}y^3$ cannot have a local maximum or a local minimum at any point.

11.26 **a.** If $f(x, y) = cx^i y^j$, find $\partial^k f / \partial x^k$.

b. If $f(x, y) = cx^i y^j$, find $\partial^{k+l} f / \partial x^k \partial y^l$.

c. If $f(x, y) = cx^i y^j$, find the value of $\partial^{k+l} f / \partial x^k \partial y^l$ at $(0, 0)$.

d. If

$$f(x, y) = \sum_{i+j=n} a_{ij} x^i y^j$$

is a homogeneous polynomial of degree n, find $(\partial^m f / \partial x^k \partial y^l)(0, 0)$, where $k + l = m$.

e. Show that the homogeneous polynomial $f(x, y)$ of part d may be written in the form

$$f(x, y) = \frac{1}{n!} \sum_{k=0}^{n} \binom{n}{k} \frac{\partial^n f}{\partial x^k \partial y^{n-k}} (0, 0) x^k y^{n-k}.$$

f. Show that an arbitrary polynomial $P(x, y)$ of degree m may be written in the form

$$P(x, y) = \sum_{n=0}^{m} \left[\frac{1}{n!} \sum_{k=0}^{n} \binom{n}{k} \frac{\partial^n P}{\partial x^k \partial y^{n-k}} (0, 0) x^k y^{n-k} \right].$$

12 Taylor's theorem; maxima and minima

Let us review the developments since the introduction of the fundamental Lemma 5.1 to see how fundamental this lemma really was. Using it as a basis, we were able to derive in turn the properties of directional derivatives and gradient, the chain rule and its various applications, the mean value theorem, the implicit function theorem, and the Lagrange multiplier method. In our discussion we had assumed only that the functions under consideration were continuously differentiable, and we had not considered the possible existence or behavior of higher order derivatives. As we have seen in Sect. 11, the values of the second derivatives at a point can lead to much more precise information about the nature of a function near that point.

The essence of Lemma 5.1 is that a continuously differentiable function may be approximated near any point by a linear function, and that the coefficients of that linear function are determined by the partial derivatives of the function at the point. The extension of this result, which we wish to prove next, is based on the observation that a linear function is simply a polynomial of first degree. It states that for a function with higher order derivatives one obtains a better approximation by using a polynomial of higher degree, whose coefficients are determined by the higher order derivatives of the function at the point. In precise form, this statement is known as Taylor's theorem.

We first recall Taylor's theorem for functions of a single variable.[8] One form is the following.

Let $f(t)$ have $n + 1$ continuous derivatives near $t = 0$. Then for all sufficiently small t, there is a t_1 between 0 and t such that

$$(12.1) \quad f(t) = f(0) + f'(0)t + \frac{f''(0)}{2} t^2 + \cdots + \frac{f^{(n)}(0)}{n!} t^n + \frac{f^{(n+1)}(t_1)}{(n+1)!} t^{n+1}.$$

For most purposes, the value of Eq. (12.1) is that it represents the function $f(t)$ in the form

$$(12.2) \qquad\qquad f(t) = P_n(t) + R_n(t),$$

where

$$P_n(t) = f(0) + f'(0)t + \cdots + \frac{f^{(n)}(0)}{n!} t^n$$

is a polynomial of degree n, and $R_n(t)$ is a remainder term that tends to zero with t more rapidly than t^n; that is, $R_n(t)/t^n \to 0$ as $t \to 0$. Conversely, one can show that if $f(t)$ has such a representation, then the coefficients of the polynomial $P_n(t)$ must be given by the successive derivatives of $f(t)$ at $t = 0$, as in Eq. (12.1).

Suppose now that $f(x, y) \in \mathscr{C}^{n+1}$ in a domain D. Given any point (x_0, y_0) in D, let C be the straight line $x = x_0 + s \cos \alpha$, $y = y_0 + s \sin \alpha$. Then we may apply Eq. (12.1) to the function $f_C(s)$:

$$f_C(s) = f_C(0) + f_C'(0)s + \frac{f_C''(0)}{2} s^2 + \cdots + \frac{f_C^{(n)}(0)}{n!} s^n + \frac{f_C^{(n+1)}(s_1)}{(n+1)!} s^{n+1}.$$

But the coefficients here are precisely the quantities we introduced in Sect. 11 as higher order directional derivatives. We may therefore rewrite this equation in the form

$$(12.3) \quad f(x, y) = f(x_0, y_0) + \nabla_\alpha f(x_0, y_0)s + \frac{\nabla_\alpha^2 f(x_0, y_0)}{2} s^2 + \cdots$$

$$+ \frac{\nabla_\alpha^n f(x_0, y_0)}{n!} s^n + \frac{\nabla_\alpha^{n+1} f(x_1, y_1)}{(n+1)!} s^{n+1},$$

where (x_1, y_1) is on the line segment between (x_0, y_0) and (x, y).

---

[8] See, for example, [3], Section 7.6 or [14], Section 9.6.

In many cases, Eq. (12.3), which is no more than Taylor's theorem for a function of one variable rewritten in different notation, is completely adequate for studying the function $f(x, y)$ near (x_0, y_0). We shall illustrate this later on. First, however, we rewrite Eq. (12.3) in several ways, using the expressions for the various directional derivatives in terms of the partial derivatives of f. By virtue of Eqs. (11.7) and (11.12), we have

(12.4) $\quad f(x, y) = f(x_0, y_0) + [f_x(x_0, y_0) \cos \alpha + f_y(x_0, y_0) \sin \alpha]s$

$$+ \tfrac{1}{2}[f_{xx}(x_0, y_0) \cos^2 \alpha + 2f_{xy}(x_0, y_0) \cos \alpha \sin \alpha$$

$$+ f_{yy}(x_0, y_0) \sin^2 \alpha]s^2 + \cdots$$

$$+ \frac{1}{n!} \left[\sum_{k=0}^{n} \binom{n}{k} \frac{\partial^n f}{\partial x^{n-k} \partial y^k} (x_0, y_0) \cos^{n-k} \alpha \sin^k \alpha \right] s^n + R_n(x, y),$$

where

(12.5) $\quad R_n(x, y)$

$$= \frac{1}{(n+1)!} \left[\sum_{k=0}^{n+1} \binom{n+1}{k} \frac{\partial^{n+1} f}{\partial x^{n+1-k} \partial y^k} (x_1, y_1) \cos^{n+1-k} \alpha \sin^k \alpha \right] s^{n+1}$$

and (x_1, y_1) is on the line joining (x_0, y_0) to (x, y). By the assumption that $f \in \mathscr{C}^{n+1}$, each of the $(n + 1)$-order partial derivatives of f is continuous, and hence bounded in a circle about (x_0, y_0). Hence

(12.6) $\qquad\qquad |R_n(x, y)| \le M s^{n+1} \qquad$ for some constant M,

inside a circle about (x_0, y_0).

If we substitute the values

$$s \cos \alpha = x - x_0, \qquad s \sin \alpha = y - y_0$$

in Eq. (12.4), we obtain the expression

(12.7) $\quad f(x, y) = f(x_0, y_0) + f_x(x_0, y_0)(x - x_0) + f_y(x_0, y_0)(y - y_0)$

$$+ \tfrac{1}{2}[f_{xx}(x_0, y_0)(x - x_0)^2 + 2f_{xy}(x_0, y_0)(x - x_0)(y - y_0)$$

$$+ f_{yy}(x_0, y_0)(y - y_0)^2] + \cdots$$

$$+ \frac{1}{n!} \left[\sum_{k=0}^{n} \binom{n}{k} \frac{\partial^n f}{\partial x^{n-k} \partial y^k} (x_0, y_0)(x - x_0)^{n-k}(y - y_0)^k \right]$$

$$+ R_n(x, y)$$

$$= P_n(x, y) + R_n(x, y),$$

where $P_n(x, y)$ is a polynomial of degree n, and $R_n(x, y)$ satisfies Eq. (12.6). The fact that a function $f(x, y) \in \mathscr{C}^{n+1}$ has an expansion of the form of Eq. (12.7) near any point (x_0, y_0) is usually referred to as *Taylor's theorem*. Formula (12.7) itself is called the *Taylor expansion* of $f(x, y)$ through terms of degree n. (See also Ex. 12.14.)

Example 12.1

Find the Taylor expansion of the function $f(x, y) = \sin xy$ at the point $(\pi, 0)$ through terms of the second degree.

We have

$$f_x = y \cos xy, \qquad f_y = x \cos xy,$$

$$f_{xx} = -y^2 \sin xy, \qquad f_{xy} = \cos xy - xy \sin xy, \qquad f_{yy} = -x^2 \sin xy.$$

Hence

$$f(\pi, 0) = 0, \qquad f_x(\pi, 0) = 0, \qquad f_y(\pi, 0) = \pi,$$

$$f_{xx}(\pi, 0) = 0, \qquad f_{xy}(\pi, 0) = 1, \qquad f_{yy}(\pi, 0) = -\pi^2.$$

By Eq. (12.7),

$$f(x, y) = \pi y + (x - \pi)y - \tfrac{1}{2}\pi^2 y^2 + R(x, y) = xy - \tfrac{1}{2}\pi^2 y^2 + R(x, y).$$

Remark If we let

$$P(x, y) = xy - \tfrac{1}{2}\pi^2 y^2,$$

then

$$P_x = y, \qquad P_y = x - \pi^2 y,$$

$$P_{xx} = 0, \qquad P_{xy} = 1, \qquad P_{yy} = -\pi^2$$

and

$$P(\pi, 0) = 0, \qquad P_x(\pi, 0) = 0, \qquad P_y(\pi, 0) = \pi.$$

Thus in Example 12.1, we have expressed $f(x, y)$ as a polynomial of degree 2 plus a remainder term, and this polynomial has the property that its value at $(\pi, 0)$ together with the values of its first and second derivatives at that point coincide with the corresponding value of f and its derivatives at $(\pi, 0)$.

The situation pointed out in the preceding Remark is completely general: *the polynomial $P_n(x, y)$ in Taylor's formula (12.7) is the unique polynomial of degree n whose value and derivatives of degree up to n at (x_0, y_0) coincide with those of $f(x, y)$.* This fact may be proved by a direct verification, writing down a general polynomial of degree n, and computing its derivatives of different orders (see Ex. 12.15).

Example 12.2

Find the Taylor expansion at the point $(1, -2)$ through terms of degree 3 of the function

$$f(x, y) = x^4 + 3x^2 y^2 - xy + 2.$$

Method 1. We find

$$f_x = 4x^3 + 6xy^2 - y, \qquad f_y = 6x^2y - x,$$

$$f_{xx} = 12x^2 + 6y^2, \qquad f_{xy} = 12xy - 1, \qquad f_{yy} = 6x^2,$$

$$f_{xxx} = 24x, \qquad f_{xxy} = 12y, \qquad f_{xyy} = 12x, \qquad f_{yyy} = 0.$$

Substitution in Eq. (12.7) with $x_0 = 1$, $y_0 = -2$, and $n = 3$ yields

$$f(x, y) = 17 + 30(x - 1) - 13(y + 2)$$
$$+ \tfrac{1}{2}[36(x - 1)^2 - 50(x - 1)(y + 2) + 6(y + 2)^2]$$
$$+ \tfrac{1}{6}[24(x - 1)^3 + 3\cdot(-24)(x - 1)^2(y + 2) + 3\cdot 12(x - 1)(y + 2)^2]$$
$$+ R(x, y)$$
$$= P_3(x, y) + R(x, y).$$

In this case the polynomial $P_3(x, y)$ is much more complicated than the original function, but it also gives a much better description of the behavior of that function near the point $(1, -2)$. Thus, the linear terms in $P_3(x, y)$ give the equation of the tangent plane to the surface $z = f(x, y)$ at $(1, -2)$ and the quadratic terms determine the relative position of the surface and its tangent plane (see Th. 12.1). Actually we can write down the remainder term $R(x, y)$ explicitly, since it is the difference of two polynomials, $f(x, y) - P_3(x, y)$.

In fact, whenever $f(x, y)$ is a polynomial, we can find its Taylor expansion, including the remainder term, much more quickly by purely algebraic methods. We illustrate the procedure as follows.

Method 2. Setting

$$x = 1 + (x - 1), \qquad y = -2 + (y + 2),$$

we have

$$x^2 = 1 + 2(x - 1) + (x - 1)^2, \qquad y^2 = 4 - 4(y + 2) + (y + 2)^2,$$

$$x^4 = 1 + 4(x - 1) + 6(x - 1)^2 + 4(x - 1)^3 + (x - 1)^4,$$
$$3x^2y^2 = 12 + 24(x - 1) + 12(x - 1)^2$$
$$- 12(y + 2) - 24(x - 1)(y + 2) - 12(x - 1)^2(y + 2)$$
$$+ 3(y + 2)^2 + 6(x - 1)(y + 2)^2 + (x - 1)^2(y + 2)^2$$
$$-xy = 2 + 2(x - 1) - (y + 2) - (x - 1)(y + 2).$$

Thus

$$x^4 + 3x^2y^2 - xy + 2$$
$$= 17 + 30(x - 1) - 13(y + 2) + 18(x - 1)^2 - 25(x - 1)(y + 2) + 3(y + 2)^2$$
$$+ 4(x - 1)^3 - 12(x - 1)^2(y + 2) + 6(x - 1)(y + 2)^2 + R(x, y),$$

where we have written down all the terms of degree up to 3, which constitute the polynomial $P_3(x, y)$, while the remaining terms give

$$R(x, y) = (x - 1)^4 + (x - 1)^2(y + 2)^2.$$

Note that this remainder term satisfies Eq. (12.6) with $n = 3$, and one can prove (just as we noted following Eq. (12.2) for functions of one variable) that there is a unique polynomial $P_n(x, y)$ of degree n such that the difference $R_n(x, y) = f(x, y) - P_n(x, y)$ satisfies Eq. (12.6).

We now formulate precisely the role of the second partial derivatives in determining the behavior of a function at a point.

Theorem 12.1 *Let $f(x, y) \in \mathscr{C}^3$ in a domain D. Suppose that*

$$f_{xx}f_{yy} - f_{xy}^2 \neq 0$$

at a point (x_0, y_0) in D. There are three possibilities.

Case 1. $f_{xx}f_{yy} - f_{xy}^2 > 0,$ $f_{xx} > 0$

Case 2. $f_{xx}f_{yy} - f_{xy}^2 > 0,$ $f_{xx} < 0$

Case 3. $f_{xx}f_{yy} - f_{xy}^2 < 0.$

In case 1—the surface $z = f(x, y)$ lies above its tangent plane at (x_0, y_0) in some neighborhood of (x_0, y_0).
In case 2—the surface lies below the tangent plane.
In case 3—the surface crosses the tangent plane (Fig. 12.1).

PROOF.   We first verify that the three cases listed comprise all the possibilities if $f_{xx}f_{yy} - f_{xy}^2 \neq 0$. Namely, either $f_{xx}f_{yy} - f_{xy}^2 < 0$, which is case 3,

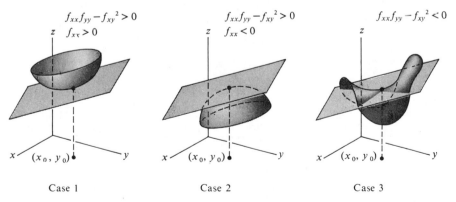

Case 1   Case 2   Case 3

FIGURE 12.1   Position of a surface relative to its tangent plane at a point where $f_{xx}f_{yy} - f_{xy}^2 \neq 0$ (Cases 1, 2, and 3 of Th. 12.1)

or else $f_{xx}f_{yy} - f_{xy}^2 > 0$. In the latter case it is impossible that $f_{xx} = 0$, since that would imply $f_{xx}f_{yy} - f_{xy}^2 = -f_{xy}^2 \leq 0$. Thus $f_{xx} > 0$ or $f_{xx} < 0$, which correspond to cases 1 and 2.

We note next that if we introduce the quadratic form

(12.8) $$Q(X, Y) = AX^2 + 2BXY + CY^2,$$

$$A = f_{xx}(x_0, y_0), \quad B = f_{xy}(x_0, y_0), \quad C = f_{yy}(x_0, y_0)$$

then the three cases in the theorem correspond precisely to

Case 1. $Q(X, Y)$ is positive definite;
Case 2. $Q(X, Y)$ is negative definite;
Case 3. $Q(X, Y)$ takes on both positive and negative values.

We may write Taylor's formula (12.7) for $f(x, y)$ in the form

(12.9) $$f(x, y) = L(x, y) + \tfrac{1}{2}Q(X, Y) + R_2(x, y),$$

$$X = x - x_0, \quad Y = y - y_0,$$

where

(12.10) $$L(x, y) = a(x - x_0) + b(y - y_0) + c,$$

$$a = f_x(x_0, y_0), \quad b = f_y(x_0, y_0), \quad c = f(x_0, y_0),$$

and $Q(X, Y)$ is defined by Eq. (12.8). Setting

$$s^2 = X^2 + Y^2$$

the remainder term $R_2(x, y)$ satisfies, in view of Eq. (12.6),

(12.11) $$R_2(x, y)/s^2 \to 0 \quad \text{as} \quad (x, y) \to (x_0, y_0).$$

Note that $z = L(x, y)$ is the equation of the tangent plane to the surface $z = f(x, y)$ at the point (x_0, y_0).

Consider now case 1. $Q(X, Y)$ is positive definite, and we have $Q(X, Y) \geq \lambda > 0$ for $X^2 + Y^2 = 1$. Then

$$Q(X, Y) \geq \lambda(X^2 + Y^2)$$

for all $(X, Y) \neq (0, 0)$. By Eq. (12.11) we have for some $s_0 > 0$, that

$$|R_2(x, y)|/s^2 \leq \lambda/3$$

for $s < s_0$. Hence

$$\frac{f(x, y) - L(x, y)}{s^2} = \frac{1}{2}\frac{Q(X, Y)}{s^2} + \frac{R_2(x, y)}{s^2} \geq \frac{\lambda}{2} - \frac{\lambda}{3} = \frac{\lambda}{6} > 0$$

for $0 < s < s_0$. But this means

$$f(x, y) - L(x, y) > 0 \quad \text{or} \quad f(x, y) > L(x, y)$$

for $0 < s < s_0$, which is precisely the statement that $z = f(x, y)$ lies above its tangent plane in a neighborhood of (x_0, y_0).

The same argument works for case 2.

Case 3 is an immediate consequence of Th. 11.2 applied to the function $f(x, y) - L(x, y)$ whose second derivatives coincide with those of $f(x, y)$. ◆

Remark   The statement of Th. 12.1 is intuitively obvious if we observe that $Q(X, Y) = \nabla_\alpha^2 f(x_0, y_0) s^2$, so that in case 1, for example, $\nabla_\alpha^2 f(x_0, y_0) > 0$ for all α. This means that the surface $z = f(x, y)$ is concave upwards over every straight line through (x_0, y_0). Similarly, for the other two cases. In fact, this intuitive idea settles case 3 completely, without using Taylor's theorem. However, it can be deceptive if one wants to use it in cases 1 and 2. Namely, one can have a function $f(x, y)$ with the following property: for each straight line through the origin, $f(x, y) \geq 0$ in some interval about the origin, but nevertheless in every neighborhood of the origin, there are points where $f(x, y) < 0$.

Example 12.3

The function

(12.12) $$f(x, y) = (x^2 - y)(2x^2 - y) = y^2 - 3x^2 y + 2x^4$$

illustrates the preceding Remark. We have $f(0, 0) = 0$, and

$$f(x, y) > 0 \quad \text{for} \quad y < x^2$$
$$f(x, y) < 0 \quad \text{for} \quad x^2 < y < 2x^2$$
$$f(x, y) > 0 \quad \text{for} \quad y > 2x^2.$$

(See Fig. 12.2.) Here the tangent plane to the surface $z = f(x, y)$ at the origin is horizontal. It is simply the x, y plane. For any straight line through the origin, the surface lies above the tangent plane for some interval about the origin. Nevertheless, every neighborhood of the origin contains points where $f(x, y) < 0$, and at these points the surface lies below the tangent plane.

Example 12.3 is one more illustration of the danger of trying to restrict the study of functions of two variables to properties of the functions along straight lines.

We return now to the question posed at the end of Sect. 6. We observed there that if $f(x, y)$ has a local maximum or minimum at a point (x_0, y_0) and if the gradient of f exists at (x_0, y_0), then

(12.13) $$\nabla f(x_0, y_0) = \mathbf{0}.$$

We noted later (after Def. 7.1) that Eq. (12.13) also holds at the summit of a pass.

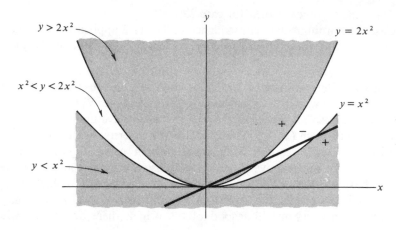

FIGURE 12.2 The set of points where $(x^2 - y)(2x^2 - y) > 0$

Definition 12.1 A point (x_0, y_0) for which Eq. (12.13) holds is called a *critical point* of the function $f(x, y)$.

The question, then, is to decide whether a given critical point corresponds to a local maximum or a local minimum or neither. The answer is provided in many cases by the following "second-derivative test."

Theorem 12.2 *Let* $f(x, y) \in \mathscr{C}^3$ *in a domain D, and suppose* $\nabla f(x_0, y_0) = 0$. *Then at the point* (x_0, y_0) *we have*

Case 1.

$f_{xx}f_{yy} - f_{xy}^2 > 0, \quad f_{xx} > 0 \Rightarrow f$ *has a local minimum;*

Case 2.

$f_{xx}f_{yy} - f_{xy}^2 > 0, \quad f_{xx} < 0 \Rightarrow f$ *has a local maximum;*

Case 3.

$f_{xx}f_{yy} - f_{xy}^2 < 0 \qquad \Rightarrow f$ *has neither a local maximum nor minimum.*

PROOF. The condition $\nabla f(x_0, y_0) = 0$ simply means that the tangent plane to $z = f(x, y)$ is the horizontal plane $z = f(x_0, y_0)$. The result follows immediately from Th. 12.1 (Fig. 12.3). ◆

Remark The name *saddle point* is sometimes used for a point at which $f_{xx}f_{yy} - f_{xy}^2 < 0$, and it is sometimes used more generally for a point at which the surface crosses the tangent plane. Note that the surface $z = x^4 - y^4$ is "saddle-shaped" at the origin, although $f_{xx}f_{yy} - f_{xy}^2 = 0$ there. Also the

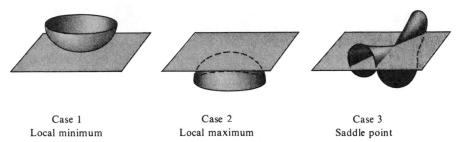

| Case 1 | Case 2 | Case 3 |
| Local minimum | Local maximum | Saddle point |

FIGURE 12.3 Picture of a surface $z = f(x, y)$ near a critical point of $f(x, y)$

function defined by Eq. (12.12) would have a saddle point in the more general sense at the origin. The main fact to observe is that a saddle point in either sense can never be a local maximum or minimum.

When looking for an absolute maximum or minimum of a function of two variables, it is often not necessary to use the second derivative test. One can try to locate all critical points and then find the maximum or minimum value of the function on the critical points.

Example 12.4

Let x, y, z be any three positive numbers whose sum is a fixed number s, that is,

(12.14) $$x + y + z = s.$$

What is the maximum value of the product xyz?
   Using Eq. (12.14), we may write the product as a function of two variables:

$$f(x, y) = xy(s - x - y) = sxy - x^2y - xy^2.$$

We are interested in the domain where each factor is positive:

$$D: x > 0, \quad y > 0, \quad s - x - y > 0.$$

Note that $f(x, y)$ must assume a maximum at some point (x_0, y_0) in the closed set (Fig. 12.4):

$$x \geq 0, \quad y \geq 0, \quad x + y \leq s.$$

But on the boundary of this set $f(x, y) = 0$, and hence (x_0, y_0) must lie inside the domain D. Since $f(x, y)$ is differentiable throughout D, we conclude that (x_0, y_0) is a critical point of f. But

$$f_x = sy - 2xy - y^2 = y(s - 2x - y),$$
$$f_y = sx - x^2 - 2xy = x(s - x - 2y).$$

Since $x \neq 0$ and $y \neq 0$ in D, at the critical point (x_0, y_0), we must have

$$2x_0 + y_0 = s$$
$$x_0 + 2y_0 = s.$$

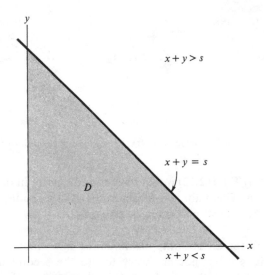

FIGURE 12.4 The set of points: $x \geq 0$, $y \geq 0$, $x + y \leq s$

Solving, we find $x_0 = s/3$, $y_0 = s/3$. (In this case we have a single critical point, which must coincide with the absolute maximum.) Thus we have

$$f(x, y) \leq f(s/3, s/3) = (s/3)^3,$$

or in terms of the original problem

(12.15)
$$xyz \leq \left(\frac{s}{3}\right)^3 = \left(\frac{x + y + z}{3}\right)^3.$$

We may note that inequality (12.15) has been proved for any three positive numbers. Omitting the middle term, we have

(12.16)
$$\sqrt[3]{xyz} \leq \frac{x + y + z}{3},$$

which is called the *inequality of the geometric and arithmetic means*. (Compare this with inequality (9.2).)

Example 12.5

Find the critical points of the function

$$f(x, y) = 5 + y - x^2 - y^2 + x^2 y,$$

and examine the behavior of the function near each point.
We have

$$f_x = -2x + 2xy = 2x(y - 2); \qquad f_y = 1 - 2y + x^2.$$
$$f_x = 0 \Leftrightarrow x = 0 \quad \text{or} \quad y = 2; \qquad f_y = 0 \Leftrightarrow 2y = 1 + x^2.$$

Thus, if $x = 0$, then $y = \frac{1}{2}$, and if $y = 2$, then $x^2 = 1$, and $x = \pm 1$. The critical points are therefore $(0, \frac{1}{2})$, $(1, 2)$, and $(-1, 2)$. Further

$$f_{xx} = -2 + 2y, \qquad f_{xy} = 2x, \qquad f_{yy} = -2,$$

and

$$\Delta = f_{xx}f_{yy} - f_{xy}^2 = 4(1 - y - x^2).$$

At $(0, \frac{1}{2})$: $\Delta = 2, f_{xx} = -1$; f has a local maximum (Case 2 of Th. 12.2).

At $(\pm 1, 1)$: $\Delta = -4$; f has a saddle point (Case 3 of Th. 12.2).

Remark It may be observed that no mention is made in Th. 12.2 of the case when $f_{xx}f_{yy} - f_{xy}^2 = 0$. The reason is that in this case the behavior of the function cannot be predicted by the second derivatives alone. The quadratic form (12.8) is then semidefinite, and we might be tempted to think that if it were positive semidefinite the function would still have a local minimum. That would be true, for example, for the function $z = x^2 + y^4$ at the origin, but it is false for $z = x^2 - y^4$, which has the same quadratic terms.

We conclude Chapter 2 by proving a theorem that at first sight may seem to contradict the preceding Remark. The theorem states in effect that even if we cannot predict the behavior of a function at a place where $f_{xx}f_{yy} - f_{xy}^2$ vanishes, we can still describe its over-all behavior in a region where this expression may vanish part of the time. We shall make no further use of this theorem in the present book, and it could easily be omitted on first reading. However, its Corollaries play a fundamental role in many more advanced areas of mathematics, and the proof provides a good illustration of some of the methods developed in this chapter.

Theorem 12.3 *Let D be a bounded domain, and let $\bar{D}$ be the set of points in D together with all its boundary points. Let $f(x, y)$ be continuous in $\bar{D}$, and $f(x, y) \in \mathscr{C}^3$ in D. Suppose that*

$$(12.17) \qquad\qquad f_{xx}f_{yy} - f_{xy}^2 \leq 0 \qquad\qquad \text{throughout } D.$$

Then the maximum and minimum values of f in $\bar{D}$ are assumed on the boundary of D.

Remark 1 For simplicity one may picture D to be the interior of the unit circle $x^2 + y^2 < 1$, in which case $\bar{D}$ is the disk $x^2 + y^2 \leq 1$. This case is sufficient for many applications.

Remark 2 If we had strict inequality in (12.17), then the result would follow immediately from Th. 11.2, since f could not have even a local maximum or minimum in D. The difficulty here is that we allow points where $f_{xx}f_{yy} - f_{xy}^2 = 0$.

PROOF. We give the proof for the maximum. The result follows then for the minimum by considering the function $-f$.

Let M be the maximum value of $f(x, y)$ in $\bar{D}$, and let m be the maximum on the boundary of D. Clearly $m \leq M$, and the assertion of the theorem is that $m = M$. We proceed by contradiction. Suppose that

$$(12.18) \qquad m < M.$$

Then if (x_0, y_0) is a point in $\bar{D}$ where $f(x_0, y_0) = M$, (x_0, y_0) cannot lie on the boundary of D, and hence must lie in D itself (Fig. 12.5). We consider next the function

$$(12.19) \qquad g(x, y) = M - \epsilon[(x - x_0)^2 + (y - y_0)^2], \qquad \epsilon > 0.$$

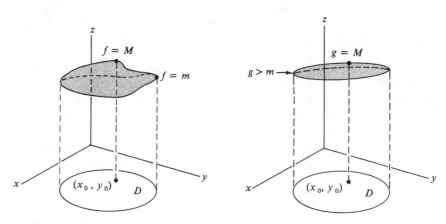

FIGURE 12.5 The surfaces $z = f(x, y)$ and $z = g(x, y)$ in the proof of Th. 12.3

The number ϵ is a fixed positive number that is to be chosen sufficiently small so that $g(x, y) > m$ on the boundary of D. That is always possible by (12.18). This means that $f(x, y) < g(x, y)$ on the boundary of D, whereas $f(x_0, y_0) = g(x_0, y_0) = M$. We introduce the difference

$$\varphi(x, y) = f(x, y) - g(x, y),$$

which is continuous in $\bar{D}$ and has a maximum δ at some point (x_1, y_1). Then

$$\delta = \varphi(x_1, y_1) \geq \varphi(x_0, y_0) = 0,$$

and since $\varphi(x, y) < 0$ on the boundary of D, (x_1, y_1) must lie somewhere inside D. Since (x_1, y_1) is a maximum of $\varphi(x, y)$, it is a local maximum along every line through (x_1, y_1), and hence for every α,

$$(12.20) \qquad \nabla_\alpha \varphi(x_1, y_1) = 0, \qquad \nabla_\alpha^2 \varphi(x_1, y_1) \leq 0.$$

But by direct computation from Eq. (12.19) we have

$$g_{xx} = -2\epsilon, \qquad g_{xy} = 0, \qquad g_{yy} = -2\epsilon,$$

and

(12.21) $\qquad \nabla_\alpha^2 g(x_1, y_1) = g_{xx} \cos^2 \alpha + g_{yy} \sin^2 \alpha = -2\epsilon.$

Hence,

$$
\begin{aligned}
(12.22) \quad \nabla_\alpha^2 f(x_1, y_1) &= f_{xx} \cos^2 \alpha + 2f_{xy} \cos \alpha \sin \alpha + f_{yy} \sin^2 \alpha \\
&= (\varphi_{xx} + g_{xx}) \cos^2 \alpha + 2(f_{xy} + g_{xy}) \cos \alpha \sin \alpha \\
&\qquad\qquad\qquad\qquad\qquad\qquad + (\varphi_{yy} + g_{yy}) \sin^2 \alpha \\
&= \nabla_\alpha^2 \varphi(x_1, y_1) + \nabla_\alpha^2 g(x_1, y_1) \\
&\le -2\epsilon,
\end{aligned}
$$

for all α by (12.20) and (12.21). This means that the quadratic form (12.22) is negative definite, and hence, $f_{xx}f_{yy} - f_{xy}^2 > 0$, $f_{xx} < 0$ at (x_1, y_1). But this contradicts (12.17). Hence the assumption (12.18) must be false, and the theorem is proved. ◆

Remark The idea of this proof is a simple geometric one. We are trying to prove that if $f(x, y)$ satisfies (12.17) throughout D, then the surface $z = f(x, y)$ cannot protrude over D above its highest point on the boundary of D. If it did, then by choosing a very flat paraboloid, $z = g(x, y)$, which is concave downward, we could lower it until it just touched the surface $z = f(x, y)$ over a point (x_1, y_1) in D. Then the two surfaces would have the same tangent plane at (x_1, y_1), and the surface $z = f(x, y)$ would lie below the surface $z = g(x, y)$. But that would mean that $z = f(x, y)$ was strictly concave downward at (x_1, y_1), which would contradict (12.17).

Corollary 1 *If a function $f(x, y)$ satisfies (12.17) in D and is constant on the boundary of D, then it is constant throughout D.*

PROOF. If $f(x, y) \equiv c$ on the boundary of D, then for all (x, y) in D we have

$$c = \min f(x, y) \le f(x, y) \le \max f(x, y) = c. \qquad ◆$$

Corollary 2 *If $f(x, y)$ is harmonic in D, then it assumes its maximum and minimum values on the boundary of D.*

PROOF. For a harmonic function $f_{xx} + f_{yy} = 0$, or $f_{xx} = -f_{yy}$, and (12.17) always holds. ◆

Corollary 3 *If $f(x, y)$ is harmonic in D and zero on the boundary, then it is zero throughout D.*

PROOF.   Same as Corollary 1.   ◆

Corollary 4 *A harmonic function in D is uniquely determined by its values on the boundary of D.*

PROOF.   Suppose $f(x, y)$ and $g(x, y)$ are both harmonic in D and take on the same value at each boundary point. Then the function $\varphi(x, y) = f(x, y) - g(x, y)$ is harmonic in D and zero on the boundary of D. By Corollary 3, $\varphi(x, y) \equiv 0$, or $f(x, y) \equiv g(x, y)$, in D.   ◆

Exercises

12.1   Find all critical points of each of the following functions, and classify them as local maxima, local minima, or saddle points.

    a. $f(x, y) = 4y^3 + x^2 - 12y^2 + 36y$
    b. $f(x, y) = x^4 + y^4 - 4xy$
    c. $f(x, y) = 2x^3 - y^3 + 12x^2 + 27y$
    d. $f(x, y) = (x - 1)(y - 1)(x + y - 1)$

12.2   Let $f(x, y)$ be the general quadratic polynomial

$$f(x, y) = Ax^2 + 2Bxy + Cy^2 + 2Dx + 2Ey + F.$$

    a. Show that if $AC - B^2 \neq 0$, then $f(x, y)$ has exactly one critical point.
    b. Find the critical point of part a in terms of the coefficients A, B, C, D, E, F.
    c. Describe the nature of the critical point in terms of the coefficients.
    d. What is the totality of critical points if A, B, and C are all zero?
    e. What is the totality of critical points if A, B, and C are not all zero, but if $AC - B^2 = 0$?
    f. Describe the nature of the critical points in part e in terms of the coefficients. (*Hint:* see Ex. 10.28.)

12.3   Find the critical points of the following functions, and try to determine whether they are local maxima, local minima, or neither, even in the cases where the "second-derivative test," Th. 12.2, fails.

    a. $f(x, y) = x^6 + y^{10}$ *e.* $f(x, y) = (x^2 - y^2)^2$
    b. $f(x, y) = x^3 + y^3$ *f.* $f(x, y) = 2x^2 - 2\sqrt{2}\,xy + y^2$
    c. $f(x, y) = e^{x^2 - y^2}$
    d. $f(x, y) = \log(2x^2 - 3xy + 3y^2 - 10x + 15y + 21)$

12.4   Show that the function $f(x, y) = \cos(xe^y)$ satisfies the equation $f_{xx}f_{yy} - f_{xy}^2 = 0$ at all critical points. (See the discussion of the local maxima and minima of this function at the end of Sect. 6.)

12.5 Let $f(x, y) = x^3 - 3xy^2$.

  a. Show that the origin is the only critical point of this function, and that $f_{xx}f_{yy} - f_{xy}^2 = 0$ at the origin.

  b. Describe the surface $z = x^3 - 3xy^2$ near the origin and explain why it is called a "monkey saddle." (*Hint:* see Ex. 3.5e.)

12.6 Show that each of the following functions $f(x, y)$ has a critical point at the origin and try to describe the surface $z = f(x, y)$ in the vicinity of the origin; state, in particular, which functions have a local maximum or minimum at the origin.

  a. $f(x, y) = x^2 y^2$
  b. $f(x, y) = x^3 - y^3$
  c. $f(x, y) = x^4 - x^2 y^2 + y^4$
  d. $f(x, y) = 4x^4 - 4x^2 y^2 + y^4$
  e. $f(x, y) = x^3 y - xy^3$
  f. $f(x, y) = (2x^2 - xy + y^2)(-x^2 + 2xy - 3y^2)$

12.7 For each of the following functions find the Taylor expansion through terms of second degree at the points indicated.

  a. $f(x, y) = 1/(x + 2y)$ at $(1, 1)$
  b. $f(x, y) = \sin(x + y)$ at $(\frac{1}{4}\pi, 0)$
  c. $f(x, y) = \log \sqrt{x^2 + y^2}$ at $(0, 1)$
  d. $f(x, y) = \operatorname{arc cot}(x/y)$ at $(0, 1)$

12.8 For each of the following functions, find the Taylor expansion through terms of the third degree at the points indicated.

  a. $f(x, y) = 2x^2 - xy + 4y - 7$,   at $(1, -2)$
  b. $f(x, y) = x^3 - 2x^2 y + xy^2 + 4y^3 + 7xy - 7$,   at $(0, 0)$
  c. $f(x, y) = x^3 - 2x^2 y + xy^2 + 4y^3 + 7xy - 7$,   at $(1, 0)$
  d. $f(x, y) = x^2 y^2 - xy^3$,   at $(2, -1)$
  e. $f(x, y) = x^2 y^2 - xy^3$,   at $(0, 0)$

12.9 For each of the following functions find the Taylor expansion through terms of degree n at the points indicated.

  a. $f(x, y) = e^{x+y}$,   at $(0, 1)$
  b. $f(x, y) = (x + y)^n$,   at $(2, -1)$
  c. $f(x, y) = \sin(x + y)$,   at $(0, 0)$
  d. $f(x, y) = \cos(x + y)$,   at $(0, 0)$
  e. $f(x, y) = e^{xy}$,   at $(0, 0)$ (*Hint:* see Ex. 11.12e.)

12.10 Let $f(x, y) = e^{25 - x^2 - y^2}$.

  a. Find the equation of the tangent plane to the surface $z = f(x, y)$ at the point $(3, 4)$, and decide whether the surface lies above or below the tangent plane near that point, or crosses it.

  b. Do the same for the point $(\frac{1}{2}, 0)$.

12.11 Show that the surface $z = 3x^2 - 2xy + 2y^2$ lies entirely above every one of its tangent planes. (*Hint:* find explicitly the remainder term after expanding through linear terms, and show that it is always positive.)

*12.12 Theorem 12.1 was proved in the text under the assumption that $f(x, y) \in \mathscr{C}^3$. Actually it is sufficient to assume $f(x, y) \in \mathscr{C}^2$. For case 3, this is a consequence of Th. 11.2. For cases 1 and 2, it can be shown by noting that the Taylor expansion through linear terms is of the form $f(x, y) = L(x, y) + R(x, y)$, where $z = L(x, y)$ is the equation the tangent plane, and $R(x, y)$ is a quadratic form whose coefficients depend on the values of f_{xx}, f_{xy}, f_{yy} at a point (x_1, y_1) near the given point (x_0, y_0). But for $f(x, y) \in \mathscr{C}^2$, the functions f_{xx} and $f_{xx}f_{yy} - f_{xy}^2$ are continuous, so that if they are positive or negative at (x_0, y_0), they have the same sign in some disk about (x_0, y_0). Using these remarks, give a complete proof of Th. 12.1 for $f(x, y) \in \mathscr{C}^2$. (Note that Ths. 12.2 and 12.3, which are consequences of Th. 12.1, then also follow for $f(x, y) \in \mathscr{C}^2$.)

12.13 Let $f(x, y) \in \mathscr{C}^2$ near (x_0, y_0) and suppose

$$f(x_0, y_0) = f_x(x_0, y_0) = f_y(x_0, y_0) = 0.$$

Show that

$$\lim_{s \to 0} \frac{f(x_0 + s \cos \alpha, y_0 + s \sin \alpha)}{s^2} = \tfrac{1}{2}\nabla_\alpha^2 f(x_0, y_0).$$

12.14 If we formally define the "product" $(\partial/\partial x)^k(\partial/\partial y)^l$ of partial differentiation operators to be $\partial^{k+l}/\partial x^k \, \partial y^l$, we may write Taylor's formula in the more compact form:

$$f(x_0 + h, y_0 + k) = P_n(h, k) + R_n(h, k),$$

where

$$P_n(h, k) = \sum_{m=0}^{n} \frac{1}{m!} \left[\left(h \frac{\partial}{\partial x} + k \frac{\partial}{\partial y} \right)^m f(x, y) \right]_{(x,y) = (x_0, y_0)}$$

and

$$R_n(h, k) = \frac{1}{(n+1)!} \left[\left(h \frac{\partial}{\partial x} + k \frac{\partial}{\partial y} \right)^{n+1} f(x, y) \right]_{(x,y) = (x_1, y_1)},$$

where $x_1 = x_0 + \lambda h, \, y_1 = y_0 + \lambda k$ for some $\lambda, \, 0 < \lambda < 1$.

Show that by formally using the binomial theorem to expand the expression $(h \, \partial/\partial x + k \, \partial/\partial y)^n$, applying the corresponding derivatives to $f(x, y)$ and evaluating the result at the point indicated, one obtains Taylor's formula in the form we have given it. (Note that the final step is to substitute $h = x - x_0, \, k = y - y_0$.)

Equivalently, we may write

$$\nabla_\alpha^n f = (\nabla_\alpha)^n f = \left(\cos \alpha \frac{\partial}{\partial x} + \sin \alpha \frac{\partial}{\partial y} \right)^n f$$

as a brief (and more easily retained) version of Eq. (11.12).

12.15 *a.* Show that if $P(x, y) = \sum c_{ij}(x - x_0)^i/(y - y_0)^j$, then

$$c_{kl} = \frac{1}{k!} \frac{1}{l!} \frac{\partial^{k+l}P}{\partial x^k \, \partial y^l}(x_0, y_0).$$

(*Hint:* see Ex. 11.26.)

b. Show that every polynomial $P(x, y)$ can be written in the form given in part a, where x_0 and y_0 are arbitrary.
(*Hint:* $x^m y^n = [x_0 + (x - x_0)]^m [y_0 + (y - y_0)]^n$.)

c. Show that if $f(x, y) \in \mathscr{C}^n$ and if $P(x, y)$ is a polynomial that has the same value as $f(x, y)$ and the same partial derivatives of all orders up to n at (x_0, y_0), then

$$P(x, y) = \sum_{0 \le k+l \le n} \frac{1}{k!} \frac{1}{l!} \frac{\partial^{k+l} f}{\partial x^k \, \partial y^l} (x_0, y_0)(x - x_0)^k (y - y_0)^l$$

$$= \sum_{m=0}^{n} \frac{1}{m!} \left[\sum_{k+l=m} \binom{m}{k} \frac{\partial^m f}{\partial x^k \, \partial y^l} (x_0, y_0)(x - x_0)^k (y - y_0)^l \right].$$

Note: a number of problems involve finding the maximum or minimum of a function of three variables subject to a relation between those variables. By using the relation to express one of the variables in terms of the other two, we reduce the problem to a maximum-minimum problem for a function of two variables, which may then be solved by looking for critical points. Exercises 12.16–12.19 are examples.

12.16 Among all rectangular boxes having a given volume V, find the dimensions that require the least material for its construction (that is, the least total surface area).

12.17 Solve Ex. 12.16 for a box without a top.

12.18 If the post office accepts parcels whose total length plus girth does not exceed 60 inches, what dimensions will maximize the volume?

12.19 What is the maximum area of a triangle having a given perimeter p? (*Hint:* the area A is related to the perimeter p by the formula

$$A^2 = s(s - x)(s - y)(s - z),$$

where x, y, z are the sides of the triangle, and $p = 2s = x + y + z$.)

Note that in the preceding four exercises the problem was to find an *absolute* maximum or minimum of a given function, where the variables are generally restricted by the nature of the problem. (For example, in Ex. 12.19, $x \ge 0$, $y \ge 0$, and $z = 2s - x - y \ge 0$, since x, y, and z represent lengths.) A critical approach to these exercises would require specifying the set of points where the function is defined, and verifying first that the extreme value sought does not lie on the boundary of this set of points. (If the function in question satisfied the hypotheses of Th. 12.3, then the extreme value would have to lie on the boundary.) Then assuming that an absolute maximum or minimum exists, it must be taken on at an interior point, and that point must be a critical point of the function. In practice, it is usual to locate a critical point and simply to assume that it provides the answer to the problem. Exercise 12.20 indicates how a careful solution of extremum problems in two variables may be given in a large number of cases.

12.20 For each of the following functions $f(x, y)$, find the absolute maximum and minimum on the given set S. (Note that since in each case S is a closed bounded set and $f(x, y)$ is continuous on S, the function $f(x, y)$ must assume its maximum and minimum somewhere in S. If at an interior point, then the gradient of f must vanish there; so the first step is to find the value of $f(x, y)$ at each critical point in the interior of S. Next, find the maximum and minimum of $f(x, y)$ on the boundary of S. This amounts to finding the extreme values of $f(x, y)$ along a curve, and may be done by one-variable methods or by using Lagrange multipliers with the precaution that the values of $f(x, y)$ at any possible "corners," where the boundary curve is not smooth, must be examined also. Note two cases in which the extreme values cannot occur at an interior point, and must occur on the boundary: if $\nabla f \neq \mathbf{0}$ on S (Lemma 6.1), or if $f_{xx}f_{yy} - f_{xy}^2 \leq 0$ on S (Th. 12.3).)

a. $f(x, y) = x + y$, $\quad S: x^2 + y^2 \leq 1$
b. $f(x, y) = x^2 - 4xy + y^2$, $\quad S: x^2 + y^2 \leq 1$
c. $f(x, y) = x^2 - 2xy + 2y^2$, $\quad S: x^2 + y^2 \leq 1$
d. $f(x, y) = x^4 - y^4$, $\quad S: x^2 + y^2 \leq 1$
e. $f(x, y) = (x - 1)^2 + y^2$, $\quad S: x^2 + y^2 \leq 4$
f. $f(x, y) = (x + 3)^2 + y^2$, $\quad S: x^2 + y^2 \leq 4$
g. $f(x, y) = x + y$, $\quad S: -2 \leq x \leq 1, -1 \leq y \leq 3$
h. $f(x, y) = (x - 2)^2 + y^2$, $\quad S: -2 \leq x \leq 1, -1 \leq y \leq 3$
i. $f(x, y) = (1 - x^2) \sin y$, $\quad S: -1 \leq x \leq 1, -\pi \leq y \leq \pi$
j. $f(x, y) = (x^2 - y^2)e^{1-x^2-y^2}$, $\quad S: x^2 + y^2 \leq 4$
k. $f(x, y) = (2x^2 + y^2)e^{1-x^2-y^2}$, $\quad S: x^2 + y^2 \leq 4$

12.21 *a.* Show that if $f(x, y) \in \mathscr{C}^2$ in a domain D, and if $f(x, y)$ has a local minimum at (x_0, y_0), then $f_{xx} \geq 0$, $f_{yy} \geq 0$, and $f_{xx}f_{yy} - f_{xy}^2 \geq 0$ at (x_0, y_0).
b. What can be said at a local maximum?

12.22 A function $f(x, y) \in \mathscr{C}^2$ is called *subharmonic* if it satisfies the inequality $f_{xx} + f_{yy} \geq 0$ in a domain.

a. Show that if $f_{xx} + f_{yy} > 0$ throughout a domain, then $f(x, y)$ cannot have a local maximum.

***b.** Using a reasoning analogous to the proof of Th. 12.3, show that if $f(x, y)$ is subharmonic in a bounded domain D and continuous in $\overline{D}$, then $f(x, y)$ takes on its maximum on the boundary of D. (*Hint:* if $f(x, y)$ is subharmonic, then

$$h(x, y) = f(x, y) + \epsilon[(x - x_0)^2 + (y - y_0)^2]$$

satisfies the hypotheses of part a for every $\epsilon > 0$. But if $f(x, y)$ had a larger value at (x_0, y_0) than on the boundary of D, ϵ could be chosen sufficiently small so that $h(x, y)$ would have a maximum at an interior point (x_1, y_1) of D, contradicting part a.)

CHAPTER THREE

Transformations

13 Pairs of functions; geometric representation

One of the major lessons learned in Chapter 2 was that for the proper study of a function of two variables it is not sufficient to consider the function with respect to each variable separately. In the present chapter we shall see that when confronted with a pair of functions it is often advantageous to treat the pair as a single entity rather than as two separate functions.

Pairs of functions of two variables appear in many guises. We give the following examples.

Example 13.1

The gradient   Given a differentiable function $f(x, y)$, its gradient associates with it a pair of functions; namely, the partial derivatives $f_x(x, y)$ and $f_y(x, y)$.

Example 13.2

Vector fields   Given a domain D, a *vector field* or *vector function* on D assigns to each point (x, y) of D a vector $\mathbf{v}(x, y)$. This is equivalent to giving a pair of functions in D; namely, the two components of the vector $\mathbf{v}(x, y)$. Example 13.1 is a special case of this, in which to each point (x, y) corresponded the gradient vector of $f(x, y)$ at that point. Many other examples arise in physics, such as the gravitational or electromagnetic force fields and the velocity field of a fluid flow.

Example 13.3

Coordinate transformations   Under a change of coordinates in the plane, the expressions for the new coordinates of a point in terms of the old constitute a pair of functions. Thus

$$x = r \cos \theta, \qquad y = r \sin \theta,$$

the relations between rectangular and polar coordinates, express x and y as functions of the two variables r and θ.

Example 13.4

Motions in the plane A motion of the plane into itself, such as a Euclidean motion or a similarity transformation, may be described by assigning to each point (x, y) the point (u, v) to which it is displaced by the motion. Each of the coordinates u, v may then be considered as a function of (x, y). More general motions, such as fluid motion, may be described in the same way, by assigning to each point (x, y) the coordinates of the point it reaches after a given interval of time.

It is important to observe that a single pair of functions may arise in all of these ways. Conversely, a given pair of functions may be interpreted in various ways. Let us illustrate these general remarks with a particular example.

Example 13.5

For a fixed value of α, consider the pair of functions

(13.1)
$$u = x \cos \alpha + y \sin \alpha$$
$$v = -x \sin \alpha + y \cos \alpha$$

of x and y.

Interpreted as a *motion*, these equations describe a clockwise rotation of the plane through an angle α (Fig. 13.1). With respect to a fixed coordinate system, we have

$$(x, y) \rightarrow (u, v) = (x \cos \alpha + y \sin \alpha, \ -x \sin \alpha + y \cos \alpha).$$

(See Sect. 15 for a detailed discussion of rotations.)

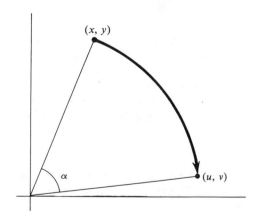

FIGURE 13.1 Rotation through angle α

As a *coordinate transformation*, Eqs. (13.1) describe the relation between the coordinates (x, y) of a point with respect to one set of axes, and the coordinates (u, v) of the same point with respect to a pair of axes making an angle α with the original pair (Fig. 13.2).

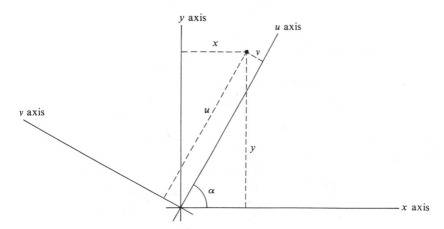

FIGURE 13.2 Rotation of coordinate axes

As a *vector field*, Eqs. (13.1) assign to each point (x, y) the vector $\langle u, v \rangle = \langle x \cos \alpha + y \sin \alpha, -x \sin \alpha + y \cos \alpha \rangle$. We note that

$$|\langle u, v \rangle| = (u^2 + v^2)^{1/2} = (x^2 + y^2)^{1/2},$$

so that the vectors $\langle u, v \rangle$ all have the same length r along the circle $x^2 + y^2 = r^2$. Furthermore, $\langle u, v \rangle$ makes an angle α with the vector $\langle x, y \rangle$, which is the radius vector to this circle. The vector field is illustrated in Fig. 13.3.

Finally, we may try to find a function $f(x, y)$ such that $\langle u, v \rangle = \nabla f$. However, for the pair of Eqs. (13.1) no such function exists, as we shall see later, by virtue of Lemma 19.1.

We defer further discussion of vector fields until Sect. 19, and of coordinate transformations until Sect. 20. For the greater part of this chapter we shall concentrate on the last of our four interpretations of a pair of functions, the one given in Example 13.4. The word "motion" will be interpreted in its wider sense, and replaced by the term "transformation" or "mapping." For ease of visualization it is most convenient to picture two separate planes. It is always possible later to consider the two planes as coinciding if we wish to picture a transformation of a plane into itself.

Definition 13.1 A *plane transformation* or *mapping* assigns to each point p of a plane domain D, a point $q = F(p)$ of another (or the same) plane. The point q is called the *image* of p under F (Fig. 13.4).

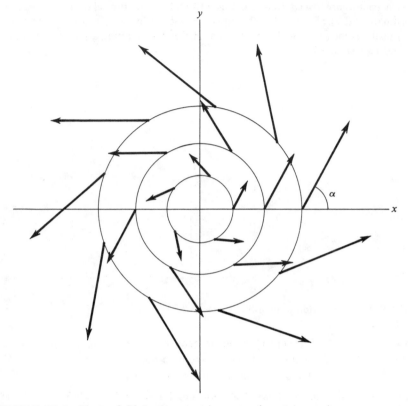

FIGURE 13.3 Vector field $\langle x \cos \alpha + y \sin \alpha, - x \sin \alpha + y \cos \alpha \rangle$

Remark The words "transformation" and "mapping" (or "map") are used interchangeably throughout most of mathematics, although in certain contexts it is more customary to use one rather than the other.

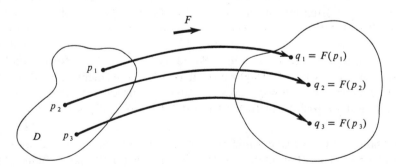

FIGURE 13.4 Plane mapping

The ability to visualize transformations is a skill that may be acquired with practice. It is useful not only in understanding the statement of theorems, but in discovering the underlying idea of their proofs. As an aid in the visualization, it is often useful to consider not only the image of individual points, but of sets of points, such as all points lying on a certain line, or on a certain figure. Then letting the line or figure vary over D, we may picture the way the image varies, and thus obtain an over-all view of the transformation.

In order to describe a transformation analytically, let us denote the coordinates of points in the first plane by (x, y) and those of the second plane by (u, v). We use the notation

$$F:(x, y) \rightarrow (u, v)$$

to indicate that the mapping F takes the point (x, y) onto the point (u, v). The mapping is given explicitly by expressing each of the coordinates u, v as a function of x and y:

$$F:\begin{cases} u = f(x, y) \\ v = g(x, y). \end{cases}$$

These general statements will be clarified by a number of concrete examples. In these examples the domain D where F is defined consists of the entire x, y plane.

Example 13.6

$$F:\begin{cases} u = x \\ v = 2y. \end{cases}$$

Under this transformation each vertical line $x = a$ maps onto the vertical line $u = a$ with the same abscissa. Each horizontal line $y = b$ maps onto the horizontal line $v = 2b$, twice as far from the horizontal axis. Thus F takes the point (a, b) into the point $(a, 2b)$. If the u, v plane is pictured as coinciding with the x, y plane, then this transformation may be described as a uniform stretching of the plane, by a factor of two, away from the x axis (Fig. 13.5).

Example 13.7

$$F:\begin{cases} u = x \\ v = -y. \end{cases}$$

In this case each horizontal line goes into a horizontal line on the opposite side of the axis. If the two planes coincide, these equations define a reflection in the x axis (Fig. 13.6).

Example 13.8

$$F:\begin{cases} u = y \\ v = x. \end{cases}$$

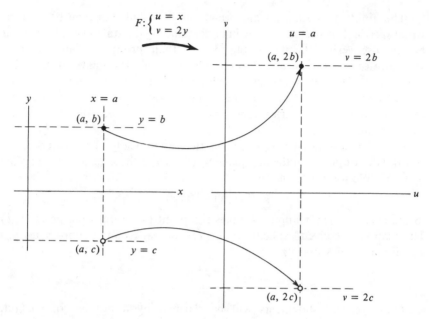

FIGURE 13.5 Map $F:(x, y) \rightarrow (x, 2y)$

Here each horizontal line goes into a vertical one, and each vertical line into a horizontal one. If the two planes coincide, then all points on the line $y = x$ remain fixed, while all other points are reflected in this line (Fig. 13.7).

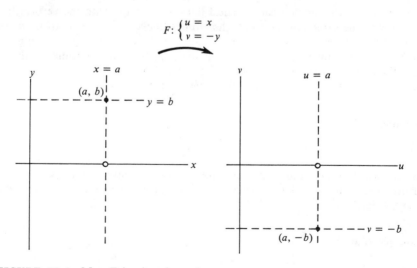

FIGURE 13.6 Map $F:(x, y) \rightarrow (x, -y)$

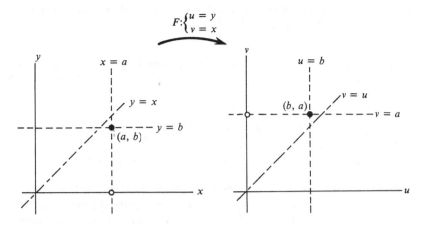

FIGURE 13.7 Map $F:(x, y) \to (y, x)$

Example 13.9

$$F:\begin{cases} u = |x| \\ v = y. \end{cases}$$

Here we have a new feature. Each *pair* of vertical lines $x = a$, $x = -a$ maps onto the same vertical line $u = |a|$.

All points on the horizontal line $y = b$ map into points of the horizontal line $v = b$, but in this case we get only half of the line $v = b$ in our image, because $u = |x| \geq 0$. Thus the whole x, y plane maps into the right half of the u, v plane. Each pair of points (a, b), $(-a, b)$ map onto a single point. If the u, v plane is pictured as coinciding with the x, y plane then this transformation can be described very simply as a "folding over." We leave the right half plane fixed, and we fold the left half plane along the y axis over onto the right half (Fig. 13.8).

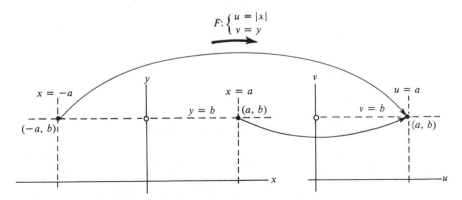

FIGURE 13.8 Map $F:(x, y) \to (|x|, y)$

At this point we interrupt our catalogue of special transformations to make some general definitions.

Definition 13.2 A mapping is called *one-to-one* or *injective* if any two distinct points map onto two distinct points. In other words, F is one-to-one if $p_1 \neq p_2$ implies $F(p_1) \neq F(p_2)$.

A mapping of a domain D into a domain E is called *onto* or *surjective* if every point q of E is the image of some point p. In other words, if to each q in E there exists at least one point p in D such that $F(p) = q$.

A mapping which is *both one-to-one and onto* is called *bijective*.

Note that under a bijective mapping F, each point q of E is the image of one and only one point p. We then have an *inverse* mapping $q \rightarrow p$, defined throughout the domain E, which assigns to each point q in E the unique point p such that $q = F(p)$.

Definition 13.3 A mapping G is called the *inverse* of a mapping F if

$$p = G(q) \Leftrightarrow q = F(p).$$

The standard notation for the inverse mapping of F is F^{-1} (Fig. 13.9).

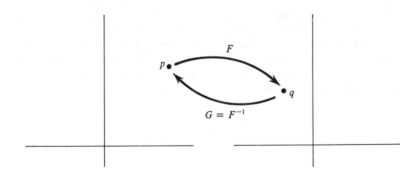

FIGURE 13.9 Inverse map

Remark For the remainder of this section, and also in the next two sections, the domains D and E consist of the entire plane.

If a mapping is given in coordinates by $u = f(x, y)$, $v = g(x, y)$, then finding the inverse means solving these equations for x and y in terms of u and v. In Examples 13.6–13.8, the mappings are all bijective, and it is easy to find the inverse.

Example 13.6a

$$F^{-1}:\begin{cases} x = u \\ y = \tfrac{1}{2}v. \end{cases}$$

Example 13.7a

$$F^{-1}:\begin{cases} x = u \\ y = -v. \end{cases}$$

Example 13.8a

$$F^{-1}:\begin{cases} x = v \\ y = u. \end{cases}$$

However, in Example 13.9, the mapping is neither one-to-one nor onto, since the point $(1, 0)$, for example, is the image of two distinct points, $(1, 0)$ and $(-1, 0)$, while the point $(-1, 0)$ is not an image at all. Thus, this mapping does not have an inverse.

We now continue our list of examples.

Example 13.10

$$F:\begin{cases} u = x^2 \\ v = y. \end{cases}$$

This mapping resembles the one in Example 13.9 in many ways. Each pair of vertical lines $x = a$, $x = -a$ maps onto a single vertical line $u = a^2$, while each horizontal line $y = b$ maps onto the right half of the horizontal line $v = b$ (Fig. 13.10). However there is a great deal of "distortion," since vertical lines near the y axis are mapped onto lines still nearer to the v axis, while those far from the y axis are mapped onto lines much further from the v axis. In some sense, though, this mapping has the same basic properties as the one in Example 13.9. Namely, each point in the right half plane $u > 0$ is the image of precisely two points in the x, y plane, while points in the left half plane $u < 0$ are not images at all.

Example 13.11

$$F:\begin{cases} u = x^2 - y^2 \\ v = 2xy. \end{cases}$$

Again we try to visualize this mapping by looking at the horizontal and vertical lines in the x, y plane, and their images in the u, v plane. Whenever $x = a$, we have $u = a^2 - y^2, v = 2ay$, or $u = a^2 - v^2/4a^2$. This is the equation of a parabola in the u, v plane with vertex at $(a^2, 0)$, and opening to the left. Similarly, if $y = b$, then $u = x^2 - b^2$, $v = 2bx$, and $u = v^2/4b^2 - b^2$, which is a parabola with vertex $(-b^2, 0)$, and opening to the right. Note also that the lines $x = -a$ and $y = -b$ map into the same two parabolas as $x = a$ and $y = b$ (Fig. 13.11). It is

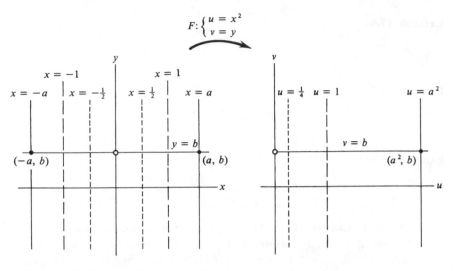

FIGURE 13.10 Map $F: (x, y) \rightarrow (x^2, y)$

clear that the points (a, b) and $(-a, -b)$ both map into the same point $(a^2 - b^2, 2ab)$, while the points $(a, -b)$ and $(-a, b)$ map into the point $(a^2 - b^2, -2ab)$.

Although we get some idea of the mapping F in Example 13.11 from the above description, it is certainly not as complete a picture as we were able to form in the previous examples. There are two very useful devices that may

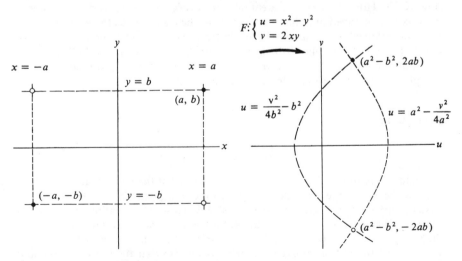

FIGURE 13.11 Map $F: (x, y) \rightarrow (x^2 - y^2, 2xy)$

aid in many cases in the visualization of a mapping. The first is that of the "inverse image."

Definition 13.4 The *inverse image* of a set S under a mapping F is the set of all points p such that $F(p)$ is in S. It is denoted by $F^{-1}(S)$.

Note that if F has an inverse map, then the inverse image of S is simply the image of S under the inverse map F^{-1}. However, the inverse image of S is a well-defined set of points for any mapping F. In particular, the inverse image of a point q consists of all points p that map into q under F.

Now it is often advantageous, when studying a particular map F, to look at the inverse image of lines in the u, v plane. Note that the inverse image of a vertical line $u = c$, is no more nor less than a level curve of the function $u(x, y)$. For the mapping of Example 13.11, the inverse image of $u = c$ is the curve $x^2 - y^2 = c$, which is a rectangular hyperbola. Similarly, the inverse image of the horizontal line $v = d$ is the curve $2xy = d$. The intersection of these two curves gives the inverse image of the point (c, d) (Fig. 13.12). See Exercise 13.9 for a more detailed analysis of Example 13.11 by the method of inverse images.

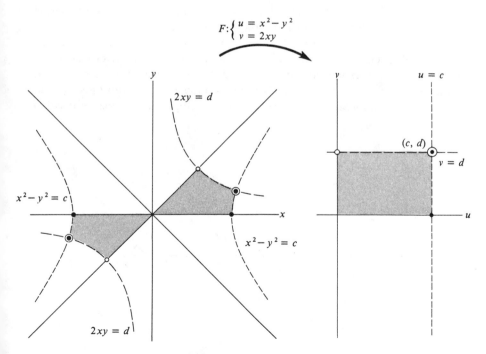

FIGURE 13.12 Map $F:(x, y) \to (x^2 - y^2, 2xy)$; inverse images

We may note that as a general rule it is much easier to construct inverse images than "forward images," because the former are directly given as a relation between x and y, whereas the latter involve a parameter that must be eliminated to get a relation between u and v.

The second device, which turns out to be the more useful in this case, is the introduction of polar coordinates. Just as in several of the examples in Sect. 3, the fact that we are dealing with homogeneous polynomials is particularly suggestive of this approach. Let us use r, θ for polar coordinates in the x, y plane, and R, φ in the u, v plane. Then

$$x = r \cos \theta, \qquad y = r \sin \theta;$$
$$u = R \cos \varphi, \qquad v = R \sin \varphi;$$

and

$$u = x^2 - y^2 = r^2(\cos^2 \theta - \sin^2 \theta) = r^2 \cos 2\theta$$
$$v = 2xy \qquad = 2r^2 \cos \theta \sin \theta \qquad = r^2 \sin 2\theta.$$

Thus the mapping F may be described in polar coordinates by

$$R = r^2$$
$$\varphi = 2\theta.$$

Each circle $x^2 + y^2 = r_1^2$ maps onto the circle $u^2 + v^2 = R_1^2 = r_1^4$, whose radius is the square of the original radius. Each ray $\theta = \theta_1$ maps onto the ray $\varphi = 2\theta_1$ (Fig. 13.13). It is now easy to visualize the total behavior of this map. Small circles about the origin are contracted into smaller circles, while large circles are expanded into still larger ones. One may give a good "kinematic" description by observing that as a ray $\theta = c$ rotates once around in the positive direction, its image is a ray rotating twice as fast, and going

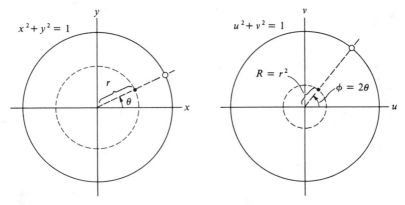

FIGURE 13.13 Map $F:(x, y) \to (x^2 - y^2, 2xy)$; polar coordinates

twice around in the positive direction. Incidentally, although it was clear from our first description that this map is not one-to-one, it was not obvious that every point in the u, v plane was the image of some point in the x, y plane. However, we now see that the origin in the u, v plane corresponds precisely to the origin the x, y plane, while every other point in the u, v plane is the image of exactly two points in the x, y plane. This map is therefore surjective.

Exercises

13.1 Try to visualize each of the following transformations. Make a sketch, and describe the transformation in words.

a. $\begin{cases} u = 3x \\ v = 3y \end{cases}$

g. $\begin{cases} u = x + y \\ v = y \end{cases}$

b. $\begin{cases} u = \frac{1}{2}y \\ v = \frac{1}{2}x \end{cases}$

h. $\begin{cases} u = x + y \\ v = |y| \end{cases}$

c. $\begin{cases} u = -x \\ v = -y \end{cases}$

i. $\begin{cases} u = |x| \\ v = |y| \end{cases}$

d. $\begin{cases} u = 2x \\ v = 3y \end{cases}$

j. $\begin{cases} u = x^2 \\ v = y^2 \end{cases}$

e. $\begin{cases} u = 2x \\ v = -3y \end{cases}$

k. $\begin{cases} u = x \\ v = y^3 \end{cases}$

f. $\begin{cases} u = x + y \\ v = -x + y \end{cases}$

l. $\begin{cases} u = e^x \\ v = y \end{cases}$

m. $\begin{cases} u = \text{arc tan } x \\ v = \text{arc tan } y \end{cases}$ (*Note*: use principal value of arc tan.)

13.2 For each of the following transformations, introduce polar coordinates r, θ in the x, y plane and R, φ in the u, v plane; express R and φ as functions of r, θ, and use these expressions to describe the transformation geometrically as in Example 13.11.

a. $\begin{cases} u = 2x \\ v = 2y \end{cases}$

e. $\begin{cases} u = x - y \\ v = x + y \end{cases}$

b. $\begin{cases} u = x/(x^2 + y^2) \\ v = y/(x^2 + y^2) \end{cases}$

f. $\begin{cases} u = x + y \\ v = x - y \end{cases}$

c. $\begin{cases} u = x \\ v = -y \end{cases}$

g. $\begin{cases} u = y \\ v = x \end{cases}$

d. $\begin{cases} u = xe^{x^2+y^2} \\ v = -ye^{x^2+y^2} \end{cases}$

*h. $\begin{cases} u = x^4 - 6x^2y^2 + y^4 \\ v = 4x^3y - 4xy^3 \end{cases}$

13.3   A transformation of the type

$$F = \begin{cases} u = x + by, \\ v = y, \end{cases}$$

where b is a nonzero constant, is called a "shear transformation."

   a. Draw the horizontal line segments $0 \le x \le 1$, $y = c$ for $c = 0$, $c = \frac{1}{2}$, and $c = 1$, and draw their images under the transformation F. (Consider separately the cases $b > 0$ and $b < 0$.)

   b. Draw the inverse images of the line segments $0 \le u \le 1$, $v = c$, where $c = 0, \frac{1}{2}, 1$. (Consider again $b > 0$ and $b < 0$.)

   c. Find the inverse of a shear transformation, and show that it is also a shear transformation.

13.4   For each of the following transformations F, find the inverse transformation F^{-1}.

a. $F: \begin{cases} u = 2x \\ v = 2y \end{cases}$
   d. $F: \begin{cases} u = x + y \\ v = x - y \end{cases}$

b. $F: \begin{cases} u = 2x \\ v = \frac{1}{2}y \end{cases}$
   e. $F: \begin{cases} u = 2x^3 - 8y^5 \\ v = 4x^3 + 16y^5 \end{cases}$

c. $F: \begin{cases} u = x + 1 \\ v = y - x \end{cases}$
   f. $F: \begin{cases} u = 6x + 5y \\ v = 7x + 6y \end{cases}$

13.5   Classify each of the following transformations as injective, surjective, bijective, or none of these. Explain your answer.

a. $\begin{cases} u = x^3 \\ v = y^5 \end{cases}$
   f. $\begin{cases} u = x + e^x \\ v = y + e^y \end{cases}$

b. $\begin{cases} u = x^3 + x \\ v = y \end{cases}$
   g. $\begin{cases} u = \sin x \\ v = \cos y \end{cases}$

c. $\begin{cases} u = x^3 - x \\ v = y^3 \end{cases}$
   h. $\begin{cases} u = 2x + 3y \\ v = 3x + 4y \end{cases}$

d. $\begin{cases} u = y^2 - y \\ v = x^3 - x \end{cases}$
   i. $\begin{cases} u = x \\ v = x^3 \end{cases}$

e. $\begin{cases} u = e^x \\ v = e^y \end{cases}$
   j. $\begin{cases} u = x + 2y \\ v = e^{x + 2y} \end{cases}$

13.6   In each of the following transformations, find the inverse image of the u axis.

a. $\begin{cases} u = x + y \\ v = -x + y \end{cases}$
   c. $\begin{cases} u = x^2 - y^2 \\ v = 2xy \end{cases}$

b. $\begin{cases} u = x^2 - y^2 \\ v = x^2 + y^2 \end{cases}$
   d. $\begin{cases} u = y \\ v = x^3 - x \end{cases}$

13.7   For each of the transformations in Ex. 13.6, find the inverse image of the line $v = u$.

13.8   For each of the following transformations, find the inverse image of the circle $u^2 + v^2 = R^2$.

a. $\begin{cases} u = x + y \\ v = -x + y \end{cases}$ *e.* $\begin{cases} u = x + 2 \\ v = y - 3 \end{cases}$

b. $\begin{cases} u = y \\ v = x \end{cases}$ *f.* $\begin{cases} u = 2x + 2 \\ v = 2y - 4 \end{cases}$

c. $\begin{cases} u = \frac{1}{2}x \\ v = \frac{1}{3}y \end{cases}$ *g.* $\begin{cases} u = x^2 - y^2 \\ v = 2xy \end{cases}$

d. $\begin{cases} u = x + 2y \\ v = y \end{cases}$ *h.* $\begin{cases} u = x^3 - 3xy^2 \\ v = 3x^2y - y^3 \end{cases}$

13.9   Study the transformation $u = x^2 - y^2$, $v = 2xy$ by a detailed use of inverse images. Carry out the following steps.

a. What is the inverse image of the u axis?

b. What is the inverse image of the v axis?

c. Describe the inverse image of the horizontal line $v = d$ for the two cases $d > 0$ and $d < 0$.

d. Describe the inverse image of the vertical line $u = c$ for the two cases $c > 0$ and $c < 0$.

e. Describe the way the curves in part d vary as the line $u = c$ moves to the right, starting with the v axis.

f. Copy Fig. 13.12. Note that the u, v plane is divided into nine parts by the lines $u = 0$, $u = c$, $v = 0$, $v = d$, and that the x, y plane is divided into eighteen parts by the inverse images of these lines. Number the nine parts in the u, v plane, and place the corresponding number in the inverse image of each of these parts.

Note: seeing how the inverse images of the different parts "fit together" is a useful device in many cases for obtaining an over-all picture of a given transformation.

13.10   Consider the transformation

$$\begin{cases} u = e^x \cos y \\ v = e^x \sin y \end{cases}$$

a. Introduce polar coordinates R, φ in the u, v plane and express R and φ as functions of x, y.

b. What is the inverse image of a circle $u^2 + v^2 = R^2$?

c. If a point moves upward along a vertical line $x = c$, describe the motion of the image point in the u, v plane.

d. What is the image of the x axis?

e. What is the image of the line $y = \frac{1}{2}\pi$?

f. What is the image of the line $y = \pi$?

g. What is the image of the line $y = 2\pi$?

h. If a point moves from left to right along a horizontal line $y = d$, describe the motion of the image point in the u, v plane.

i. Picture the line $y = d$ moving upward, and describe the way the image of this line varies in the u, v plane.

j. Is this transformation surjective? Explain.

13.11 Consider the transformation

$$F: \begin{cases} u = x^2 \\ v = y. \end{cases}$$

a. If p is the point $(-1, 1)$, what is $F(p)$?

b. If p is the point $(-1, 1)$, show that $F^{-1}(F(p)) \neq p$.

c. Find a set S of points in the u, v plane such that the image under F of the set $F^{-1}(S)$ does not coincide with S.

13.12 Let F be a transformation of a domain D into a domain E.

a. Show that F is injective if and only if there exists a transformation G of E into D satisfying $G(F(p)) = p$ for all p in D.

b. Show that F is surjective if and only if there exists a transformation G of E into D satisfying $F(G(q)) = q$ for all q in E.

c. Show that F is bijective if and only if there exists a transformation G of E into D satisfying both $G(F(p)) = p$ for all p in D and $F(G(q)) = q$ for all q in E; show that this transformation G is F^{-1}.

Note: the fact that D and E are plane domains is irrelevant in Ex. 13.12. The three parts hold without change if F is a mapping of any set D into any set E. The transformation G defined in part a is called a *left inverse* of F and the transformation G of part b is called a *right inverse* of F. Thus Ex. 13.12 may be summed up as follows: F has a left inverse $\Leftrightarrow F$ is injective; F has a right inverse $\Leftrightarrow F$ is surjective; F has an inverse $\Leftrightarrow F$ has a simultaneous left and right inverse $\Leftrightarrow F$ is bijective.

13.13 If F is a mapping of a domain D into a domain E, the image of the domain D is called the *range* of F. In other words, the range of F is the set S of points q in E such that $q = F(p)$ for some p in D. Thus F is surjective $\Leftrightarrow S = E$. It is always possible to consider F as a mapping of D into S. If F is injective, it will have an inverse when considered as a map of D into the range S.

For each of the following mappings, find the range S. If the mapping is injective, find the inverse mapping defined on the range.

a. $F: \begin{cases} u = x^3 \\ v = y^4 \end{cases}$

b. $F: \begin{cases} u = \sin xy \\ v = \cos (x + y) \end{cases}$

c. $F: \begin{cases} u = e^x \\ v = -e^y \end{cases}$

d. $F: \begin{cases} u = 1/(1 + x^2) \\ v = y \end{cases}$

e. $F: \begin{cases} u = \arctan x \\ v = \arctan y \end{cases}$ (*Note*: use principal value of arc tan.)

f. $F: \begin{cases} u = x + y \\ v = (x + y)^2 \end{cases}$

g. $F: \begin{cases} u = x^2 + y^2 \\ v = (x^2 + y^2)^2 \end{cases}$

13.14   In many cases the simplest way to show that a given transformation $F: u = f(x, y), v = g(x, y)$ is bijective is to solve explicitly for x and y as functions of u and v. Consider, for example, the transformation

$$u = x + y^2$$
$$v = x + y + y^2.$$

a. Solve these equations for x and y as functions of u and v.

b. Explain carefully why your solution demonstrates that the image of the x, y plane is the entire u, v plane.

c. Explain carefully why the solution implies that two different points in the x, y plane cannot have the same image in the u, v plane.

14 Linear transformations

Definition 14.1   A *linear transformation* is a transformation of the form

(14.1) $$F: \begin{cases} u = ax + by \\ v = cx + dy, \end{cases}$$

where a, b, c, d are constants. The transformation is uniquely determined by the array of coefficients

(14.2) $$\begin{pmatrix} a & b \\ c & d \end{pmatrix},$$

which is called the *matrix* of the transformation.

The study of linear transformations from an algebraic point of view forms a basic part of "linear algebra" or "matrix theory."

In this section, we investigate the geometric properties of linear transformations. Although it may seem that linear transformations are quite special, we shall see later that many questions concerning general transformations can be answered by referring to related linear transformations.

The first thing to observe is that solving the simultaneous linear equations

(14.3) $$\begin{aligned} ax + by &= e \\ cx + dy &= f \end{aligned}$$

is precisely equivalent to finding those points in the x, y plane which map onto the point (e, f) in the u, v plane under the transformation (14.1).

We use the following standard notation

(14.4) $$\Delta = \det \begin{pmatrix} a & b \\ c & d \end{pmatrix} = ad - bc.$$

This number Δ is called the *determinant* of the transformation (14.1), as well as of the matrix (14.2) and of the system (14.3). Two cases must be distinguished.

Case 1. $\Delta \neq 0$. In this case, for each choice of the right-hand side of (14.3), there is a unique solution, namely

(14.5)
$$x = (de - bf)/\Delta$$
$$y = (af - ce)/\Delta.$$

In terms of the transformation (14.1), this means that each point (e, f) in the u, v plane is the image of one and only one point in the x, y plane, or equivalently, that F maps the x, y plane one-to-one onto the u, v plane. Equations (14.5) are in fact the equations of the inverse mapping of F, which we may write as

(14.6)
$$F^{-1}: \begin{cases} x = \dfrac{d}{\Delta} u - \dfrac{b}{\Delta} v \\ y = -\dfrac{c}{\Delta} u + \dfrac{a}{\Delta} v. \end{cases}$$

Note that the map F^{-1} is a linear transformation of the u, v plane into the x, y plane. Its matrix is

(14.7)
$$\begin{pmatrix} \dfrac{d}{\Delta} & -\dfrac{b}{\Delta} \\ -\dfrac{c}{\Delta} & \dfrac{a}{\Delta} \end{pmatrix}$$

and its determinant is

(14.8)
$$\frac{d}{\Delta} \cdot \frac{a}{\Delta} - \left(-\frac{b}{\Delta}\right) \cdot \left(-\frac{c}{\Delta}\right) = \frac{ad - bc}{\Delta^2} = \frac{\Delta}{\Delta^2} = \frac{1}{\Delta}.$$

Case 2. $\Delta = 0$. Within case 2 there are two possibilities:

a. a, b, c, d are all zero. Then the transformation (14.1) simply maps the whole x, y plane onto the origin of the u, v plane.

b. Not all coefficients are zero. Suppose that either a or c is different from zero. Multiply the first equation of (14.1) by c, the second by a, and subtract. Using $\Delta = 0$, we find

$$cu - av = 0,$$

which is the equation of a straight line through the origin. Thus, the image of every point (x, y) lies on this straight line. (In terms of the system of equations (14.3), a solution exists for a given right-hand side if and only if the point (e, f) lies on this line; that is, if and only if $ce - af = 0$.) If either b or d is different from zero, we find similarly that the whole plane maps onto the line $du - bv = 0$.

Definition 14.2 The linear transformation (14.1) (or the matrix (14.2)) is called *singular* if $\Delta = 0$ and *nonsingular* if $\Delta \neq 0$.

We may summarize the above discussion as follows. Under a singular linear transformation the whole plane maps into a straight line (or in the extreme case when all the coefficients are zero, into a single point). The mapping is *neither* one-to-one *nor* onto. A nonsingular linear transformation is *both* one-to-one *and* onto. It has an inverse, which is again a nonsingular linear transformation.

Example 14.1

$$F = \begin{cases} u = x \\ v = 2y. \end{cases}$$

Matrix is

$$\begin{pmatrix} 1 & 0 \\ 0 & 2 \end{pmatrix},$$

$\Delta = 2$. F is nonsingular.

Example 14.2

$$F = \begin{cases} u = x \\ v = -y. \end{cases}$$

Matrix is

$$\begin{pmatrix} 1 & 0 \\ 0 & -1 \end{pmatrix},$$

$\Delta = -1$. F is nonsingular.

Example 14.3

$$F = \begin{cases} u = y \\ v = x. \end{cases}$$

Matrix

$$\begin{pmatrix} 0 & 1 \\ 1 & 0 \end{pmatrix},$$

$\Delta = -1$. F is nonsingular.

Example 14.4

$$F = \begin{cases} u = 2x - 4y \\ v = -3x + 6y. \end{cases}$$

Matrix

$$\begin{pmatrix} 2 & -4 \\ -3 & 6 \end{pmatrix},$$

$\Delta = 0$. F is singular. For all x, y, we have $3u + 2v = 0$.

Nonsingular Transformations

For the remainder of this section we restrict our attention to the case of nonsingular transformations. The introduction of polar coordinates once again proves advantageous. Let us set

$$x = r \cos \theta, \qquad y = r \sin \theta;$$
$$u = R \cos \varphi, \qquad v = R \sin \varphi.$$

Then Eqs. (14.1) become

(14.9)
$$F = \begin{cases} u = r(a \cos \theta + b \sin \theta) = R \cos \varphi \\ v = r(c \cos \theta + d \sin \theta) = R \sin \varphi. \end{cases}$$

Thus

(14.10)
$$\tan \varphi = \frac{v}{u} = \frac{c \cos \theta + d \sin \theta}{a \cos \theta + b \sin \theta} = \frac{c + d \tan \theta}{a + b \tan \theta}.$$

If we hold θ fixed and let r vary, it follows from Eq. (14.10) that φ is constant, which means geometrically that each ray starting at the origin in the x, y plane maps into a ray from the origin in the u, v plane (Fig. 14.1). We ask the question: "if the ray in the x, y plane is rotated about the origin in the positive direction what happens to the image ray in the u, v plane?" More precisely, we may ask: "how does the angle φ depend on θ?" The answer is obtained simply by differentiating Eq. (14.10) with respect to θ.

$$\sec^2 \varphi \, \frac{d\varphi}{d\theta} = \frac{(a + b \tan \theta)(d \sec^2 \theta) - (c + d \tan \theta)(b \sec^2 \theta)}{(a + b \tan \theta)^2}$$

$$= (ad - bc) \frac{\sec^2 \theta}{(a + b \tan \theta)^2},$$

or

(14.11)
$$\frac{d\varphi}{d\theta} = \Delta \frac{\sec^2 \theta}{\sec^2 \varphi} \frac{1}{(a + b \tan \theta)^2}.$$

It follows that $d\varphi/d\theta$ has the same sign as the determinant Δ. When Δ is positive we have $d\varphi/d\theta > 0$ and φ is an increasing function of θ, while for Δ negative, φ is a decreasing function of θ. We may summarize the situation geometrically as follows.

Case 1. $\Delta > 0$. As a ray in the x, y plane is rotated in the counterclockwise direction (θ increasing), the image ray in the u, v plane rotates in the same direction. The transformation is called *orientation preserving* (Fig. 14.2).

Case 2. $\Delta < 0$. As a ray in the x, y plane rotates in the counterclockwise direction, the image ray in the u, v plane rotates in the opposite direction. The transformation is called *orientation reversing* (Fig. 14.3).

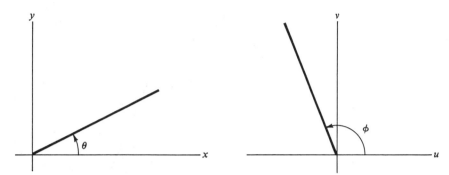

FIGURE 14.1   Correspondence between polar angles

Example 14.1a

Since $\Delta > 0$ in Example 14.1 the transformation is orientation preserving.

Example 14.2a

The transformation in Example 14.2 is orientation reversing. In fact, Eq. (14.10) becomes $\tan \varphi = -\tan \theta$. It follows that, in this example, $d\varphi/d\theta = -1$, so that the image ray rotates at the same rate in the opposite direction as the ray in the x, y plane. The reason is clear geometrically if we recall from Example 13.7 that this transformation corresponds to a reflection in the horizontal axis.

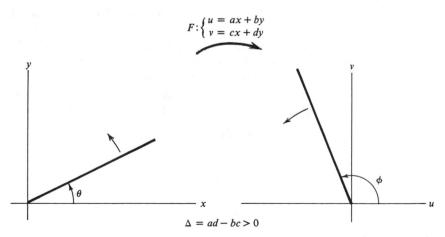

FIGURE 14.2   Orientation-preserving transformation

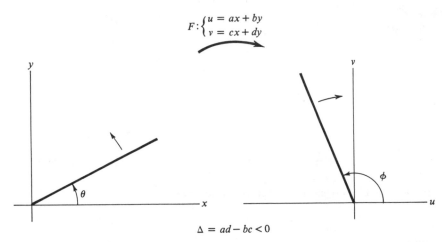

$$\Delta = ad - bc < 0$$

FIGURE 14.3   Orientation-reversing transformation

Example 14.3a

The transformation in Example 14.3 is also orientation reversing. Equation (14.10) becomes $\tan \varphi = \cot \theta$, and again $d\varphi/d\theta = -1$. As in Example 14.2a the reason is clear geometrically if we recall from Example 13.8 that this transformation corresponds to a reflection across the diagonal.

We should remark that actually every straight line through the origin in the x, y plane maps into a straight line through the origin in the u, v plane. We have simply restricted our discussion to rays, or half-lines, in order to single out a direction along each line.

We now return to Eqs. (14.9) and ask what happens if we hold r fixed and let θ vary. Geometrically, we are looking for the image of a circle about the origin in the x, y plane. Although this question can be answered directly by examining Eqs. (14.9), it turns out to be easier to use the device mentioned in Sect. 13, and ask what is the *inverse image* of a circle in the u, v plane. If we hold R fixed, we want the set of points in the x, y plane satisfying $u^2 + v^2 = R^2$. Returning to Eqs. (14.1), we find that x and y must satisfy

(14.12) $(a^2 + c^2)x^2 + 2(ab + cd)xy + (b^2 + d^2)y^2 = R^2$

or

(14.13) $Ax^2 + 2Bxy + Cy^2 = 1,$

where

(14.14) $A = \dfrac{a^2 + c^2}{R^2}, \qquad B = \dfrac{ab + cd}{R^2}, \qquad C = \dfrac{b^2 + d^2}{R^2}.$

Equation (14.13) defines a curve whose nature depends on the sign of the quantity $AC - B^2$ (see Corollary 3 of Th. 10.2). But from (14.14)

$$(14.15) \qquad AC - B^2 = \frac{a^2d^2 + c^2b^2 - 2abcd}{R^4} = \frac{(ad - bc)^2}{R^4}$$

$$= \left(\frac{\Delta}{R^2}\right)^2 > 0.$$

Also $A + C = (a^2 + b^2 + c^2 + d^2)/R^2 > 0$. Thus $Ax^2 + 2Bxy + Cy^2$ is a positive definite quadratic form, and Eq. (14.13) represents an ellipse. Furthermore, we know from Corollary 4 of Th. 10.2 that the area $\mathscr{A}$ inside the ellipse is

$$\mathscr{A} = \frac{\pi}{\sqrt{AC - B^2}} = \frac{\pi R^2}{|\Delta|},$$

so that

$$(14.16) \qquad |\Delta| = \pi R^2/\mathscr{A}.$$

We may summarize as follows.

Under a nonsingular linear transformation the inverse image of each circle with center at the origin in the u, v plane is an ellipse centered at the origin of the x, y plane. The absolute value of the determinant Δ is equal to the ratio of the area inside the circle to the area inside the ellipse (Fig. 14.4). Note in particular that this ratio of the areas does not depend on R; it is the same for all circles.

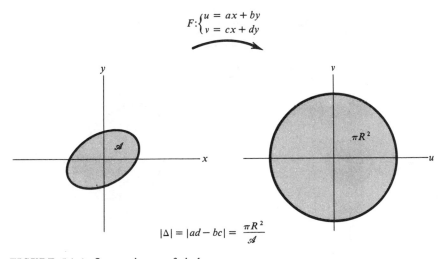

$$F: \begin{cases} u = ax + by \\ v = cx + dy \end{cases}$$

$$|\Delta| = |ad - bc| = \frac{\pi R^2}{\mathscr{A}}$$

FIGURE 14.4   Inverse image of circle

We now may return to our original question concerning the image of a circle in the x, y plane. The answer is provided by the simple observation that the inverse mapping $G = F^{-1}$ is also a nonsingular transformation, and the image of the circle $x^2 + y^2 = r^2$ under F is the same as its inverse image under G (Fig. 14.5). Applying the above reasoning to the linear transformation G, we see that the circle $x^2 + y^2 = r^2$ corresponds to an ellipse about the origin in the u, v plane, and that the area $\mathscr{A}'$ inside the ellipse satisfies

$$|\Delta'| = \pi r^2 / \mathscr{A}',$$

where Δ' is the determinant of G. But Eq. (14.8) relates the determinant of the inverse transformation to that of the original. We find therefore, that

(14.17) $$|\Delta| = \frac{1}{|\Delta'|} = \frac{\mathscr{A}'}{\pi r^2}.$$

Comparing Eqs. (14.16) and (14.17), we see that in both cases *the magnitude of Δ represents the ratio of the area in the u, v plane to the area in the x, y plane.* This interpretation of the absolute value of Δ together with our earlier discussion of the significance of the sign of Δ, allows us to describe the value of the determinant Δ in a purely geometric fashion. This description is fundamental to the study of differentiable mappings in the same way as the interpretation of the magnitude and sign of the slope of a straight line is to the study of differentiable functions of a single variable.

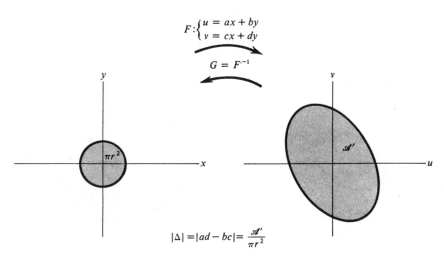

FIGURE 14.5 Image of a circle

Example 14.5

$$F:\begin{cases} u = x \\ v = 2y. \end{cases}$$

A circle of radius r in the x, y plane may be written in parametric form as $x = r \cos t$, $y = r \sin t$. Its image is then $u = r \cos t$, $v = 2r \sin t$, which is the parametric form of the ellipse (Fig. 14.6)

$$\frac{u^2}{r^2} + \frac{v^2}{(2r)^2} = 1.$$

The area inside this ellipse is $\pi r \cdot (2r) = 2\pi r^2$, which is twice the area inside the original circle $x^2 + y^2 = r^2$. The ratio of areas is therefore the same as the value of the determinant $\Delta = 2$. Similarly, the inverse image of the circle $u^2 + v^2 = R^2$ is given by the equation

$$x^2 + (2y)^2 = R^2 \qquad \text{or} \qquad \frac{x^2}{R^2} + \frac{y^2}{(R/2)^2} = 1,$$

which is an ellipse bounding an area of $\pi \cdot R \cdot R/2 = \tfrac{1}{2}\pi R^2$. The ratio of areas is again equal to 2 (Fig. 14.7).

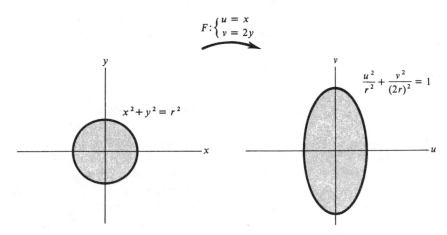

FIGURE 14.6 Map $F:(x, y) \rightarrow (x, 2y)$; image of circle

Translations

There is another class of elementary transformations that is often considered together with the linear transformations. Namely, the transformation

(14.18) $$F:\begin{cases} u = x + h \\ v = y + k, \end{cases}$$

where h and k are constants. If we picture the x, y plane and u, v plane as coinciding, then the effect of the transformation (14.18) is to move each point (x, y) a distance h horizontally and a distance k vertically. Equivalently,

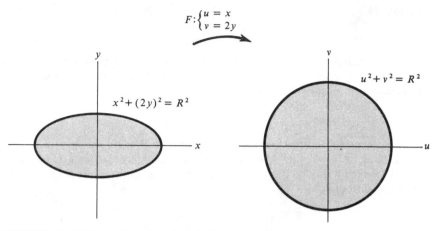

FIGURE 14.7 Map $F:(x, y) \rightarrow (x, 2y)$; inverse image of circle

each point undergoes a displacement represented by the fixed vector $\langle h, k \rangle$. Such a transformation is called a *parallel translation*, or simply a *translation*.

Combining a translation with a linear transformation, yields a transformation of the form

$$(14.19) \qquad F: \begin{cases} u = ax + by + h \\ v = cx + dy + k. \end{cases}$$

A transformation of this form is called an *affine transformation*.

There is a certain ambiguity in the use of the word "linear," which is sometimes used to mean "purely linear," that is to say, homogeneous of first degree, and sometimes in the more general sense of a polynomial of degree one, which may include constant terms. Thus, in the case of functions, "linear" may mean $ax + by$ or $ax + by + c$. However, in the case of transformations it has become standard to reserve the expression "linear transformation" for the homogeneous case, that is, Eqs. (14.1), and to use the word "affine" for the more general transformations of the form (14.19).

An affine transformation can be most readily visualized as the result of two successive transformations. This is a special case of the following general situation, which we encounter frequently.

Definition 14.3 Let F_1 be an arbitrary transformation of a domain D_1 into a domain D_2, and let F_2 be a transformation of D_2 into a domain D_3. Then the transformation F of D_1 into D_3 obtained by applying successively the transformations F_1 and F_2 (see Fig. 14.8) is called the *composition* of F_1 and F_2, and is denoted by

$$F = F_2 \circ F_1.$$

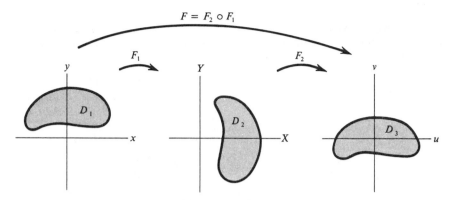

FIGURE 14.8   Composed map

Note that the transformation written on the *right* is the one that is applied *first*. The reason for this convention is that if

$$q = F_1(p) \quad \text{and} \quad r = F_2(q),$$

then

$$r = F(p) = F_2(F_1(p)),$$

so that the transformation applied last appears on the left.

In terms of coordinates, if

$$F_1 : (x, y) \rightarrow (X, Y) \quad \text{and} \quad F_2 : (X, Y) \rightarrow (u, v)$$

then

$$F = F_2 \circ F_1 : (x, y) \rightarrow (u, v).$$

Thus, the general affine transformation (14.19) is obtained as the composition of

$$F_1 : \begin{cases} X = ax + by \\ Y = cx + dy \end{cases} \quad \text{and} \quad F_2 : \begin{cases} u = X + h \\ v = Y + k, \end{cases}$$

where F_1 is a general linear transformation and F_2 is a translation. The equations for $F = F_2 \circ F_1$ are obtained by direct substitution into the equations for F_2 of the expressions for X and Y given by F_1.

In Sect. 15 we shall study in detail the composition of linear transformations, and in Sect. 17 we shall treat the composition of more general transformations.

Exercises

14.1   Write down the matrices associated with the following linear transformations; compute the determinants, and decide whether the transformations are singular or nonsingular.

a. $\begin{cases} u = x \\ v = y \end{cases}$

g. $\begin{cases} u = x\cos\alpha + y\sin\alpha \\ v = -x\sin\alpha + y\cos\alpha, \ \alpha \text{ constant} \end{cases}$

b. $\begin{cases} u = -x \\ v = -y \end{cases}$

h. $\begin{cases} u = ax + by \\ v = bx - ay, \ a \text{ and } b \text{ not both zero} \end{cases}$

c. $\begin{cases} u = ay \\ v = ax, \quad a \neq 0 \end{cases}$

i. $\begin{cases} u = ax + \lambda ay \\ v = bx + \lambda by \end{cases}$

d. $\begin{cases} u = x + y \\ v = x - y \end{cases}$

j. $\begin{cases} u = ax + by \\ v = \lambda ax + \lambda by \end{cases}$

e. $\begin{cases} u = x - 3y \\ v = -2x + 6y \end{cases}$

k. $\begin{cases} u = 7x \\ v = -11x \end{cases}$

f. $\begin{cases} u = x + by \\ v = y \end{cases}$

l. $\begin{cases} u = 2x - 3y \\ v = 0 \end{cases}$

14.2   Show that each of the following linear transformations has nonzero determinant Δ. Find the inverse transformation, compute its determinant Δ', and show that $\Delta' = 1/\Delta$.

a. $\begin{cases} u = 2y \\ v = 2x \end{cases}$

c. $\begin{cases} u = 3x + y \\ v = x - 3y \end{cases}$

b. $\begin{cases} u = x - y \\ v = x + y \end{cases}$

d. $\begin{cases} u = 3x + 5y \\ v = 4x + 7y \end{cases}$

14.3   Show that each of the following linear transformations has zero determinant. Find the line in the u, v plane that is the image of the whole plane under the transformation.

a. $\begin{cases} u = y \\ v = -y \end{cases}$

c. $\begin{cases} u = -x + 4y \\ v = 3x - 12y \end{cases}$

b. $\begin{cases} u = 0 \\ v = 3x + 5y \end{cases}$

d. $\begin{cases} u = x - 3y \\ v = x - 3y \end{cases}$

14.4 Consider the transformation

$$F: \begin{cases} u = x + y \\ v = 2x + 2y. \end{cases}$$

   a. What is the inverse image under F of a point on the line $v = 2u$?
   b. Show that F maps the line $y = x$ in a one-to-one fashion onto the line $v = 2u$.

c. Show that each line perpendicular to $y = x$ is mapped by F onto a single point.

d. Show that two different lines perpendicular to $y = x$ are mapped by F onto two different points.

e. Deduce that the transformation F may be described geometrically in the following fashion. There is a family of parallel lines in the x, y plane such that F maps each line into a point on the line $v = 2u$; as a variable line sweeps out the family of lines, the image under F describes the entire line $v = 2u$.

14.5 Let F be a singular linear transformation,

$$F: \begin{cases} u = ax + by \\ v = cx + dy, \end{cases}$$

where a and b are not both zero.

a. Show that there is a number λ such that F may be written in the form

$$F: \begin{cases} u = ax + by \\ v = \lambda ax + \lambda by. \end{cases}$$

(*Hint*: see Ex. 1.9a.)

b. What is the inverse image of a point on the line $v = \lambda u$?

c. Show that an arbitrary singular linear transformation F may be visualized geometrically in a fashion analogous to that described in Ex. 14.4e, unless F maps every point into the origin. Namely, there is a family of parallel lines in the x, y plane, such that F maps each line onto a point, F maps two different lines onto two different points, and as a variable line sweeps through the family, the image under F describes a line in the u, v plane. (*Hint*: consider separately the case where a and b are both zero and the case where a and b are not both zero.)

14.6 For each of the following linear transformations, introduce polar coordinates r, θ in the x, y plane and R, φ in the u, v plane. Find φ explicitly as a function of θ, and verify that φ is an increasing function of θ when the determinant is positive, and a decreasing function of θ when the determinant is negative.

a. $\begin{cases} u = x \\ v = -y \end{cases}$
  d. $\begin{cases} u = 3x \\ v = 3y \end{cases}$

b. $\begin{cases} u = y \\ v = x \end{cases}$
  e. $\begin{cases} u = x + y \\ v = -x + y \end{cases}$

c. $\begin{cases} u = -x \\ v = -y \end{cases}$

14.7 Let F be a nonsingular linear transformation.

a. Show that if F is orientation-preserving, then so is F^{-1}.

b. Show that if F is orientation-reversing, then so is F^{-1}.

14.8 For each of the transformations in Ex. 14.6, find the inverse image of $u^2 + v^2 \leq R^2$, find its area A, and show that the ratio of $\pi R^2/A$ is equal to the absolute value of the determinant.

14.9 Let F be a linear transformation

$$F: \begin{cases} u = ax + by \\ v = cx + dy. \end{cases}$$

a. Show that if $F(x_0, y_0) = (u_0, v_0)$ and $F(x_1, y_1) = (u_1, v_1)$, then F maps the line segment joining (x_0, y_0) to (x_1, y_1) onto the line segment joining (u_0, v_0) to (u_1, v_1). (*Hint*: write the first line segment in parametric form, and substitute in the equations for F to get the equations of the image.)

b. Show that F maps every triangle onto a triangle, where the image triangle may be degenerate.

c. If T_1 is the triangle whose vertices are $(0, 0)$, $(1, 0)$, and $(0, 1)$, what are the vertices of the triangle T_2, the image of T_1 under F?

d. If A_1, A_2 are the areas of T_1, T_2, respectively, show that $A_2/A_1 = |\Delta| = |ad - bc|$. (*Hint*: see Ex. 1.29d.)

e. Show that the unit square, with vertices $(0, 0)$, $(1, 0)$, $(1, 1)$, $(0, 1)$, maps onto a parallelogram whose area is $|ad - bc|$.

14.10 Let F be the transformation

$$F: \begin{cases} u = x + 3y \\ v = 3x + y. \end{cases}$$

a. What is the shortest distance to the origin of points $F(x, y)$, where (x, y) lies on the circle $x^2 + y^2 = 1$?
(*Hint*: if d is the shortest distance, then d^2 is the minimum of $u^2 + v^2$ subject to the condition $x^2 + y^2 = 1$.)

b. What is the furthest distance to the origin of points $F(x, y)$, where $x^2 + y^2 = 1$?

c. Verify that the product of the answers to parts a and b is equal to the absolute value of the determinant of F, and explain why that is so.

*14.11 Let F be an arbitrary linear transformation

$$F: \begin{cases} u = ax + by \\ v = cx + dy. \end{cases}$$

For any two points (x_0, y_0), (x_1, y_1), let $(u_0, v_0) = F(x_0, y_0)$ and $(u_1, v_1) = F(x_1, y_1)$. Consider the displacement vectors

$$\langle X, Y \rangle = \langle x_1 - x_0, y_1 - y_0 \rangle, \qquad \langle U, V \rangle = \langle u_1 - u_0, v_1 - v_0 \rangle.$$

a. Show that $F(X, Y) = (U, V)$.

b. Show that $|\langle U, V \rangle|^2 = AX^2 + 2BXY + CY^2$, and determine the coefficients A, B, C.

c. Show that the quadratic form in part b is positive definite if and only if F is nonsingular, and positive semidefinite if and only if F is singular.

d. Using part c, give a direct proof that for a nonsingular linear transformation F, no two distinct points can map onto a single point (that is, F is injective), whereas for F singular, certain pairs of points map into a single point (that is, F is not injective).

e. Show that as (x_0, y_0) and (x_1, y_1) range over all pairs of distinct points in the x, y plane, the ratio $|\langle U, V \rangle| / |\langle X, Y \rangle|$ of the distance between the image points and the distance between the original points ranges between a minimum m and a maximum M, where the product mM is the absolute value of the determinant of F. In particular, if F is nonsingular, $m > 0$.

f. Show that for the transformation in Ex. 14.10, the quantities m and M of part e are $m = 2$, $M = 4$.

14.12   Let F be an arbitrary linear transformation.

a. Show that F always maps the origin into the origin.

b. Show that if (x_1, y_1) is any point other than the origin, then specifying $F(x_1, y_1)$ determines $F(x, y)$ on the entire line through the origin and (x_1, y_1).

c. Show that if (x_1, y_1) and (x_2, y_2) are any two points that do not lie on a straight line through the origin, then specifying $F(x_1, y_1)$ and $F(x_2, y_2)$ determines $F(x, y)$ completely.

d. Show that if (x_1, y_1) and (x_2, y_2) are any two points that do not lie on a straight line through the origin, and if (u_1, v_1), (u_2, v_2) are any two points at all, then there exists one and only one linear transformation F such that $F(x_1, y_1) = (u_1, v_1)$ and $F(x_2, y_2) = (u_2, v_2)$.

e. Is the statement in part d still correct if (x_1, y_1) and (x_2, y_2) lie on a line through the origin? Explain.

14.13   Prove the following statements.

a. Every parallel translation has an inverse, which is again a parallel translation.

b. The composition of two parallel translations is again a parallel translation.

c. If F_1 and F_2 are parallel translations, then $F_1 \circ F_2 = F_2 \circ F_1$.

14.14   Consider the linear transformations

$$F_1 : (x, y) \to (x, -y)$$
$$F_2 : (x, y) \to (y, x).$$

a. What is the image of $(0, 1)$ under $F_2 \circ F_1$?

b. What is the image of $(0, 1)$ under $F_1 \circ F_2$?

c. Show that $F_1 \circ F_2 \neq F_2 \circ F_1$.

d. Indicate with a sketch how the answers to parts a and b may be obtained geometrically, using the interpretations of F_1 and F_2 as transformations of a plane into itself.

*14.15   Let F_1 and F_2 be linear transformations such that $F_2 \circ F_1$ leaves every point unchanged.

a. Show that F_1 and F_2 are both nonsingular.

b. Show that $F_2 = F_1^{-1}$.

14.16   A transformation F such that $F \circ F$ leaves every point unchanged is called an *involution*.

a. Show that if a linear transformation F is an involution, then the determinant of F is ± 1. (*Hint*: see Ex. 14.15.)

b. Which of the following transformations are involutions:

$$F:(x, y) \to (x, -y); \qquad F:(x, y) \to (2x, \tfrac{1}{2}y); \qquad F:(x, y) \to (2y, \tfrac{1}{2}x)?$$

*14.17 Let F be a linear transformation

$$F:\begin{cases} u = ax + by \\ v = cx + dy \end{cases}$$

where a, b, c, d are integers. Find a necessary and sufficient condition that the inverse transformation F^{-1} will have integer coefficients. Prove your assertion.

14.18 Let F be an affine transformation

$$F:\begin{cases} u = ax + by + h \\ v = cx + dy + k. \end{cases}$$

a. Show that F maps every straight line segment onto a straight line segment or a point.

b. Under what condition does F have an inverse?

c. Write down the equations for F^{-1} when it exists.

14.19 Let F_1 map D_1 into D_2, and let F_2 map D_2 into D_3. Show that if F_1 and F_2 have inverses, then so does $F = F_2 \circ F_1$, and $F^{-1} = F_1^{-1} \circ F_2^{-1}$.

Remark Given a matrix

$$\begin{pmatrix} a & b \\ c & d \end{pmatrix},$$

we may associate with it not only the transformation F of the plane defined by Eqs. (14.1), but also a corresponding transformation of vectors, which we may denote by **F**. Namely,

$$\mathbf{F}(\langle X, Y \rangle) = \langle U, V \rangle, \qquad \text{where} \qquad \begin{aligned} U &= aX + bY \\ V &= cX + dY. \end{aligned}$$

(By Ex. 14.11a, **F** is precisely the transformation of displacement vectors induced by the transformation F of the plane.) In fact, it is more usual to consider linear transformations as acting upon vectors. The following five exercises explore linear transformations from this point of view. (Note that as transformations of number pairs into number pairs, F and **F** are identical. The only difference is in the geometric interpretation we give to these number pairs, and this difference only becomes essential when we consider allowable coordinate changes corresponding to the geometric interpretation.)

14.20 Show that the transformation **F** defined above has the following properties.

a. $\mathbf{F}(\lambda \mathbf{w}) = \lambda \mathbf{F}(\mathbf{w})$ for any vector $\mathbf{w}$ and any scalar λ.

b. $\mathbf{F}(\mathbf{w}_1 + \mathbf{w}_2) = \mathbf{F}(\mathbf{w}_1) + \mathbf{F}(\mathbf{w}_2)$ for any vectors $\mathbf{w}_1, \mathbf{w}_2$.

14.21 Show that the properties in Ex. 14.20a, b are together equivalent to the single property

$$F(\lambda_1 w_1 + \lambda_2 w_2) = \lambda_1 F(w_1) + \lambda_2 F(w_2)$$

for any vectors w_1, w_2 and scalars λ_1, λ_2.

14.22 Let F be an arbitrary transformation that assigns to every vector w a vector F(w). Suppose that F satisfies Ex. 14.20a, b (or equivalently, the property in Ex. 14.21).

    a. Show that F is completely determined by its values on $\langle 0, 1 \rangle$ and on $\langle 1, 0 \rangle$.

    b. Show that F is a linear transformation.

14.23 *a.* Show that $F(0) = 0$.

    b. Show that $F(w) = 0$ for some $w \neq 0$ if and only if $\det \begin{pmatrix} a & b \\ c & d \end{pmatrix} = 0$.

*14.24 Definition: a nonzero vector w that is carried by F into a multiple of itself is called a *characteristic vector* of the transformation F and of the matrix $\begin{pmatrix} a & b \\ c & d \end{pmatrix}$. (The terms *proper vector* and *eigenvector* are also used.) Thus, if w is a characteristic vector, $F(w) = \lambda w$ for some scalar λ. This scalar λ is then called a *characteristic value* of F and of the matrix $\begin{pmatrix} a & b \\ c & d \end{pmatrix}$. (Again, the expressions *proper value* and *eigenvalue* are used.)

    a. Show that if λ is a characteristic value of the matrix $\begin{pmatrix} a & b \\ c & d \end{pmatrix}$, then λ satisfies the equation

$$\det \begin{pmatrix} a - \lambda & b \\ c & d - \lambda \end{pmatrix} = 0$$

or

$$\lambda^2 - (a + d)\lambda + (ad - bc) = 0.$$

(*Hint*: write out the equation $F(w) - \lambda w = 0$ in components, and use the fact that there exists a nonzero solution w.)

    b. Show that a 2×2 matrix can have at most two characteristic values.

    c. Show that the matrix $\begin{pmatrix} 1 & 1 \\ -1 & 1 \end{pmatrix}$ has no (real) characteristic values and hence no (real) characteristic vectors. Can you give a reason for this in terms of the geometric interpretation of the linear transformation

$$u = x + y \qquad v = -x + y?$$

(*Hint*: see Ex. 14.6e.)

    d. Show that if $\det \begin{pmatrix} a & b \\ c & d \end{pmatrix} < 0$, then there always exist real nonzero characteristic values and characteristic vectors. Can you give a reason for this in terms of the geometric interpretation of the sign of the determinant?

e. Show that the maximum and minimum of the quadratic form $AX^2 + 2BXY + CY^2$, subject to the condition $X^2 + Y^2 = 1$, are the characteristic values of the matrix $\begin{pmatrix} A & B \\ B & C \end{pmatrix}$; furthermore if (X_0, Y_0), (X_1, Y_1) are points where these maximum and minimum values are attained, then $\langle X_0, Y_0 \rangle$ and $\langle X_1, Y_1 \rangle$ are characteristic vectors of this matrix.

14.25 Let C be the curve $x = a \cos t + b \sin t$, $y = c \cos t + d \sin t$, $0 \le t \le 2\pi$.

a. Show that if $ad - bc \ne 0$, then C is an ellipse.

b. Describe C for the case $ad - bc = 0$.

15 Special linear transformations; composition

Let us consider a linear transformation F of the x, y plane into the X, Y plane:

$$(15.1) \qquad F: \begin{cases} X = ax + by \\ Y = cx + dy \end{cases}$$

followed by a linear transformation G of the X, Y plane into the u, v plane,

$$(15.2) \qquad G: \begin{cases} u = AX + BY \\ v = CX + DY. \end{cases}$$

The composition $G \circ F$ of F and G is a transformation H of the x, y plane into the u, v plane (Fig. 15.1). The equations of H are found by substituting the expressions for X and Y from Eqs. (15.1) into Eqs. (15.2). Explicitly,

$$(15.3) \qquad H: \begin{cases} u = \alpha x + \beta y \\ v = \gamma x + \delta y, \end{cases}$$

where

$$(15.4) \qquad \begin{pmatrix} \alpha & \beta \\ \gamma & \delta \end{pmatrix} = \begin{pmatrix} Aa + Bc & Ab + Bd \\ Ca + Dc & Cb + Dd \end{pmatrix}.$$

The main conclusions are first, that the composed transformation H is again a linear transformation, and, second, that the determinant of the transformation H is given by

$$(15.5) \qquad \alpha\delta - \beta\gamma = (AD - BC)(ad - bc).$$

Equation (15.5) is obtained by direct computation from Eq. (15.4).

This result is sufficiently important to be stated as a separate theorem.

Theorem 15.1 *If F and G are any two linear transformations, then their composition $G \circ F$ is again a linear transformation, and*

$$(15.5a) \qquad \det (G \circ F) = (\det F)(\det G).$$

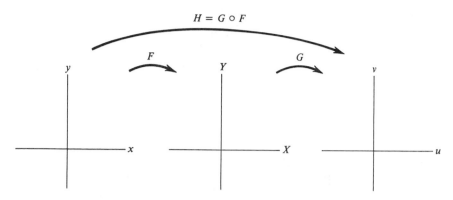

FIGURE 15.1   Composition of mappings

It follows that the result of applying any number of successive linear transformations is again a linear transformation.

One way to arrive at an understanding of general linear transformations is to start with some special types of transformations that are easy to visualize, and then build up more complicated ones by a succession of simple ones. We consider now a number of special classes of linear transformations.

Rotations

Let us picture momentarily the x, y plane and the u, v plane as coinciding, and ask for the equations of the transformation that consists of a rotation of the plane about the origin through an angle α in the positive direction (Fig. 15.2). If, in polar coordinates, we set

$$x = r \cos \theta, \qquad y = r \sin \theta,$$

then

(15.6)
$$u = r \cos (\theta + \alpha) = r(\cos \theta \cos \alpha - \sin \theta \sin \alpha) = x \cos \alpha - y \sin \alpha$$
$$v = r \sin (\theta + \alpha) = r(\sin \theta \cos \alpha + \cos \theta \sin \alpha) = x \sin \alpha + y \cos \alpha.$$

Thus this rotation is a linear transformation, and its matrix is

(15.7)
$$\begin{pmatrix} \cos \alpha & -\sin \alpha \\ \sin \alpha & \cos \alpha \end{pmatrix}.$$

If we solve Eqs. (15.6) for x and y in terms of u and v, we find that the equations of the inverse transformation are

(15.8)
$$x = u \cos \alpha + v \sin \alpha$$
$$y = -u \sin \alpha + v \cos \alpha$$

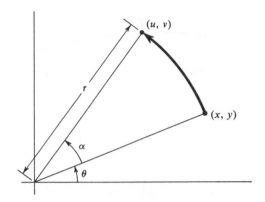

FIGURE 15.2 Rotation through angle α

and therefore that the matrix of the inverse transformation is

$$(15.9) \qquad \begin{pmatrix} \cos\alpha & \sin\alpha \\ -\sin\alpha & \cos\alpha \end{pmatrix}.$$

Noting that $\cos(-\alpha) = \cos\alpha$, $\sin(-\alpha) = -\sin\alpha$, we observe that the matrix (15.9) may be derived from (15.7) simply by changing α to $-\alpha$. With the wisdom of hindsight, we deduce that the inverse transformation (15.8) could have been written down immediately by observing that the inverse of a rotation through an angle α is a rotation through the angle $-\alpha$.

Reflections

We have already noted in Example 13.7 that the equations $u = x$, $v = -y$ define a reflection in the horizontal axis. If we wish to find the equations for a reflection in any line through the origin, we may proceed in steps: first, make a rotation so that the line becomes the horizontal axis, then reflect in this axis, and then rotate back again. Specifically, if we want to reflect in a line that makes an angle β with the x axis, we perform the following transformations in succession (Fig. 15.3):

$$F: \begin{cases} X = & x\cos\beta + y\sin\beta \\ Y = & -x\sin\beta + y\cos\beta \end{cases}; \qquad \text{rotation through angle } -\beta.$$

$$G: \begin{cases} U = & X \\ V = & -Y \end{cases}; \qquad \text{reflection in horizontal axis.}$$

$$H: \begin{cases} u = U\cos\beta - V\sin\beta \\ v = U\sin\beta + V\cos\beta \end{cases}; \qquad \text{rotation through angle } \beta.$$

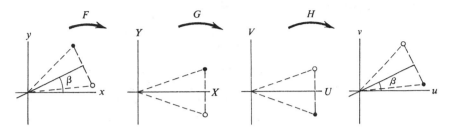

FIGURE 15.3   Reflection in line $y = x \tan \beta$

The resulting transformation is obtained by successively substituting each set of variables into the equations of the following transformation. We find

$$u = U \cos \beta - V \sin \beta = X \cos \beta + Y \sin \beta$$
$$= x(\cos^2 \beta - \sin^2 \beta) + 2y \sin \beta \cos \beta$$
$$v = U \sin \beta + V \cos \beta = X \sin \beta - Y \cos \beta$$
$$= 2x \sin \beta \cos \beta + y(\sin^2 \beta - \cos^2 \beta).$$

If we set $\alpha = 2\beta$, these equations take the form

(15.10)
$$u = x \cos \alpha + y \sin \alpha$$
$$v = x \sin \alpha - y \cos \alpha.$$

Thus a reflection is a linear transformation with matrix

(15.11)
$$\begin{pmatrix} \cos \alpha & \sin \alpha \\ \sin \alpha & -\cos \alpha \end{pmatrix}.$$

The resemblance between this matrix and matrices (15.7) and (15.9) corresponding to a rotation is certainly striking. Note, however, the important difference that the determinant of matrix (15.11) is equal to -1, whereas for both (15.7) and (15.9) the determinant is $+1$. This is the algebraic confirmation of the geometrically obvious fact that rotations preserve orientation, and reflections reverse orientation.

Dilations

By a dilation we mean a uniform stretching or contraction. Analytically, for some fixed number $\lambda > 0$ the image of an arbitrary point (x, y) is the point (u, v) (Fig. 15.4), where

$$\langle u, v \rangle = \lambda \langle x, y \rangle.$$

Thus the equations of a dilation are

(15.12)
$$u = \lambda x$$
$$v = \lambda y$$

and its matrix is

(15.13)
$$\begin{pmatrix} \lambda & 0 \\ 0 & \lambda \end{pmatrix}.$$

A basic property of dilations is that not only distances to the origin, but all distances are multiplied by the same constant λ. In other words, if the distance between two points in the x, y plane is d, then the distance between their image points in the u, v plane will be λd. This follows immediately from Eq. (15.12) and the formula for the distance between two points. In general, the following terminology is used.

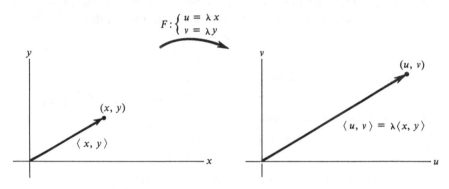

FIGURE 15.4 Dilation

Definition 15.1 A *similarity transformation* is a linear transformation under which all distances are multiplied by a fixed constant $\lambda > 0$.

The reason for this terminology is that such a transformation transforms every triangle into a similar triangle.

Now as we have just seen, a dilation is always a similarity transformation. Furthermore, both rotations and reflections are similarity transformations in which $\lambda = 1$. That is, both of these types of transformation preserve all distances, as is clear geometrically, and can be verified algebraically by direct substitution in Eqs. (15.6) and (15.10). Finally, since the composition of two similarity transformations is clearly again a similarity transformation, any transformation obtained by composing rotations, reflections, and dilations is a similarity transformation. An important fact is that the converse is also true.

Theorem 15.2 *Every similarity transformation consists of a dilation followed by a rotation or reflection.*

PROOF.   We note first of all that a similarity transformation is always nonsingular. Namely, two distinct points are mapped onto two distinct points, whereas a singular transformation is never one-to-one. Consequently the unit circle $x^2 + y^2 = 1$ is mapped onto an ellipse. But since every point on the unit circle is unit distance from the origin, every point on the image must be a distance λ from the origin. Hence the image ellipse is just the circle $u^2 + v^2 = \lambda^2$.

Suppose now that the equations of the transformation are

(15.14)
$$u = ax + by$$
$$v = cx + dy.$$

Then

$$u^2 + v^2 = (a^2 + c^2)x^2 + 2(ab + cd)xy + (b^2 + d^2)y^2$$

and the inverse image of the circle $u^2 + v^2 = \lambda^2$ has the equation

$$(a^2 + c^2)x^2 + 2(ab + cd)xy + (b^2 + d^2)y^2 = \lambda^2.$$

But this equation must represent the circle $x^2 + y^2 = 1$. This means that we must have

(15.15) $ab + cd = 0,$ $a^2 + c^2 = b^2 + d^2 = \lambda^2.$

If we introduce the vectors $\langle a, c \rangle$, $\langle b, d \rangle$, these equations reduce to

(15.16) $\langle a, c \rangle \perp \langle b, d \rangle,$ $|\langle a, c \rangle| = |\langle b, d \rangle| = \lambda.$

From the first of these equations we deduce that

$$\langle b, d \rangle = \mu \langle -c, a \rangle$$

for some number μ, by Lemma 1.1. But since $|\langle -c, a \rangle| = |\langle a, c \rangle| = (a^2 + c^2)^{1/2}$, we deduce further from the second of Eqs. (15.16) that

$$\lambda = |\langle b, d \rangle| = |\mu| |\langle -c, a \rangle| = |\mu| \langle a, c \rangle = |\mu| \lambda,$$

and hence $|\mu| = 1$ or $\mu = \pm 1$. Thus either

$$\langle b, d \rangle = \langle -c, a \rangle \qquad \text{or} \qquad \langle b, d \rangle = \langle c, -a \rangle.$$

In terms of components, we have

(15.17) $b = -c, \quad d = a$ or else $b = c, \quad d = -a.$

Finally, again using the second of Eqs. (15.16), we note that $\langle a, c \rangle / \lambda$ is a unit vector and hence is of the form $\langle \cos \alpha, \sin \alpha \rangle$ for some α. In other words,

(15.18) $a = \lambda \cos \alpha,$ $c = \lambda \sin \alpha.$

Case 1.   If the first alternative in (15.17) holds, then

$$b = -\lambda \sin \alpha, \qquad d = \lambda \cos \alpha.$$

Substituting in Eqs. (15.14), the equations of the transformation take the form

$$u = \lambda x \cos \alpha - \lambda y \sin \alpha$$
$$v = \lambda x \sin \alpha + \lambda y \cos \alpha.$$

But these equations may be obtained by setting

$$X = \lambda x \qquad u = X \cos \alpha - Y \sin \alpha$$
$$Y = \lambda y \qquad v = X \sin \alpha + Y \cos \alpha,$$

the first of which represents a dilation, and the second a rotation through the angle α (see (15.7)).

Case 2. If the second alternative in (15.17) holds, then by (15.18),

$$b = \lambda \sin \alpha, \qquad d = -\lambda \cos \alpha.$$

Substituting in (15.14) yields

$$u = \lambda x \cos \alpha + \lambda y \sin \alpha$$
$$v = \lambda x \sin \alpha - \lambda y \cos \alpha,$$

which is the composition of

$$X = \lambda x \qquad \text{and} \qquad u = X \cos \alpha + Y \sin \alpha$$
$$Y = \lambda y \qquad\qquad\qquad v = X \sin \alpha - Y \cos \alpha.$$

The first of these is again a dilation, but the second is in this case a reflection (see (15.11)).

This concludes the proof of the theorem. ◆

Remark An examination of the proof reveals that it falls into two distinct parts. In the first part we showed that if the linear transformation (15.14) is a similarity transformation, then its coefficients must satisfy (15.15). In the second part we deduced that if Eqs. (15.15) hold, then the transformation must consist of a dilation followed by a rotation or a reflection, and hence it must be a similarity transformation. Further, we showed that Eqs. (15.15) imply (15.17), and the converse is obvious. We have therefore the following conclusion.

Corollary *The algebraic condition that the general linear transformation (15.14) be a similarity transformation is that any of the equivalent relations (15.15), (15.16), or (15.17) on the coefficients should hold.*

We are now in a position to characterize two basic classes of linear transformations—those which preserve angles and those which preserve distances. We start with the former.

Under a linear transformation every pair of straight lines in the x, y plane maps into a pair of straight lines in the u, v plane. We say that the transformation *preserves angles* if the angle between the original pair of lines equals the angle between the image pair of lines. We restrict ourselves to nonsingular transformations, since otherwise certain lines map into a single point, and the angle is not defined.

Theorem 15.3 *For a nonsingular linear transformation the following are equivalent. The transformation*

1. *preserves all angles*
2. *preserves all right angles*
3. *is a similarity transformation.*

PROOF. Clearly, if all angles are preserved then all right angles are preserved. Hence $1 \Rightarrow 2$. To show that $2 \Rightarrow 3$, suppose that every pair of perpendicular lines maps onto a pair of perpendicular lines. Let the transformation be given by

$$F: \begin{cases} u = ax + by \\ v = cx + dy. \end{cases}$$

Then F maps the origin into the origin, $(1, 0)$ onto (a, c), and $(0, 1)$ onto (b, d). Thus the image of the x axis is the line through $(0, 0)$ and (a, c), and the image of the y axis is the line through $(0, 0)$ and (b, d). By our assumption the two image lines must be perpendicular. This means that the vectors $\langle a, c \rangle$ and $\langle b, d \rangle$ are perpendicular: $\langle a, c \rangle \perp \langle b, d \rangle$ (Fig. 15.5). But this is precisely the first of Eqs. (15.16). Next consider the images of the two diagonal lines $y = x$ and $y = -x$. Since F maps $(1, 1)$ onto $(a + b, c + d)$ and $(1, -1)$ onto $(a - b, c - d)$, we have $\langle a + b, c + d \rangle \cdot \langle a - b, c - d \rangle = 0$, or $a^2 - b^2 + c^2 - d^2 = 0$. Thus $a^2 + c^2 = b^2 + d^2$, and setting $\lambda = |\langle a, c \rangle| = |\langle b, d \rangle|$ we see that (15.16) is satisfied. By the Corollary to Th. 15.2, F is a similarity transformation. Finally, the fact that $3 \Rightarrow 1$ is obvious, since a similarity transformation takes every triangle into a similar triangle, and hence all angles are preserved. Thus conditions 1, 2, and 3 are all equivalent, and the theorem is proved. ♦

We consider next the class of transformations that preserve all distances.

Theorem 15.4 *The linear transformations that preserve distances are precisely the rotations and reflections. Algebraically they are characterized by the conditions*

(15.19) $\qquad ab + cd = 0, \qquad a^2 + c^2 = b^2 + d^2 = 1,$

where we assume that the equations of the transformation are in the standard form (15.14).

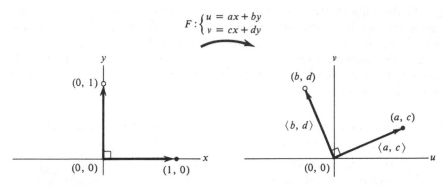

FIGURE 15.5 Distance-preserving transformation

PROOF. That a transformation preserves distances means that the images of any pair of points are the same distance apart as the original pair of points. But such a transformation is simply a similarity transformation with $\lambda = 1$. The result is then an immediate consequence of Th. 15.2 and its Corollary. ◆

Although this theorem may appear to some as intuitively clear, it has simple consequences that are not so obvious. We give one example.

Corollary *The composed transformation obtained by a reflection in some axis followed by a rotation may also be obtained by making a single reflection in a suitable axis.*

PROOF. Since both the reflection and rotation preserve distances, so does the composed transformation, which must itself be a reflection or rotation by Th. 15.4. But the determinant of a reflection is -1, and that of a rotation is $+1$. Hence the composed transformation has determinant -1 (Th. 15.1), and it must be a reflection. ◆

The above results on linear transformations are all that we need in our ensuing discussion of differentiable transformations. For the sake of completeness, however, we note the following additional facts.

Consider a transformation of the form

(15.20)
$$u = \lambda_1 x$$
$$v = \lambda_2 y$$

with matrix

$$\begin{pmatrix} \lambda_1 & 0 \\ 0 & \lambda_2 \end{pmatrix}.$$

Under such a transformation horizontal distances are multiplied by $|\lambda_1|$ and vertical distances by $|\lambda_2|$. The over-all effect is easy to visualize. Furthermore, we can then visualize the most general linear transformation by using the following result.

Theorem 15.5   *Every linear transformation may be obtained by performing in succession first a rotation, then a transformation of the form* (15.20), *and then another rotation.*

The idea of the proof is the following. The unit circle in the x, y plane maps onto an ellipse in the u, v plane. The lines in the x, y plane which map onto the major and minor axes of the ellipse must be perpendicular, since they correspond to directions in which a quadratic form takes on its maximum and minimum values. Making a rotation in the x, y plane so that these lines go into the coordinate axes, and a rotation in the u, v plane, which takes the axes of the ellipse into the coordinate axes, the linear transformation reduces to the form (15.20), where the numbers λ_1 and λ_2 are equal in magnitude to the semimajor and semiminor axes of the ellipse. (See Ex. 15.18 for further details.)

As an application of this result, consider an arbitrary triangle in the x, y plane and its image in the u, v plane under a given linear transformation. Under a transformation of the form (15.20), the area of this triangle is multiplied by $|\lambda_1 \lambda_2|$, since if we divide the triangle into two parts by a horizontal line through a vertex, each part has the base multiplied by $|\lambda_1|$ and the altitude by $|\lambda_2|$. Now let F be an arbitrary linear transformation and denote its determinant by Δ. If we compose F out of rotations and a transformation of the form (15.20), then the determinant of F is the product of the determinants of these transformations. But the determinant of a rotation is equal to 1, and the determinant of the transformation (15.20) is just $\lambda_1 \lambda_2$. Hence $\Delta = \lambda_1 \lambda_2$. Finally, since rotations preserve areas, we come to the following conclusion.

Corollary   *Under an arbitrary linear transformation the area of every triangle is multiplied by a fixed constant, $|\Delta|$, that is, the absolute value of the determinant of the transformation.*

This statement is restricted to triangles, since their area is computed in an elementary fashion. After we have discussed the concept of area for more general figures in the following chapter, it will be clear that the above result holds for such figures (see the Corollary to Th. 26.2).

As a final consequence, we have the result: *a linear transformation preserves areas if and only if its determinant Δ satisfies $|\Delta| = 1$, that is, $\Delta = \pm 1$.*

Exercises

15.1 Find the matrix of the transformation $G \circ F$ if the matrices of F and G, respectively, are equal to

a. $\begin{pmatrix} 1 & 0 \\ 0 & 1 \end{pmatrix}, \begin{pmatrix} 1 & 0 \\ 0 & 1 \end{pmatrix}$

e. $\begin{pmatrix} a & b \\ c & d \end{pmatrix}, \begin{pmatrix} 0 & 1 \\ 1 & 0 \end{pmatrix}$

b. $\begin{pmatrix} 0 & 1 \\ 1 & 0 \end{pmatrix}, \begin{pmatrix} 0 & 1 \\ 1 & 0 \end{pmatrix}$

f. $\begin{pmatrix} 0 & 1 \\ 1 & 0 \end{pmatrix}, \begin{pmatrix} a & b \\ c & d \end{pmatrix}$

c. $\begin{pmatrix} a & b \\ c & d \end{pmatrix}, \begin{pmatrix} 1 & 0 \\ 0 & 1 \end{pmatrix}$

g. $\begin{pmatrix} 1 & 1 \\ -1 & 1 \end{pmatrix}, \begin{pmatrix} 1 & 1 \\ -1 & 1 \end{pmatrix}$

d. $\begin{pmatrix} 1 & 0 \\ 0 & 1 \end{pmatrix}, \begin{pmatrix} a & b \\ c & d \end{pmatrix}$

h. $\begin{pmatrix} 3 & 4 \\ 5 & 7 \end{pmatrix}, \begin{pmatrix} 7 & -4 \\ -5 & 3 \end{pmatrix}$

15.2 Carry out the computation leading from Eq. (15.4) to Eq. (15.5).

15.3 Verify Eq. (15.5a) explicitly for each of the pairs F, G in Ex. 15.1.

15.4 Prove that the composition of any number of nonsingular linear transformations is again nonsingular.

15.5 Prove that the composition of any number of nonsingular linear transformations together with one singular transformation is always singular.

15.6 Show that for any two linear transformations F, G

$$\det (F \circ G) = \det (G \circ F).$$

15.7 Find the equations of each of the following linear transformations.

    a. A counterclockwise rotation through an angle of $\frac{1}{4}\pi$.
    b. A clockwise rotation through an angle of $\frac{1}{2}\pi$.
    c. A reflection in the line $x + y = 0$.
    d. A reflection in the line of positive slope making an angle of $\frac{1}{6}\pi$ with the x axis.

15.8 Show that each of the following transformations is a reflection, and find the line in which the plane is reflected.

    a. $F_1 : (x, y) \rightarrow (y, x)$
    b. $F_2 : (x, y) \rightarrow (x, -y)$
    c. $F_3 : (x, y) \rightarrow (-x, y)$

15.9 Show that the effect of a counterclockwise rotation through 45° followed by a reflection in the line $y = x$ is the same as a single reflection in a certain line, and specify the line.

15.10   **a.** Using the notation of Ex. 15.8 show that $F_2 \circ F_1$ is a rotation through a certain angle, and determine the angle.

    b. Do the same for $F_3 \circ F_2$.

    c. Find the equations of the transformation $F_3 \circ F_2 \circ F_1$ and describe this transformation geometrically.

15.11 Show that the result of performing any even number of successive reflections in different lines may also be obtained by a single rotation.

15.12 Let F be a similarity transformation that multiplies all distances by λ. Show that the determinant of F is $\pm \lambda^2$.

15.13 Show that the inverse of a similarity transformation is also a similarity transformation.

15.14 Show that a linear transformation is a similarity transformation if and only if it maps circles onto circles.

15.15 Show that the following transformations are similarity transformations by referring to Eqs. (15.17). Then express each as the composition of a dilation and a rotation or reflection.

$$a. \begin{cases} u = & x + y \\ v = & -x + y \end{cases} \qquad c. \begin{cases} u = & -3x - 4y \\ v = & -4x + 3y \end{cases}$$

$$b. \begin{cases} u = & 2y \\ v = & -2x \end{cases} \qquad d. \begin{cases} u = & x - 2y \\ v = & 2x + y \end{cases}$$

15.16 Show that if F is a dilation and G is an arbitrary linear transformation, then $F \circ G = G \circ F$.

15.17 Give a geometric description of transformations of the form of (15.20) in each of the following cases.

$a.$ $\lambda_1 > 1$, $\lambda_2 = 1$ $\qquad$ $c.$ $\lambda_1 = 0$, $\lambda_2 = -1$

$b.$ $0 < \lambda_1 < 1$, $\lambda_2 = -1$ $\qquad$ $d.$ $-1 < \lambda_1 < 0$, $\lambda_2 < -1$

*15.18 Complete the proof of Th. 15.5 in the following steps.

$a.$ Show that if a linear transformation F maps the circle $x^2 + y^2 = 1$ onto a circle, then F is already in the form of (15.20), with $\lambda_1 = \lambda_2$.

$b.$ Verify the statement that if F is nonsingular, and if the circle $x^2 + y^2 = 1$ maps onto a certain ellipse, then the inverse images of the axes of that ellipse are perpendicular lines in the x, y plane. (*Hint*: use Corollary 2 to Th. 10.2.)

$c.$ Show that if a nonsingular linear transformation F maps the x axis into the u axis, the y axis into the v axis, and the circle $x^2 + y^2 = 1$ onto the ellipse $x^2/\lambda_1^2 + y^2/\lambda_2^2 = 1$, then F is of the form $u = \pm \lambda_1 x$, $v = \pm \lambda_2 y$. (*Hint*: observe that the coefficients of a linear transformation F are determined by the action of F on the points (1, 0) and (0, 1).)

$d.$ Show that Th. 15.5 holds for a singular transformation, in which case either λ_1 or λ_2 (or both) will be zero. (*Hint*: use Ex. 14.5c, and describe the rotations that must be made in the x, y plane and in the u, v plane.)

15.19 Let F be an arbitrary linear transformation:

$$F: \begin{cases} u = ax + by \\ v = cx + dy. \end{cases}$$

Give a direct proof of the Corollary to Th. 15.5 by carrying out the following steps.

Let (x_0, y_0), (x_1, y_1), (x_2, y_2) be the vertices of an arbitrary triangle T in the x, y plane, and let $(u_k, v_k) = F(x_k, y_k)$, $k = 0, 1, 2$. Let

$$\langle X_k, Y_k \rangle = \langle x_k - x_0, y_k - y_0 \rangle,$$
$$\langle U_k, V_k \rangle = \langle u_k - u_0, v_k - v_0 \rangle, \qquad k = 1, 2.$$

a. Express U_k, V_k in terms of X_k, Y_k.

b. Express the area A of the triangle T in terms of X_1, Y_1, X_2, Y_2 (*Hint*: see Ex. 1.29d.)

c. Let T' be the triangle with vertices $(u_0, v_0), (u_1, v_1), (u_2, v_2)$. Express the area A' of T' in terms of U_1, V_1, U_2, V_2.

d. By a direct computation, using parts a, b, c, above, show that $A' = |ad - bc|A$.

15.20 Let F_1, F_2 be any two linear transformations, and let Δ_1, Δ_2 be their determinants. Let Δ be the determinant of $F_2 \circ F_1$.

a. Using the Corollary to Th. 15.5, give a geometric proof of the fact that

$$|\Delta| = |\Delta_1| \cdot |\Delta_2|.$$

b. Using the geometric interpretation of the sign of the determinant of a linear transformation, show that if F_1 and F_2 are both nonsingular, then

$$\frac{\Delta}{|\Delta|} = \frac{\Delta_1}{|\Delta_1|} \cdot \frac{\Delta_2}{|\Delta_2|}.$$

c. Using parts a and b (or part a alone, in case either F_1 or F_2 is singular) show that we always have

$$\Delta = \Delta_1 \cdot \Delta_2.$$

(Note that this gives both a purely geometric proof and also an important geometric interpretation of Eq. (15.5a).)

*15.21 Using the notation of Exs. 14.11 and 14.20–14.24, we associate with an arbitrary linear transformation F, the corresponding linear transformation $\mathbf{F}$ of displacement vectors.

a. Show that F is a dilation with constant λ if and only if $\mathbf{F}(\mathbf{w}) = \lambda\mathbf{w}$ for all vectors $\mathbf{w}$.

b. Show that F preserves distances if and only if $|\mathbf{F}(\mathbf{w})| = |\mathbf{w}|$ for all vectors $\mathbf{w}$.

c. Show that F is a similarity transformation with factor λ if and only if $|\mathbf{F}(\mathbf{w})| = \lambda|\mathbf{w}|$ for all vectors $\mathbf{w}$.

d. Show that F preserves distances if and only if $\mathbf{F}$ preserves dot products, that is, $\mathbf{F}(\mathbf{w}_1) \cdot \mathbf{F}(\mathbf{w}_2) = \mathbf{w}_1 \cdot \mathbf{w}_2$ for all vectors $\mathbf{w}_1$, $\mathbf{w}_2$. (*Hint*: if $\mathbf{F}$ preserves dot products, set $\mathbf{w}_1 = \mathbf{w}_2$ and use part b. If F preserves distances, use part b and the dentity

$$|\mathbf{w}_1 + \mathbf{w}_2|^2 = |\mathbf{w}_1|^2 + |\mathbf{w}_2|^2 + 2\mathbf{w}_1 \cdot \mathbf{w}_2$$

applied to any two vectors $\mathbf{w}_1$, $\mathbf{w}_2$ and to $\mathbf{F}(\mathbf{w}_1) + \mathbf{F}(\mathbf{w}_2) = \mathbf{F}(\mathbf{w}_1 + \mathbf{w}_2)$.)

e. Show that F is a similarity transformation with factor λ if and only if $F(w_1) \cdot F(w_2) = \lambda^2 w_1 \cdot w_2$ for all vectors w_1, w_2. (*Hint*: use Th. 15.2 together with parts a and d.)

f. Give a purely analytic proof that if F is a similarity transformation, then **F** preserves all angles between vectors. (*Hint*: use parts c and e together with the formula $\cos \alpha = w_1 \cdot w_2 / |w_1| |w_2|$, for the angle α between the vectors w_1, w_2.)

Remark   Given matrices

$$M_1 = \begin{pmatrix} a & b \\ c & d \end{pmatrix}, \qquad M_2 = \begin{pmatrix} A & B \\ C & D \end{pmatrix}$$

associated with a pair of linear transformations F_1, F_2, respectively, the *product* of M_1 and M_2 is defined to be the matrix associated with the composition of F_1 and F_2. As is clear, for example by Ex. 15.1e, f, this product depends on the order in which the transformations are performed. Specifically, we define $M_2 M_1$ to be the matrix of $F_2 \circ F_1$. Then by Eq. (15.4), we have

$$\begin{pmatrix} A & B \\ C & D \end{pmatrix} \begin{pmatrix} a & b \\ c & d \end{pmatrix} = \begin{pmatrix} Aa + Bc & Ab + Bd \\ Ca + Dc & Cb + Dd \end{pmatrix}.$$

The matrix

$$I = \begin{pmatrix} 1 & 0 \\ 0 & 1 \end{pmatrix}$$

is called the *identity* matrix and is associated with the *identity transformation F*, which leaves every point fixed. If $M = \begin{pmatrix} a & b \\ c & d \end{pmatrix}$ is associated with a nonsingular transformation F, then

$$M^{-1} = \begin{pmatrix} \dfrac{d}{\Delta} & -\dfrac{b}{\Delta} \\ -\dfrac{c}{\Delta} & \dfrac{a}{\Delta} \end{pmatrix}, \qquad \text{where} \quad \Delta = \det \begin{pmatrix} a & b \\ c & d \end{pmatrix},$$

is associated with the inverse transformation F^{-1} by Eq. (14.7). This matrix M^{-1} is called the *inverse* matrix of M. Equation (14.8) takes the form

$$\det (M^{-1}) = (\det M)^{-1},$$

where the right-hand side is simply the reciprocal of $\det M$.

We may also define other algebraic operations on matrices in terms of elementary operations on the corresponding linear transformations. Thus

$$\begin{pmatrix} A & B \\ C & D \end{pmatrix} + \begin{pmatrix} a & b \\ c & d \end{pmatrix} = \begin{pmatrix} A + a & B + b \\ C + c & D + d \end{pmatrix}$$

and

$$\lambda \begin{pmatrix} a & b \\ c & d \end{pmatrix} = \begin{pmatrix} \lambda a & \lambda b \\ \lambda c & \lambda d \end{pmatrix}.$$

The study of matrices and their properties under these operations is known as *matrix algebra*. It may be carried out quite independently of any considerations of the associated linear transformations. One advantage of this approach is that matrices have other applications (such as to quadratic forms, in the case of symmetric matrices, as we have seen earlier) and it is not desirable to tie their study to any one interpretation.

The following exercises are designed to illustrate some of the elementary facts of matrix algebra. They are to be carried out by direct application of the matrix operations defined above.

15.22 Show that if I is the identity matrix, then

$$MI = IM = M$$

for every matrix M.

15.23 Show that for any matrix M with nonzero determinant,

$$MM^{-1} = M^{-1}M = I.$$

15.24 Let

$$M_k = \lambda_k \begin{pmatrix} \cos \alpha_k & \sin \alpha_k \\ -\sin \alpha_k & \cos \alpha_k \end{pmatrix}, \qquad k = 1, 2.$$

Show that

$$M_1 M_2 = \lambda_1 \lambda_2 \begin{pmatrix} \cos (\alpha_1 + \alpha_2) & \sin (\alpha_1 + \alpha_2) \\ -\sin (\alpha_1 + \alpha_2) & \cos (\alpha_1 + \alpha_2) \end{pmatrix}.$$

15.25 Let M_1, M_2 be given matrices with det $M_1 \neq 0$. Show that if λ_1, λ_2 are the roots of the equation

$$\det (M_1 + \lambda M_2) = 0,$$

then

$$\lambda_1 \lambda_2 = (\det M_2)/(\det M_1).$$

15.26 Show that the answer to Ex. 10.14, can be written in matrix form as

$$\begin{pmatrix} A & B \\ B & C \end{pmatrix} = \begin{pmatrix} \cos \theta & \sin \theta \\ -\sin \theta & \cos \theta \end{pmatrix} \begin{pmatrix} a & b \\ b & c \end{pmatrix} \begin{pmatrix} \cos \theta & -\sin \theta \\ \sin \theta & \cos \theta \end{pmatrix}.$$

Show how this provides a simple answer to Ex. 10.16.

16 Differentiable transformations

Having studied a few special transformations in some detail, we now examine more general transformations.

Definition 16.1 A transformation

(16.1) $$F: \begin{cases} u = f(x, y) \\ v = g(x, y) \end{cases}$$

defined in a domain D is *of class* $\mathscr{C}^k$ in D if $f(x, y) \in \mathscr{C}^k$ and $g(x, y) \in \mathscr{C}^k$ in D.

By a *differentiable transformation* we mean a transformation that is of class $\mathscr{C}^k$ for some $k \geq 1$.

Examples 13.6–13.11 were all differentiable transformations with the exception of Example 13.9. (The function $u = |x|$ is not differentiable at $x = 0$.) In all those examples the domain D was the entire plane. If we set

$$u = \sqrt{1 - x^2 - y^2},$$
$$v = x^{7/3},$$

we obtain a transformation of class $\mathscr{C}^2$ in the domain $D: x^2 + y^2 < 1$.

As in our study of a single function of two variables, we start our investigation of a pair of functions by asking what happens along an arbitrary curve C in the domain D. We use our standard notation

$$C: x(t),\ y(t), \qquad\qquad a \leq t \leq b,$$

where $x(t)$ and $y(t)$ are continuously differentiable functions for $a \leq t \leq b$. The image of the curve C under the transformation F is a curve Γ in the u, v plane:

$$\Gamma: u(t),\ v(t), \qquad\qquad a \leq t \leq b;$$
$$u(t) = f(x(t), y(t)), \qquad v(t) = g(x(t), y(t)).$$

(Figure 16.1.)

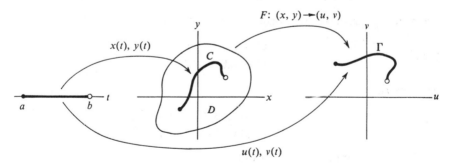

FIGURE 16.1   Image of a curve under a mapping

If F is a differentiable transformation, then the chain rule yields

$$\frac{du}{dt} = f_x \frac{dx}{dt} + f_y \frac{dy}{dt}$$

(16.2)

$$\frac{dv}{dt} = g_x \frac{dx}{dt} + g_y \frac{dy}{dt}.$$

More precisely, if $x_0 = x(t_0)$; $y_0 = y(t_0)$, then

(16.2a)
$$u'(t_0) = f_x(x_0, y_0)x'(t_0) + f_y(x_0, y_0)y'(t_0)$$
$$v'(t_0) = g_x(x_0, y_0)x'(t_0) + g_y(x_0, y_0)y'(t_0).$$

We use these equations to study the transformation F near an arbitrary point (x_0, y_0) in D. Let (u_0, v_0) be the image of (x_0, y_0) under F, and let C be any curve passing through (x_0, y_0), where $x_0 = x(t_0)$, $y_0 = y(t_0)$. Note first that the coefficients f_x, f_y, g_x, g_y are all evaluated at the point (x_0, y_0), and do not depend on the choice of the curve C through the point. Furthermore, the right-hand side of Eqs. (16.2a) depends only on the tangent vector $\langle x'(t_0), y'(t_0)\rangle$. As for the left-hand side, it defines precisely the tangent vector $\langle u'(t_0), v'(t_0)\rangle$ to the image curve Γ. Our first conclusion is therefore the following.

Lemma 16.1 *Let F be a differentiable transformation near (x_0, y_0). Then all curves through (x_0, y_0) having the same tangent vector at (x_0, y_0) map onto curves having the same tangent vector at the image of (x_0, y_0).*

This lemma is often expressed in the form: *a differentiable transformation defines at each point an induced transformation of tangent vectors.*

In order to see more clearly the nature of this induced transformation, let us introduce the notation:

$$X = x'(t_0), \quad Y = y'(t_0); \qquad U = u'(t_0), \quad V = v'(t_0).$$

Then Eq. (16.2a) takes the form of a linear transformation

(16.3)
$$U = aX + bY$$
$$V = cX + dY$$

with matrix

(16.4)
$$\begin{pmatrix} a & b \\ c & d \end{pmatrix} = \begin{pmatrix} f_x(x_0, y_0) & f_y(x_0, y_0) \\ g_x(x_0, y_0) & g_y(x_0, y_0) \end{pmatrix}.$$

We have therefore the further result.

Lemma 16.2 *Under a differentiable transformation F, the induced correspondence of tangent vectors at each point is a linear transformation.*

Definition 16.2 The linear transformation (16.3) with matrix (16.4), associated with the transformation F at the point (x_0, y_0) is called the *differential of F* and denoted by dF.

Remark   In using the notation dF it is always understood that a particular point has been chosen. If we wish to indicate the choice of this point we use the notation $dF_{(x_0,y_0)}$. Thus

$$dF_{(x_0,y_0)} : (X, Y) \to (U, V)$$

is the linear transformation defined by (16.3) and (16.4).

Definition 16.3   The transformation F is called *singular* at (x_0, y_0) if $dF_{(x_0,y_0)}$ is a singular linear transformation, and *regular* at (x_0, y_0) if $dF_{(x_0,y_0)}$ is nonsingular.

We recall that a linear transformation is singular if its determinant vanishes.

Definition 16.4   The matrix (16.4) is called the *Jacobian matrix* at (x_0, y_0) of the transformation F. We denote the Jacobian matrix by

$$(16.5) \qquad J_F = \begin{pmatrix} u_x & u_y \\ v_x & v_y \end{pmatrix},$$

where u and v are given as functions of x and y by (16.1), and it is understood that this matrix is associated with F at a given point.

Definition 16.5   The *Jacobian* of F at a point is the determinant of the Jacobian matrix at that point.

The standard notation for the Jacobian is

$$(16.6) \qquad \frac{\partial(u, v)}{\partial(x, y)} = \det J_F = u_x v_y - u_x v_y.$$

We note that the Jacobian of F is defined at each point of the domain D and is therefore itself a function in D.

Theorem 16.1   *Let F be a differentiable transformation in D, and let F map a point (x_0, y_0) of D into (u_0, v_0).*

Case 1. F is singular at (x_0, y_0). Then all differentiable curves C through (x_0, y_0) map into curves through (u_0, v_0) whose tangent vectors lie along a single line.

Case 2. F is regular at (x_0, y_0). Then there is a one-to-one correspondence between tangent vectors to curves at (x_0, y_0) and tangent vectors to curves at (u_0, v_0).

PROOF. In case 1, $dF_{(x_0, y_0)}$ is singular and maps all vectors (X, Y) into some line. We include in this case the possibility that all four partial derivatives u_x, u_y, v_x, v_y vanish at (x_0, y_0), in which case each curve C through (x_0, y_0) maps onto a curve whose tangent vector $\langle u'(t_0), v'(t_0) \rangle$ is zero.

In case 2, $dF_{(x_0, y_0)}$ is nonsingular, and hence defines a one-to-one correspondence between $\langle X, Y \rangle$ and $\langle U, V \rangle$. ◆

Example 16.1

Let F be the mapping $u = x^2$, $v = y$. Then

$$u_x = 2x, \qquad u_y = 0, \qquad v_x = 0, \qquad v_y = 1.$$

Hence

$$J_F = \begin{pmatrix} 2x & 0 \\ 0 & 1 \end{pmatrix}; \qquad \frac{\partial(u, v)}{\partial(x, y)} = \det J_F = 2x.$$

Thus F is regular at all points except along the y axis; on the y axis, where $x = 0$, F is singular.

Case 1. Let us look at a singular point, say $(0, 1)$. Then $F: (0, 1) \to (0, 1)$. We wish to look at curves in the x, y plane through $(0, 1)$ and their images under F. Since the tangent vector of the image depends only on the tangent vector to the original curve, we get a complete picture by considering all straight lines through $(0, 1)$. Thus, let C be the line

$$x = t \cos \alpha, \qquad y = 1 + t \sin \alpha,$$

for any fixed α. The image Γ is given by

$$u = t^2 \cos^2 \alpha, \qquad v = 1 + t \sin \alpha.$$

The point $(0, 1)$ corresponds to $t = 0$. The tangent vector to C is

$$\langle X, Y \rangle = \langle x'(0), y'(0) \rangle = \langle \cos \alpha, \sin \alpha \rangle.$$

The tangent vector to Γ is

$$\langle U, V \rangle = \langle u'(0), v'(0) \rangle = \langle 0, \sin \alpha \rangle.$$

Hence the differential $dF_{(0, 1)}$ is the map

$$U = 0, \qquad V = Y.$$

This means that the tangent vector to the image curve is always vertical. We can verify this explicitly by writing the curve Γ in nonparametric form. We have $v - 1 = t \sin \alpha$, and

$$u = t^2 \cos^2 \alpha = \left(\frac{v - 1}{\sin \alpha} \right)^2 \cos^2 \alpha = (v - 1)^2 \cot^2 \alpha.$$

But this is just a parabola with vertex at $(0, 1)$ (Fig. 16.2). Thus all straight lines through $(0, 1)$ map onto parabolas with vertical tangent at $(0, 1)$.

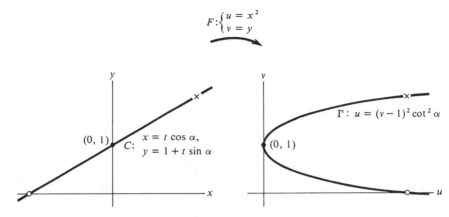

FIGURE 16.2 Map $F:(x, y) \rightarrow (x^2, y)$ near a singular point

Case 2. Let us examine a regular point, say $(1, 0)$. Again we may restrict our attention to straight lines through $(1, 0)$. Every such line has the equation

$$x = 1 + t \cos \alpha, \qquad y = t \sin \alpha.$$

The image is

$$u = 1 + 2t \cos \alpha + t^2 \cos^2 \alpha, \qquad v = t \sin \alpha.$$

This is again a parabola

$$u = 1 + 2v \cot \alpha + v^2 \cot^2 \alpha = (1 + v \cot \alpha)^2.$$

The derivative is

$$\frac{du}{dv} = 2 \cot \alpha + 2v \cot^2 \alpha.$$

At the point $(1, 0)$ we have $du/dv = 2 \cot \alpha$, and hence the slope is $dv/du = \frac{1}{2} \tan \alpha$. As α varies, we obtain as the image of lines through $(1, 0)$ in the x, y plane, a family of parabolas through $(1, 0)$ in the u, v plane. The correspondence between tangents is one-to-one (Fig. 16.3).

Definition 16.6 A transformation is called *orientation-preserving* at a point if its Jacobian is positive and *orientation-reversing* if its Jacobian is negative.

Recalling that the Jacobian of the transformation F is equal to the determinant of the linear transformation dF, we see from our discussion of the sign of the determinant in Sect. 14 that if we rotate a curve about a point (x_0, y_0) in the positive direction, then the image curve rotates in the same direction or the opposite direction, depending on whether the transformation F preserves or reverses orientation.

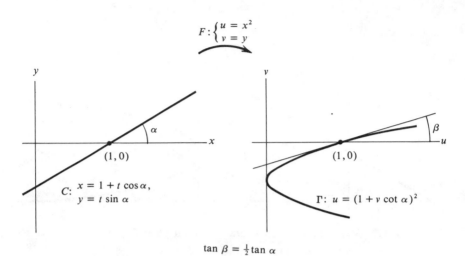

$$\tan \beta = \tfrac{1}{2}\tan \alpha$$

FIGURE 16.3 Map $F:(x, y) \to (x^2, y)$ near a regular point

In Example 16.1, the Jacobian was equal to $2x$, hence positive for $x > 0$ and negative for $x < 0$. Thus the transformation $u = x^2$, $v = y$ is orientation-preserving in the right half-plane, and orientation-reversing in the left half-plane. If we recall that this transformation may be described roughly as "folding over the left half-plane onto the right" then the intuitive content of reversing orientation becomes quite evident.

We turn next to a somewhat different way of viewing the differential dF of a transformation. Applying the fundamental lemma of Chapter 2 (Lemma 5.1) to each of the functions $u(x, y)$, $v(x, y)$, we may write

$$(16.7) \qquad \begin{aligned} u - u_0 &= a(x - x_0) + b(y - y_0) + h_1(x, y) \\ v - v_0 &= c(x - x_0) + d(y - y_0) + h_2(x, y), \end{aligned}$$

where $F(x_0, y_0) = (u_0, v_0)$, $F(x, y) = (u, v)$ and

$$(16.8) \qquad \begin{pmatrix} a & b \\ c & d \end{pmatrix} = \begin{pmatrix} u_x(x_0, y_0) & u_y(x_0, y_0) \\ v_x(x_0, y_0) & v_y(x_0, y_0) \end{pmatrix} = J_F(x_0, y_0).$$

The remainder terms $h_1(x, y)$, $h_2(x, y)$ satisfy

$$(16.9) \qquad \frac{h_1(x, y)}{d(x, y)} \to 0, \qquad \frac{h_2(x, y)}{d(x, y)} \to 0,$$

$$\text{as} \quad d(x, y) = \sqrt{(x - x_0)^2 + (y - y_0)^2} \to 0.$$

If we temporarily ignore the remainder terms in Eq. (16.7), then the right-hand side is simply a linear transformation, namely $dF_{(x_0, y_0)}$, applied to the displacement vector $\langle x - x_0, y - y_0 \rangle$ rather than to the tangent vector $\langle x'(t_0), y'(t_0) \rangle$, and the left-hand side defines the corresponding displacement vector $\langle u - u_0, v - v_0 \rangle$. We may therefore describe the content of the Eqs. (16.7) as follows.

Theorem 16.2   *The differential $dF_{(x_0,y_0)}$ applied to the displacement vector $\langle x - x_0, y - y_0 \rangle$ yields the displacement vector $\langle u - u_0, v - v_0 \rangle$ up to remainder terms satisfying (16.9) (see Fig. 16.4).*

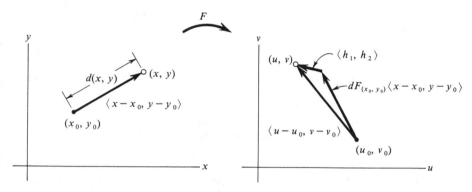

FIGURE 16.4   The differential $dF_{(x_0,y_0)}$

In other words, every differentiable transformation F may be approximated near each point by a linear transformation, just as a differentiable function may be approximated by a linear function (namely, the tangent line to the graph of the function). It is for this reason that linear transformations play such a fundamental role in the study of arbitrary differentiable transformations.

Example 16.2

Let F be the transformation

$$u = x^2 - y^2, \qquad v = 2xy.$$

Then

$$u_x = 2x, \qquad u_y = -2y,$$
$$v_x = 2x, \qquad v_y = 2y.$$

If we wish to study F near the point $(1, 1)$, then we note that at that point

$$u = 0, \quad v = 2, \qquad u_x = 2, \quad u_y = -2, \qquad v_x = 2, \quad v_y = 2.$$

Equations (16.7) take the form

$$u = 2(x - 1) - 2(y - 1) + h_1(x, y)$$
$$v - 2 = 2(x - 1) + 2(y - 1) + h_2(x, y).$$

Thus the displacement vector $\langle u, v - 2 \rangle$ is obtained from $\langle x - 1, y - 1 \rangle$, up to remainder terms, by a linear transformation with matrix

$$\begin{pmatrix} 2 & -2 \\ 2 & 2 \end{pmatrix}.$$

But we have seen in Sect. 15 that such a matrix defines a similarity transformation. In fact, if we divide each entry by $2\sqrt{2}$, we are left with the matrix

$$\begin{pmatrix} \dfrac{1}{\sqrt{2}} & -\dfrac{1}{\sqrt{2}} \\ \dfrac{1}{\sqrt{2}} & \dfrac{1}{\sqrt{2}} \end{pmatrix} = \begin{pmatrix} \cos \tfrac{1}{4}\pi & -\sin \tfrac{1}{4}\pi \\ \sin \tfrac{1}{4}\pi & \cos \tfrac{1}{4}\pi \end{pmatrix}$$

corresponding to a rotation of 45°. The approximate behavior of the transformation F near the point (1, 1) can then be described as follows (Fig. 16.5): take a neighborhood of (1, 1) and move it by a translation so that the point (1, 1) goes into (0, 2); then make a dilation about the point (0, 2) with a factor of $2\sqrt{2}$; then make a rotation about (0, 2) through 45° in the positive direction. We note once again, that the linear transformation $dF_{(x_0, y_0)}$, which when applied in this way to displacement vectors at (x_0, y_0) gives the approximate behavior of F near (x_0, y_0), is the same one, which when applied to tangent vectors at (x_0, y_0), gives the precise tangent vector of the image. Thus Fig. 16.5 is an exact description of the behavior of tangent vectors to curves through (1, 1) under the transformation F.

In conclusion, we make the following observation. When trying to visualize an arbitrary differentiable transformation F near a point (x_0, y_0), it is often helpful to recall that the linear map $dF_{(x_0, y_0)}$ takes circles into ellipses, and since this map approximates F near (x_0, y_0), we should picture a small circle about (x_0, y_0) being mapped by F onto a curve that is approximately an ellipse. Furthermore, as we noted in Sect. 14, under the map dF the ratio of the area inside the image ellipse to the area inside the circle about (x_0, y_0) is equal to the absolute value of the determinant of dF, which is the Jacobian of F at (x_0, y_0). We saw earlier that the sign of the Jacobian had the geometric significance of distinguishing between orientation-preserving and -reversing transformations. We now have the important geometric interpretation of the magnitude of the Jacobian: *the absolute value of the Jacobian of F at (x_0, y_0) is approximately equal to the ratio of areas of a small circle about (x_0, y_0) and its image under F.* It should be remarked that this approximate statement can be replaced by an exact one, using a limiting procedure. We shall do so in Sect. 27. As an aid to our intuition, we anticipate to the extent of stating the following result (see Corollary to Th. 27.2): *the Jacobian of F at (x_0, y_0) is precisely equal to the limiting value of the ratio of areas, where we choose a sequence of circles about (x_0, y_0) whose radii tend to zero.*

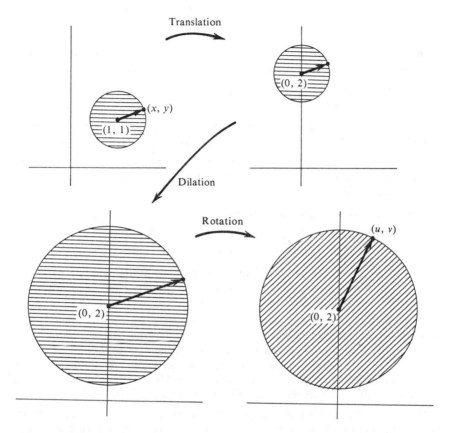

FIGURE 16.5 $dF_{(1,1)}\langle x - 1, y - 1\rangle$; where $F:(x, y) \rightarrow (x^2 - y^2, 2xy)$

Exercises

16.1   Find the Jacobian matrix of each of the following transformations.

a. $\begin{cases} u = x \\ v = y \end{cases}$

b. $\begin{cases} u = 2x - y \\ v = x + 5y \end{cases}$

c. $\begin{cases} u = x^3 - 3xy^2 \\ v = 3x^2y - y^3 \end{cases}$

d. $\begin{cases} u = xy^2 \\ v = x^2y \end{cases}$

e. $\begin{cases} u = e^{x+y} \\ v = e^{x-y} \end{cases}$

f. $\begin{cases} u = e^x \cos y \\ v = e^x \sin y \end{cases}$

g. $\begin{cases} u = e^{xy} \\ v = e^{x^2} \end{cases}$

h. $\begin{cases} u = x^2 + 2xy + y^2 + 1 \\ v = \sin(x + y) \end{cases}$

$$i. \begin{cases} u = \sin x \sin y \\ v = \cos x \cos y \end{cases}$$

$$*k. \begin{cases} u = \int_0^{xy} e^{-t^2}\, dt \\ v = \int_0^{xy} \sin t^2\, dt \end{cases}$$

$$j. \begin{cases} u = -y/(1 + x^2) \\ v = x + \tfrac{1}{3}x^3 \end{cases}$$

16.2 For each transformation in Ex. 16.1, compute the Jacobian and determine the set of points in the x, y plane where the transformation is singular.

16.3 Show that the Jacobian of a linear transformation F is constant, and coincides with the determinant of F.

16.4 Let F be a differentiable transformation whose Jacobian is constant. Does F have to be linear? What if u and v are known to be polynomials in x and y?

16.5 Let F_1 and F_2 be differentiable transformations having the same Jacobian matrix at every point of a domain D. Show that F_2 is the composition of F_1 with a translation.

16.6 Let F be an affine transformation

$$F: \begin{cases} u = ax + by + h \\ v = cx + dy + k. \end{cases}$$

  a. Find the differential dF at an arbitrary point (x_0, y_0).

  b. What does your answer to part a imply about the remainder terms in Eq. (16.7) in the case of an affine transformation?

16.7   **a.** In Example 16.1, find $dF_{(0,1)}$ by direct evaluation of the Jacobian matrix J_F at the point $(0, 1)$, and verify that the result is the same as that obtained in the text by examination of tangent vectors and their images.

  b. In the same example, evaluate J_F at the point $(1, 0)$ and describe the corresponding linear transformation geometrically. Use this description to deduce the action of $dF_{(1,0)}$ on tangent vectors as described in the text.

16.8 At which points is the transformation $u = x^2$, $v = y^2$ orientation-preserving, and at which points is it orientation-reversing? Try to picture this geometrically.

16.9 Find the differential of the transformation $F: u = x^3 - 3xy^2$, $v = 3x^2y - y^3$ at the point $(1, \sqrt{3})$, and use it to describe the behavior of F near that point, as in Example 16.2.

16.10 Show that for the transformation $F: u = e^x \cos y$, $v = e^x \sin y$, the differential dF at an arbitrary point is a similarity transformation.

***16.11** In Example 16.1 we showed that every line through the point $(0, 1)$ mapped onto a parabola with vertical tangent at $(0, 1)$. However, we assumed implicitly that the line was not horizontal; that is, $\alpha \neq 0$. Examine carefully the case of a horizontal line. Find its image, and find the tangent vector to the image at the point $(0, 1)$. Explain away any apparent contradictions.

16.12 Let F be a differential transformation of the form $u = f(x)$, $v = g(y)$.

  a. What is the Jacobian matrix of F?

b. What is the Jacobian of F?

c. Under what conditions is F orientation-preserving at (x_0, y_0), and when is it orientation-reversing?

d. Give a geometric explanation of the answer to part c.

e. Give an intuitive explanation of why you would expect the absolute value of the Jacobian in this case to represent the "area magnification" at the point.

16.13　Let F be a differentiable transformation: $u = f(x, y)$, $v = g(x, y)$. Let C_1 be the curve $x = t$, $y = y_0$ (that is, the horizontal line $y = y_0$ in parametric form), and let C_2 be the curve $x = x_0$, $y = t$. Let Γ_1, Γ_2 be the images of C_1, C_2 under F. Let $(u_0, v_0) = F(x_0, y_0)$.

a. Show that the Jacobian matrix J_F at (x_0, y_0) has the following geometric interpretation. If the columns of J_F are considered as the components of vectors; namely

$$\mathbf{T}_1 = \langle u_x(x_0, y_0), v_x(x_0, y_0) \rangle$$
$$\mathbf{T}_2 = \langle u_y(x_0, y_0), v_y(x_0, y_0) \rangle,$$

then $\mathbf{T}_1$ and $\mathbf{T}_2$ are tangent vectors at (u_0, v_0) to the curves Γ_1 and Γ_2, respectively.

b. Indicate the situation described in part a by a sketch.

c. Show that $\mathbf{T}_1$ and $\mathbf{T}_2$ are the images under $dF_{(x_0, y_0)}$ of $\langle 1, 0 \rangle$ and $\langle 0, 1 \rangle$, respectively.

d. Let $x = x_0 + \lambda t$, $y = y_0 + \mu t$ be the equations of an arbitrary straight ine C through (x_0, y_0), and let Γ be the image of C under F. Show that the tangent vector to C at (x_0, y_0) is $\lambda \langle 1, 0 \rangle + \mu \langle 0, 1 \rangle$, and that the tangent vector to Γ at (u_0, v_0) is $\lambda \mathbf{T}_1 + \mu \mathbf{T}_2$.

e. Show that $|\partial(u, v)/\partial(x, y)|_{(x_0, y_0)}$ is equal to the area of the parallelogram spanned by $\mathbf{T}_1$ and $\mathbf{T}_2$.

*16.14　Let F be defined by

$$u = x \int_1^x \sin\left(\tfrac{1}{2}\pi y t^2\right) dt$$

$$v = y \int_1^y e^{-t^2 x^2} dt.$$

Find $\partial(u, v)/\partial(x, y)$ at the points

a. $(1, 1)$
b. $(1, 2)$
c. $(2, 1)$

16.15　Let F be the transformation

$$u = ax^2 + 2bxy + cy^2$$
$$v = ex^2 + 2fxy + gy^2.$$

Show that every circle about the origin maps into an ellipse (which may degenerate into a line segment or a point).

(*Hint:* use Exs. 10.29a and 14.25a.)

16.16 Let a differentiable transformation F map a regular curve C onto a curve Γ. Show that if F is regular at every point of C, then Γ is a regular curve.

Remark The definition we have given for the differential of a transformation is one that has evolved gradually out of the formal differential notation that was used throughout the nineteenth century and is still presented in many elementary texts. In that notation the differential at the point (x_0, y_0) of a transformation $F:(x, y) \to (u, v)$ is given by

$$du = u_x(x_0, y_0) \, dx + u_y(x_0, y_0) \, dy$$
$$dv = v_x(x_0, y_0) \, dx + v_y(x_0, y_0) \, dy.$$

In some presentations the "dx" and "dy" are interpreted as independent variables, in which case the right-hand sides of these equations simply define the linear equations that represent $dF_{(x_0, y_0)}$. In other treatments, these equations are considered as purely formal expressions that may be operated on in certain specified ways. (For example, dividing through by dt and interpreting the quotients du/dt, dx/dt, etc., as derivatives, we arrive at meaningful and correct equations; namely, the chain rule.) In either case, the choice of notation is such that the expressions dx and dy tend to take on a life of their own. This is especially true when these same expressions appear later on under integral signs. (See, for example, Sect. 21 on line integrals.) As a consequence, many students come away from a calculus course with the sensation that there is something mystical about differentials. We have tried to avoid this by dispensing with the dx, dy notation altogether. In fact, we did not mention differentials at all until discussing transformations, although we could have easily given a similar treatment of the differential of a function $f(x, y)$. Namely, $df_{(x_0, y_0)}$ would be the linear function whose coefficients are $f_x(x_0, y_0)$ and $f_y(x_0, y_0)$. But it is precisely this function which enters into the definition of the tangent plane, and nothing is gained by the introduction of new terminology.

There are, on the other hand, several advantages worth mentioning to the classical notation. First of all, it is sometimes easier to manipulate in computations. Secondly, the suggestive nature of the notation may lead one to certain formulas that can then be proved rigorously. Finally, by thinking of differentials as "small quantities," applied mathematicians are often able to derive mathematical models for the objects of their study.

The student who wishes to familiarize himself with the manipulation of differentials in the classical notation should verify the formulas in Ex. 16.17.

16.17 Let $u = f(x, y)$, $v = g(x, y)$, $w = \varphi(u, v)$. Fix a point (x_0, y_0) at which differentials are evaluated. Using the notation

$$df = du = f_x \, dx + f_y \, dy$$
$$dg = dv = g_x \, dx + g_y \, dy$$
$$d\varphi = dw = \varphi_u \, du + \varphi_v \, dv,$$

show that the following statements are valid.

a. If $h(x, y) = f(x, y)g(x, y)$, then $dh = f \, dg + g \, df$.

b. If $h(x, y) = f(x, y)/g(x, y)$ and $g(x_0, y_0) \neq 0$, then

$$dh = \frac{g \, df - f \, dg}{g^2}.$$

c. If $f(x, y)^2 - g(x, y)^3 \equiv 1$, then $2f \, df = 3g \, dg$.

d. $w_u \, du + w_v \, dv = w_x \, dx + w_y \, dy$.

(*Hint*: use, if necessary, Eq. (17.2a).)

e. If we introduce the notation $\mathbf{r} = \langle x, y \rangle$, $d\mathbf{r} = \langle dx, dy \rangle$, then

$$df = \nabla f \cdot d\mathbf{r}.$$

17 Composition; inverse mappings

Our next aim is to develop what may be called a "chain rule" for transformations. We start with the following observation.

Let F be a differentiable transformation of a domain D into a domain E,

$$F: \begin{cases} u(x, y) \\ v(x, y). \end{cases}$$

Let $f(u, v)$ be a continuously differentiable function in E. We may then consider the composition of the transformation F and the function f. Let us write

$$\tilde{f} = f \circ F,$$

where

$$\tilde{f}(x, y) = f(u(x, y), v(x, y)).$$

Example 17.1

Suppose

$$F: \begin{cases} u = x - y \\ v = x + y \end{cases}$$

$$f(u, v) = u^2 + 2uv + v^2.$$

Then

$$\begin{aligned} \tilde{f}(x, y) &= (x - y)^2 + 2(x - y)(x + y) + (x + y)^2 \\ &= x^2 - 2xy + y^2 + 2(x^2 - y^2) + x^2 + 2xy + y^2 \\ &= 4x^2. \end{aligned}$$

Remark Although the notation at first appears to cloud the issue, the process of forming the function $\tilde{f}(x, y)$ from $f(u, v)$ is completely elementary; $\tilde{f}(x, y)$ is obtained by substituting the functions $u(x, y)$, $v(x, y)$, which define the transformation F, into $f(u, v)$. The function $\tilde{f}(x, y)$ obtained in this manner is sometimes referred to as "the function $f(u, v)$ pulled back to the x, y plane by the transformation F." The same process will be considered more extensively in a somewhat different light in Sect. 20, where we discuss changes of coordinates.

We next wish to compute the partial derivatives of the function $\tilde{f}(x, y)$. Since partial differentiation amounts to ordinary differentiation with one variable held constant, the result is obtained by a direct application of the chain rule, Th. 7.1. (See Ex. 17.5 for a more detailed comparison of the notation of Th. 7.1 with that of the present case.) We find

$$\tilde{f}_x(x_0, y_0) = f_u(u_0, v_0)u_x(x_0, y_0) + f_v(u_0, v_0)v_x(x_0, y_0)$$
$$\tilde{f}_y(x_0, y_0) = f_u(u_0, v_0)u_y(x_0, y_0) + f_v(u_0, v_0)v_y(x_0, y_0),$$

where

$$u_0 = u(x_0, y_0), \qquad v_0 = v(x_0, y_0).$$

Example 17.2

If, in Example 17.1 we wish to compute $\tilde{f}_x$ and $\tilde{f}_y$ at the point $(2, 3)$, we first obtain the corresponding values of u and v

$$u = 2 - 3 = -1, \qquad v = 2 + 3 = 5;$$

we then find

$$f_u = 2u + 2v, \qquad f_v = 2u + 2v,$$

and using the chain rule, we have

$$\tilde{f}_x(2, 3) = f_u(-1, 5)u_x(2, 3) + f_v(-1, 5)v_x(2, 3) = 8 \cdot 1 + 8 \cdot 1 = 16$$
$$\tilde{f}_y(2, 3) = f_u(-1, 5)u_y(2, 3) + f_v(-1, 5)v_y(2, 3) = 8 \cdot (-1) + 8 \cdot 1 = 0.$$

These values are the same as those obtained by direct differentiation of the function $\tilde{f}(x, y) = 4x^2$, and evaluation at $(2, 3)$.

It may help to write the above equations in a slightly different notation. Setting

$$(u, v) = F(x, y), \qquad z = f(u, v),$$

we have

(17.1)
$$z_x(x_0, y_0) = z_u(u_0, v_0)u_x(x_0, y_0) + z_v(u_0, v_0)v_x(x_0, y_0)$$
$$z_y(x_0, y_0) = z_u(u_0, v_0)u_y(x_0, y_0) + z_v(u_0, v_0)v_y(x_0, y_0),$$

where

$$(u_0, v_0) = F(x_0, y_0).$$

In even more compact form,

(17.1a)
$$z_x = z_u u_x + z_v v_x$$
$$z_y = z_u u_y + z_v v_y.$$

This form of the equations is undoubtedly the most perspicuous, but it must be remembered that the various quantities on the right-hand side are evaluated at different points; u_x, v_x, u_y, and v_y (as well as z_x and z_y) are all evaluated at a point (x_0, y_0), whereas z_u and z_v are evaluated at the corresponding point (u_0, v_0) in the u, v plane.

We now return to our original problem. We are given a pair of differentiable transformations, say

$$F:\begin{cases} u(x, y) \\ v(x, y) \end{cases}$$

and

$$G:\begin{cases} z(u, v) \\ w(u, v). \end{cases}$$

We wish to study their composition

$$H = G \circ F,$$

where

$$H:(x, y) \to (z, w)$$

(see Fig. 17.1). Throughout this section we assume that F maps some domain D in the x, y plane into a domain E in the u, v plane, and that G is defined in the domain E. Then the composed transformation H is defined throughout D. If F and G are both differentiable, then so is H. Differentiating the functions $z(u(x, y), v(x, y))$ and $w(u(x, y), v(x, y))$, which define H, we obtain Eqs. (17.1), and the analogous equations

(17.2)
$$w_x(x_0, y_0) = w_u(u_0, v_0)u_x(x_0, y_0) + w_v(u_0, v_0)v_x(x_0, y_0)$$
$$w_y(x_0, y_0) = w_u(u_0, v_0)u_y(x_0, y_0) + w_v(u_0, v_0)v_y(x_0, y_0),$$

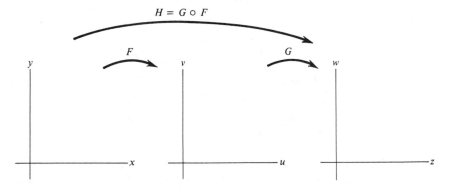

FIGURE 17.1 Composed mapping

or, in abbreviated form,

(17.2a)
$$w_x = w_u u_x + w_v v_x$$
$$w_y = w_u u_y + w_v v_y.$$

Equations (17.1a) and (17.2a) combined, yield the expression

(17.3)
$$J_H = \begin{pmatrix} z_x & z_y \\ w_x & w_y \end{pmatrix} = \begin{pmatrix} z_u u_x + z_v v_x & z_u u_y + z_v v_y \\ w_u u_x + w_v v_x & w_u u_y + w_v v_y \end{pmatrix}$$

for the Jacobian matrix of the transformation H. Note that the right-hand side of Eq. (17.3) depends only on the entries in the Jacobian matrices

$$J_F = \begin{pmatrix} u_x & u_y \\ v_x & v_y \end{pmatrix} \quad \text{and} \quad J_G = \begin{pmatrix} z_u & z_v \\ w_u & w_v \end{pmatrix}.$$

Equation (17.3) leads to the following basic result.

Theorem 17.1 *Let $F:(x, y) \to (u, v)$ and $G:(u, v) \to (z, w)$ be differentiable transformations, and let*

$$H = G \circ F:(x, y) \to (z, w).$$

Let

$$F:(x_0, y_0) \to (u_0, v_0).$$

Then

(17.4)
$$dH_{(x_0,y_0)} = dG_{(u_0,v_0)} \circ dF_{(x_0,y_0)}.$$

PROOF. We recall that dF, dG, and dH are the linear transformations

$$dF_{(x_0,y_0)} : \begin{cases} U = aX + bY \\ V = cX + dY \end{cases}, \qquad dG_{(u_0,v_0)} : \begin{cases} Z = AU + BV \\ W = CU + DV \end{cases},$$

$$dH_{(x_0,y_0)} : \begin{cases} Z = \alpha X + \beta Y \\ W = \gamma X + \delta Y \end{cases},$$

where

$$\begin{pmatrix} a & b \\ c & d \end{pmatrix} = \begin{pmatrix} u_x(x_0, y_0) & u_y(x_0, y_0) \\ v_x(x_0, y_0) & v_y(x_0, y_0) \end{pmatrix},$$

$$\begin{pmatrix} A & B \\ C & D \end{pmatrix} = \begin{pmatrix} z_u(u_0, v_0) & z_v(u_0, v_0) \\ w_v(u_0, v_0) & w_v(u_0, v_0) \end{pmatrix},$$

$$\begin{pmatrix} \alpha & \beta \\ \gamma & \delta \end{pmatrix} = \begin{pmatrix} z_x(x_0, y_0) & z_y(x_0, y_0) \\ w_x(x_0, y_0) & w_y(x_0, y_0) \end{pmatrix}.$$

Recalling Eq. (15.4), which shows how the matrices of linear transformations combine under composition, we see that Eq. (17.3) is equivalent to the statement of the theorem. ◆

Remark For those familiar with matrix multiplication, Eq. (17.3) takes the simple form

$$J_H = J_G J_F.$$

(See the Remark following Ex. 15.21.)

Corollary *Under the same hypotheses we have*

(17.5)
$$\left.\frac{\partial(z, w)}{\partial(x, y)}\right|_{(x_0, y_0)} = \left.\frac{\partial(z, w)}{\partial(u, v)}\right|_{(u_0, v_0)} \left.\frac{\partial(u, v)}{\partial(x, y)}\right|_{(x_0, y_0)}.$$

PROOF. By the definition of the Jacobian (Eq. (16.6)), Eq. (17.5) simply says

$$\det dH = (\det dG)(\det dF),$$

and this follows immediately from (17.4) and Th. 15.1. ◆

Equation (17.5) may be considered the "chain rule for Jacobians." It is one indication that the role of the Jacobian of a transformation is closely analogous to that of the derivative for a function of one variable. Many later results will confirm this observation.

It is worth pointing out that Eq. (17.5) has a simple geometric interpretation in terms of the relation of the Jacobian to "area magnification at a point," described at the end of the previous section. Namely, if we denote by A_1, the area of a small circle about (x_0, y_0), by A_2 and A_3 the area of the image of this circle under F and H, respectively, then we have

$$\frac{A_3}{A_1} = \frac{A_3}{A_2} \cdot \frac{A_2}{A_1}.$$

But as we remarked at the end of the previous section, these ratios tend to the magnitudes of the Jacobians of the maps H, G, and F, respectively. Similarly, the way the sign of the Jacobian behaves under composition is easily deduced by geometric considerations (see Ex. 17.9).

We next apply Th. 17.1 and its Corollary to the important special case where the transformation G is the inverse of F. In this case $z = x$, $w = y$, and we may identify the z, w plane with the x, y plane (Fig. 17.2).

Theorem 17.2 *If a differentiable transformation $F: (x, y) \to (u, v)$ has a differentiable inverse, then the Jacobian of F cannot equal zero, and*

(17.6)
$$\frac{\partial(x, y)}{\partial(u, v)} = \det J_{F^{-1}} = \frac{1}{\det J_F} = \frac{1}{\partial(u, v)/\partial(x, y)}.$$

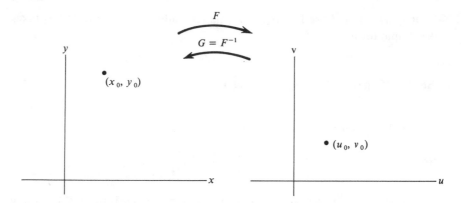

FIGURE 17.2 Inverse mapping

PROOF.   Setting $z = x$, $w = y$, we find $z_x = 1$, $z_y = 0$, $w_x = 0$, $w_y = 1$, and substituting in Eq. (17.5),

$$\frac{\partial(x, y)}{\partial(u, v)} \cdot \frac{\partial(u, v)}{\partial(x, y)} = 1.$$

But if the product of two numbers is equal to 1, neither number can equal zero. This proves the theorem.   ◆

Remark   It is important to note that the right-hand side of Eq. (17.6) is evaluated at some point (x_0, y_0) and the left-hand side at the point (u_0, v_0), which is the image of (x_0, y_0) under F.

We shall illustrate this theorem by a number of examples. First, however, let us make several general remarks.

We start by recalling that if a transformation F maps two distinct points onto the same point (u_0, v_0), then F does not have an inverse; in other words, for F to have an inverse it must be one-to-one. Now it may happen that F is not one-to-one in its entire domain of definition, but that if we restrict our attention to some neighborhood of a point (x_0, y_0), then F may define a one-to-one mapping of that neighborhood (see Example 17.3 below). On the other hand, it follows from Th. 17.2 that if the Jacobian of F is equal to zero at (x_0, y_0), then no matter how small a neighborhood of (x_0, y_0) we may choose, the map F of that neighborhood cannot have a differentiable inverse. We may formulate this consequence of Th. 17.2 as follows.

Corollary   *If F is a differentiable transformation, and if the Jacobian of F vanishes at a point, then F cannot have a differentiable inverse in any neighborhood of that point.*

Next we note that if F maps (x_0, y_0) onto (u_0, v_0), then in order for F to have an inverse defined in some neighborhood of (u_0, v_0), it must map a neighborhood of (x_0, y_0) onto an entire neighborhood of (u_0, v_0). This may not be the case if the Jacobian of F vanishes at (x_0, y_0) (see Example 17.3 below).

Finally, even if F maps some neighborhood of (x_0, y_0) onto some neighborhood of (u_0, v_0) and is one-to-one, then F has an inverse transformation G, but G need not be differentiable (see Example 17.5 below).

Example 17.3

Let

$$F: \begin{cases} u = x^2 \\ v = 2y. \end{cases}$$

Then

$$J_F = \begin{pmatrix} 2x & 0 \\ 0 & 2 \end{pmatrix},$$

and

$$\frac{\partial(u, v)}{\partial(x, y)} = 4x.$$

Thus the Jacobian vanishes at every point of the y axis, and, by the above Corollary, F cannot have an inverse in any neighborhood of a point on the y axis. Take, for example, the point $(0, 1)$. It maps onto the point $(0, 2)$ in the u, v plane. Every circle about $(0, 1)$ in the x, y plane maps into the right half of the u, v plane, since $u = x^2 \geq 0$ (Fig. 17.3). Thus the image of F never covers a full neighborhood of $(0, 2)$ in the u, v plane, and we cannot hope to find an inverse map G defined in a neighborhood of $(0, 2)$. If, on the other hand, we take a point such as $(2, 1)$ in the x, y plane, where the Jacobian of F is different from zero, then we may be able to find an inverse. If we choose a large circle about $(2, 1)$, which includes points in the left-half plane, then the map F is not one-to-one, and it cannot have an inverse. However, by choosing a smaller circle, of radius $r < 2$, F is one-to-one inside this circle, since F is a one-to-one map of the whole right half-plane $x > 0$ onto the whole right half-plane $u > 0$, and F has a differentiable inverse

$$G = F^{-1}: \begin{cases} x = \sqrt{u} \\ y = \tfrac{1}{2}v, \end{cases}$$

which maps the right half-plane $u > 0$ onto $x > 0$ (Fig. 17.4).

Example 17.4

$$F: \begin{cases} u = x^2 - y^2 \\ v = 2xy. \end{cases}$$

Here

$$\frac{\partial(u, v)}{\partial(x, y)} = \det \begin{pmatrix} 2x & -2y \\ 2y & 2x \end{pmatrix} = 4(x^2 + y^2).$$

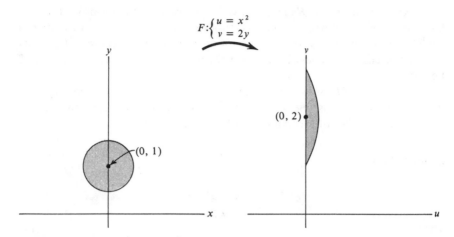

FIGURE 17.3 Map $F:(x, y) \rightarrow (x^2, 2y)$ near the point $(0, 1)$

Thus the Jacobian of F vanishes at the origin. From our discussion of this transformation at the end of Sect. 13, it is clear that F maps the interior of every circle about the origin in the x, y plane onto the full interior of a circle about the

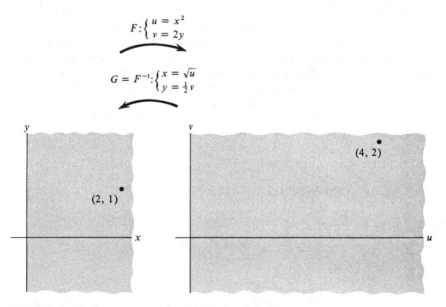

FIGURE 17.4 Inverse mapping of $F:(x, y) \rightarrow (x^2, 2y)$

origin in the u, v plane. However, no matter how small a circle we choose, the map is not one-to-one. Hence there cannot be an inverse in any neighborhood of the origin.

Example 17.5

$$F: \begin{cases} u = x^3 \\ v = y^3. \end{cases}$$

In this case F is a differentiable transformation that maps the entire x, y plane one-to-one onto the u, v plane. There is therefore an inverse mapping; namely

$$G: \begin{cases} x = \sqrt[3]{u} \\ y = \sqrt[3]{v}. \end{cases}$$

The map G is differentiable everywhere except on the u and v axes, where one of the derivatives $x_u = \frac{1}{3}u^{-2/3}$ or $y_v = \frac{1}{3}v^{-2/3}$ becomes infinite. This corresponds to the fact that the u and v axes are the image under F of the x and y axes, and that along the latter the Jacobian of F vanishes;

$$\frac{\partial(u, v)}{\partial(x, y)} = \det \begin{pmatrix} 3x^2 & 0 \\ 0 & 3y^2 \end{pmatrix} = 9x^2y^2 = 0$$

$$\Leftrightarrow x = 0 \quad \text{or} \quad y = 0.$$

In the light of Examples 17.3–17.5 it is easy to see why the terminology "singular point" is used for a point at which the Jacobian vanishes. At such a point the transformation misbehaves in one way or another. We may restate the Corollary to Th. 17.2 as follows: *if F is a differentiable transformation near (x_0, y_0) and if F has a differentiable inverse transformation in some neighborhood of (x_0, y_0), then (x_0, y_0) is a regular point of F.*

The converse of this statement is known as the *inverse mapping theorem*: *if (x_0, y_0) is a regular point of a differentiable transformation F, then there exists some neighborhood of (x_0, y_0) in which F has a differentiable inverse.*

We omit the proof of this theorem, since it follows most naturally as a consequence of general results for functions of more than two variables.[1] Stated together with its converse, the result is the following.

Theorem 17.3 Inverse Mapping Theorem *A differentiable mapping F has a differentiable inverse in some neighborhood of (x_0, y_0) if and only if the Jacobian of F at (x_0, y_0) is different from zero.*

---

[1] See, for example, Section 12.4 of [36]. A proof which is purely two-dimensional is given in Ex. 26.29 of this book. Other proofs may be found in Chapter III, Section 3 of [11], and in Section 8.4 of [19].

This theorem is of great importance in the theory of differentiable transformations, and it is worth spending some time on the precise meaning of the statement. Consider a mapping

(17.7)
$$F:\begin{cases} u = f(x, y) \\ v = g(x, y) \end{cases}$$

that is differentiable in a domain D, and choose some point (x_0, y_0) in D. We ask whether Eqs. (17.7) can be "solved" for x and y in terms of u and v, if not in the whole domain D, at least in some smaller domain D' containing (x_0, y_0). More precisely, we may state the question as follows. "Does there exist some domain D' included in D, such that (x_0, y_0) is in D', F maps D' one-to-one onto a domain E', and if the correspondence thus defined between points of D' and E' is written in the form

$$\begin{cases} x = \varphi(u, v) \\ y = \psi(u, v), \end{cases}$$

then φ and ψ are continuously differentiable functions in E'?" (See Fig. 17.5.) This question and its answer are summarized briefly in Th. 17.3: such a domain D' exists if and only if the Jacobian of F is different from zero at (x_0, y_0). A striking feature of Th. 17.3 is that such a complicated question can be answered by a simple computation with derivatives.

If we consider D in Example 17.3 above to be the full plane, and (x_0, y_0) any point in the right half-plane, then we may choose D' to be the right half-plane $x > 0$, in which case E' is the right half-plane $u > 0$, and we are able

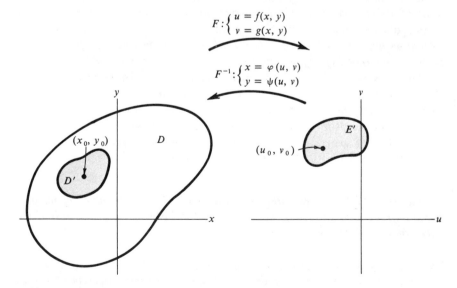

FIGURE 17.5 Inverse mapping near a point

to write down the explicit equations for the inverse mapping $\varphi(u, v) = \sqrt{u}$, $\psi(u, v) = \frac{1}{2}v$. Similarly, in Example 17.5, if (x_0, y_0) is any point not on the x or y axis, then we could choose for D' the entire quadrant containing (x_0, y_0) and we again have explicit equations for the inverse mapping.

In general, it is not possible to write down explicitly the functions $\varphi(u, v)$ and $\psi(u, v)$. Theorem 17.3 simply asserts that these functions must exist. The situation is precisely analogous to that of the implicit function theorem (Th. 8.2) and the statement that "we can solve for x and y in terms of u and v" is to be interpreted in the manner discussed in detail at that point. An important fact to notice is that again, as in the case of the implicit function theorem, even if we cannot find the functions φ and ψ explicitly, we can use implicit differentiation to compute their derivatives. Namely, let us assume that u and v have been given explicitly as functions of x and y, so that the partial derivatives u_x, u_y, v_x, v_y evaluated at any point are known constants. If we apply the chain rule, Eq. (17.1a), first to the case where $z = \varphi(u, v) = x$, and then to the case $z = \psi(u, v) = y$, we find

$$(17.8) \qquad \begin{array}{ll} \varphi_u u_x + \varphi_v v_x = 1 & \psi_u u_x + \psi_v v_x = 0 \\ \varphi_u u_y + \varphi_v v_y = 0 & \psi_u u_y + \psi_v v_y = 1. \end{array}$$

At a regular point, if we set

$$(17.9) \qquad \Delta = \frac{\partial(u, v)}{\partial(x, y)} = u_x v_y - v_x u_y,$$

then $\Delta \neq 0$, and each pair of equations in (17.8) is a pair of simultaneous linear equations with nonvanishing determinant Δ. We may therefore solve these pairs of equations, and we find

$$(17.10) \qquad \varphi_u = \frac{v_y}{\Delta}, \quad \varphi_v = -\frac{u_y}{\Delta}; \qquad \psi_u = -\frac{v_x}{\Delta}, \quad \psi_v = \frac{u_x}{\Delta}.$$

In other words, we have for the Jacobian matrix of the inverse mapping

$$(17.11) \qquad J_{F^{-1}} = \begin{pmatrix} x_u & x_v \\ y_u & y_v \end{pmatrix} = \begin{pmatrix} v_y/\Delta & -u_y/\Delta \\ -v_x/\Delta & u_x/\Delta \end{pmatrix}, \qquad \Delta = \frac{\partial(u, v)}{\partial(x, y)}.$$

Note that the right-hand side of (17.11) is evaluated at some point (x_0, y_0), and the left-hand side is to be evaluated at the point (u_0, v_0), which is the image under F of (x_0, y_0).

Remark In the terminology of elementary matrix theory, the matrix on the right-hand side of (17.11) is simply the *inverse* of the matrix J_F. We may therefore write (17.11) in the form

$$(17.12) \qquad J_{F^{-1}} = (J_F)^{-1}.$$

This also follows immediately from Eq. (17.4) applied to the case where $G = F^{-1}$, in which case dH is the identity map.

Example 17.6

If the equations

(17.13) $$u = x^2 - y^2, \qquad v = 2xy$$

are solved for x and y in terms of u and v near the point $(x_0, y_0) = (3, -4)$, what are the partial derivatives x_u, x_v, y_u, y_v at the corresponding point in the u, v plane?

We find

$$J_F = \begin{pmatrix} u_x & u_y \\ v_x & v_y \end{pmatrix} = \begin{pmatrix} 2x & -2y \\ 2y & 2x \end{pmatrix};$$

$$\frac{\partial(u, v)}{\partial(x, y)} = 4(x^2 + y^2).$$

By Eq. (17.11),

$$J_{F^{-1}} = \begin{pmatrix} x_u & x_v \\ y_u & y_v \end{pmatrix} = \begin{pmatrix} \dfrac{x}{2(x^2 + y^2)} & \dfrac{y}{2(x^2 + y^2)} \\ \dfrac{-y}{2(x^2 + y^2)} & \dfrac{x}{2(x^2 + y^2)} \end{pmatrix}.$$

If $x = 3$, $y = -4$, then $u = -7$, $v = -24$, and

$$\begin{pmatrix} x_u & x_v \\ y_u & y_v \end{pmatrix} = \begin{pmatrix} \frac{3}{50} & -\frac{4}{50} \\ \frac{4}{50} & \frac{3}{50} \end{pmatrix}, \qquad \text{at } (-7, -24).$$

In conclusion, we remark that in working a given problem, rather than referring to Eq. (17.11), it may be just as easy to use the method that led to this equation. Thus, in order to find x_u in Example 17.6, we may simply differentiate both Eqs. (17.13), using the chain rule. We find

$$2xx_u - 2yy_u = 1$$
$$2yx_u + 2xy_u = 0.$$

Substituting $x = 3$, $x = -4$, we obtain

$$6x_u + 8y_u = 1$$
$$-8x_u + 6y_u = 0,$$

which, when solved, yields $x_u = \frac{3}{50}$, $y_u = \frac{4}{50}$.

Exercises

17.1　In each of the following cases, find the partial derivatives z_x and z_y at the point indicated, first by using the chain rule (Eqs. (17.1)) and then by finding z explicitly in terms of x and y and differentiating.

a. $z = u^2 + v^2$;　$u = x \cos y$, $v = x \sin y$;　$(x_0, y_0) = (1, 0)$
b. $z = u^2 + 2v^2$;　$u = x - y$, $v = x + y$;　$(x_0, y_0) = (1, -2)$
c. $z = e^{u \cos v}$;　$u = e^x \cos y$, $v = e^x \sin y$;　$(x_0, y_0) = (0, 0)$
d. $z = \log(2u^2 + 3v^2 + 6)$;　$u = x^2 - y^2$, $v = 2xy$;　$(x_0, y_0) = (1, 2)$

17.2 Let $f(u, v)$ be continuously differentiable, and let F be the transformation $u = cx - y$, $v = cx + y$, where c is a nonzero constant. Let $\tilde{f} = f \circ F$.

    a. Find $\tilde{f}_x - c\tilde{f}_y$.
    b. Use your answer to part a to construct functions $\tilde{f}(x, y)$ satisfying the equation $\tilde{f}_x = c\tilde{f}_y$.

17.3 Let $f(u, v)$ be continuously differentiable, and let F be a differentiable transformation, $F:(x, y) \to (u, v)$. Let $\tilde{f} = f \circ F$ and $(u_0, v_0) = F(x_0, y_0)$.

    a. Show that if $\nabla f(u_0, v_0) = 0$, then $\nabla \tilde{f}(x_0, y_0) = 0$.
    b. Show that if F is regular at (x_0, y_0) and if $\nabla \tilde{f}(x_0, y_0) = 0$, then $\nabla f(u_0, v_0) = 0$.
    *c. Show that if F is singular at (x_0, y_0), then it is possible to find a function f such that $\nabla f(u_0, v_0) \neq 0$, but $\nabla \tilde{f}(x_0, y_0) = 0$.

17.4 Given a differentiable mapping $F:(x, y) \to (u, v)$, and given a function $f(u, v)$, form a geometric picture of the relation between f and $\tilde{f} = f \circ F$ in the following fashion. Draw two sets of rectangular coordinates, x, y, z and u, v, z. Sketch the surface $z = f(u, v)$. Then to obtain the corresponding surface $z = \tilde{f}(x, y)$, picture the map $F:(x, y) \to (u, v)$, and to each point (x_0, y_0), place a point directly above it at the same height as f at $(u_0, v_0) = F(x_0, y_0)$. Thus the surface $z = \tilde{f}(x, y)$ is obtained by a process of distorting the surface $z = f(u, v)$ in a manner that preserves the height of each point, although the horizontal distortion may be quite arbitrary. One would not expect many geometric properties of a surface to be preserved under such a distortion, but some of them are. For example, interpret the statement in Ex. 17.3a geometrically in terms of the surfaces $z = f(u, v)$ and $z = \tilde{f}(x, y)$, and try to visualize why the statement is true. Note also that if $f(u, v)$ has a local minimum or maximum at (u_0, v_0), then $\tilde{f}(x, y)$ has the same at (x_0, y_0) (see also Ex. 20.17f).

17.5 Let F be a differentiable transformation of a domain D in the x, y plane into a domain E in the u, v plane:

$$F: \begin{cases} u = \varphi(x, y) \\ v = \psi(x, y). \end{cases}$$

    a. Show that for any point (x_0, y_0) in D, there are numbers a, b such that $a < x_0 < b$, and such that $u = \varphi(x, y_0)$, $v = \psi(x, y_0)$, $a \leq x \leq b$, defines a differentiable curve C in E. (C would be the image under F of the line segment $a \leq x \leq b$, $y = y_0$, where x is the parameter along the curve.)
    b. Show that the vector $\langle u_x(x_0, y_0), v_y(x_0, y_0) \rangle$ is the tangent vector to the curve C of part a at the point $(u_0, v_0) = F(x_0, y_0)$.
    c. Let $f(u, v)$ be a continuously differentiable function in E, and let $\tilde{f} = f \circ F$. Using the notation of part a, show that $f_C(x) = \tilde{f}(x, y_0)$, $a \leq x \leq b$.
    d. Deduce from parts b and c that

$$\tilde{f}_x(x_0, y_0) = f_C'(x_0) = f_u(u_0, v_0)u_x(x_0, y_0) + f_v(u_0, v_0)u_y(x_0, y_0).$$

17.6 For each of the following pairs of transformations F, G, find the Jacobian matrices J_F and J_G and use Eq. (17.3) to find the Jacobian matrix of the composed transformation $H = G \circ F$ at the point indicated. Then find H explicitly, compute J_H, and check your answer.

a. $F: \begin{cases} u = 2x + y \\ v = 3x + 2y \end{cases}$ $\quad G: \begin{cases} z = 2u - v \\ w = -3u + v \end{cases}$ $\quad (x_0, y_0) = (5, 7)$

b. $F: \begin{cases} u = x^2 - 4xy + 5 \\ v = y^3 - yx \end{cases}$ $\quad G: \begin{cases} z = \sin \frac{1}{2}\pi u^2 \\ w = \sin \frac{1}{2}\pi uv \end{cases}$ $\quad (x_0, y_0) = (3, 1)$

17.7   Find $\partial(z, w)/\partial(x, y)$ at the point (x_0, y_0) for each of the following pairs of transformations.

    a. Use Ex. 17.6a.
    b. Use Ex. 17.6b.

c. $F: \begin{cases} u = xe^y \\ v = \sin(x + y) \end{cases}$ $\quad G: \begin{cases} z = ue^v \\ w = \sin(u + v) \end{cases}$ $\quad (x_0, y_0) = (0, 1)$

    d. Use part c, but let $(x_0, y_0) = (\pi, 0)$.
    e. Use part c, but let $(x_0, y_0) = (\frac{1}{2}\pi, \log 2)$.

17.8   Show that if (x_0, y_0) is a singular point of a differentiable transformation F, then it is a singular point of $G \circ F$ for every differentiable transformation G.

17.9   Let F be a differentiable transformation that is regular at (x_0, y_0), and let $(u_0, v_0) = F(x_0, y_0)$. Let G be a differentiable transformation that is regular at (u_0, v_0). Let $H = G \circ F$.

    a. Give a geometric argument that shows that H is orientation-preserving at (x_0, y_0) if and only if F and G are both orientation-preserving or are both orientation-reversing at the points (x_0, y_0) and (u_0, v_0), respectively.

    b. For any real number $\lambda \neq 0$, let

$$sg\lambda = \begin{cases} 1 & \text{if } \lambda > 0 \\ -1 & \text{if } \lambda < 0 \end{cases}$$

(that is, $sg\lambda = \lambda/|\lambda|$). Use a geometric argument to show that

$$sg \left.\frac{\partial(z, w)}{\partial(x, y)}\right|_{(x_0, y_0)} = sg \left.\frac{\partial(z, w)}{\partial(u, v)}\right|_{(u_0, v_0)} sg \left.\frac{\partial(u, v)}{\partial(x, y)}\right|_{(x_0, y_0)}.$$

17.10   Let F be a differentiable transformation that is an involution (see Ex. 14.16).

    a. Show that F has no singular points.
    b. Show that if $F(x_0, y_0) = (u_0, v_0)$, then the Jacobian of F at (u_0, v_0) is the reciprocal of the Jacobian of F at (x_0, y_0).
    c. Show that the transformation

$$F: \begin{cases} u = \dfrac{x}{x^2 + y^2} \\[2mm] v = \dfrac{y}{x^2 + y^2} \end{cases}$$

is an involution, and verify part b directly by computing the Jacobian.

17.11   For each of the following transformations, F, check by Th. 17.3 that F has a differentiable inverse in some neighborhood of $(0, 0)$; use implicit differentiation to find the partial derivatives of the inverse transformation at the point $F(0, 0)$, and verify Eq. (17.6).

a. $F:\begin{cases} u = e^{x+y} \\ v = e^{x-y} \end{cases}$
  e. $F:\begin{cases} u = e^x \cos y \\ v = e^x \sin y \end{cases}$

b. $F:\begin{cases} u = \tan x \\ v = \tan y \end{cases}$
  f. $F:\begin{cases} u = x^2 + 3y + 1 \\ v = x^3 - x - 2y + 5 \end{cases}$

c. $F:\begin{cases} u = x + y^3 \\ v = y \end{cases}$
  g. $F:\begin{cases} u = \int_0^x e^{t^2}\, dt + \sin y \\ v = ye^{-x^2} \end{cases}$

d. $F:\begin{cases} u = (x - 2)^3 \\ v = (y + 1)^5 \end{cases}$

17.12   In Ex. 17.11a–e, solve explicitly for x and y as functions of u and v, and thus write down the inverse transformation in explicit form. Specify carefully some domain containing the point $F(0, 0)$ in which F^{-1} is defined and differentiable and check your answers to Ex. 17.11 by computing the derivatives of F^{-1} at $F(0, 0)$.

17.13   Let us call a "generalized shear transformation," a transformation of the form

$$F:\begin{cases} u = x + g(y) \\ v = y. \end{cases}$$

   a. Describe the map F geometrically by examining the behavior of F on each horizontal line $y = y_0$.
   b. Show that F is a bijective map from the whole x, y plane to the whole u, v plane, and find the explicit equations of the inverse.
   c. Note that $\partial(u, v)/\partial(x, y) \equiv 1$. Interpret this fact in terms of the vectors $\langle u_x, v_x \rangle$ and $\langle u_y, v_y \rangle$.

***17.14*   *a.* State and prove an analog of Th. 17.3 if the differentiable mapping F is replaced by a differentiable function of one variable, and if the Jacobian of F is replaced by the derivative of f.
   b. Discuss the various ways in which a function f of one variable can fail to have a differentiable inverse near a point where the derivative vanishes, and give examples analogous to those given in the text for a mapping F near a point where the Jacobian vanishes.

17.15   Let F be a differentiable transformation, $F:(x, y) \to (u, v)$. Let $P(u, v)$ and $Q(u, v)$ be continuously differentiable, and let

$$\tilde{P}(x, y) = P \circ F, \qquad \tilde{Q}(x, y) = Q \circ F.$$

Define

$$p(x, y) = \tilde{P}u_x + \tilde{Q}v_x, \qquad q(x, y) = \tilde{P}u_y + \tilde{Q}v_y.$$

a. Show that

$$p_y - q_x = (P_v - Q_u) \frac{\partial(u, v)}{\partial(x, y)}.$$

b. Show that if the transformation F satisfies the Cauchy-Riemann equations $u_x = v_y$, $v_x = -u_y$, then

$$p_x + q_y = (P_u + Q_v) \frac{\partial(u, v)}{\partial(x, y)}.$$

c. Show that if there exists a function $f(u, v)$ such that $P = f_u$, $Q = f_v$, then $p = \tilde{f}_v$ and $q = \tilde{f}_y$, where $\tilde{f}(x, y) = f \circ F$.

17.16 *a.* Given a matrix

$$M = \begin{pmatrix} a & b \\ c & d \end{pmatrix},$$

show that for every $\epsilon > 0$ there exists a $\delta > 0$ such that if

$$|a - a'| < \delta, \qquad |b - b'| < \delta, \qquad |c - c'| < \delta, \quad \text{and} \quad |d - d'| < \delta,$$

then

$$|ad - bc - (a'd' - b'c')| < \epsilon.$$

(*Note*: this says that the determinant of the matrix M is a continuous function of the four variables a, b, c, d.)
(*Hint*: let m be the maximum of $|a|, |b|, |c|, |d|$. Suppose that $|a - a'| < \delta$, etc. Writing $ad - a'd' = a(d - d') + d'(a - a')$, show that $|ad - a'd'| \le 2m\delta + \delta^2$, and similarly $|bc - b'c'| \le 2m\delta + \delta^2$. If $m = 0$, choose $\delta < \sqrt{\epsilon/2}$; if $m \neq 0$, choose $\delta < \sqrt{\epsilon/2}$ and $\delta < \epsilon/8m$.)
b. Using the notation of part a, show that if $ad - bc \neq 0$, then also $a'd' - b'c' \neq 0$ for δ sufficiently small.

17.17 Prove the following statement, which is part of Theorem 17.3.
If F is a differentiable mapping in a domain D, and if the Jacobian of F at (x_0, y_0) is different from zero, then F is one-to-one in some neighborhood of (x_0, y_0).
(*Hint*: if F is given by $u = f(x, y)$, $v = g(x, y)$, then by hypothesis,

$$\det M \neq 0, \qquad \text{where} \quad M = \begin{pmatrix} f_x(x_0, y_0) & f_y(x_0, y_0) \\ g_x(x_0, y_0) & g_y(x_0, y_0) \end{pmatrix}.$$

By Ex. 17.16b, it follows that there exists $\delta > 0$ such that

$$\det M' \neq 0, \qquad \text{where} \quad M' = \begin{pmatrix} f_x(x_3, y_3) & f_y(x_3, y_3) \\ g_x(x_4, y_4) & g_y(x_4, y_4) \end{pmatrix},$$

provided

$$|f_x(x_3, y_3) - f_x(x_0, y_0)| < \delta, \qquad |f_y(x_3, y_3) - f_y(x_0, y_0)| < \delta,$$
$$|g_x(x_4, y_4) - g_x(x_0, y_0)| < \delta, \qquad |g_y(x_4, y_4) - g_y(x_0, y_0)| < \delta.$$

By continuity of the partial derivatives f_x, f_y, g_x, g_y, these four inequalities hold, provided (x_3, y_3) and (x_4, y_4) lie in some disk $D':(x - x_0)^2 + (y - y_0)^2 < r^2$. Fix $r > 0$ by this requirement, and let (x_1, y_1), (x_2, y_2) be any two distinct points in D'. We wish to show that $F(x_1, y_1) \neq F(x_2, y_2)$, so that F is one-to-one in D'. By the mean-value theorem applied to the functions f and g,

$$f(x_2, y_2) - f(x_1, y_1) = (x_2 - x_1)f_x(x_3, y_3) + (y_2 - y_1)f_y(x_3, y_3)$$

$$g(x_2, y_2) - g(x_1, y_1) = (x_2 - x_1)g_x(x_4, y_4) + (y_2 - y_1)g_y(x_4, y_4),$$

where (x_3, y_3) and (x_4, y_4) are points which lie on the line segment joining (x_1, y_1) to (x_2, y_2), and which are therefore both in D'. It follows by our choice of r that the matrix M' of this system has nonzero determinant, so that the left-hand sides of these equations cannot be zero unless $x_2 - x_1 = 0$ and $y_2 - y_1 = 0$.)

18 Functional dependence; conformal mapping

We start once more with a differentiable transformation,

$$(18.1) \qquad F:\begin{cases} u = f(x, y) \\ v = g(x, y) \end{cases}$$

defined in some domain D. In the previous section we concentrated chiefly on the case where the Jacobian was different from zero at all points in D. We wish now to consider the other extreme, namely those transformations F for which the Jacobian is zero at every point of D. Let us start with the special case in which F is linear,

$$F:\begin{cases} u = ax + by \\ v = cx + by. \end{cases}$$

Then

$$J_F = \begin{pmatrix} u_x & u_y \\ v_x & v_y \end{pmatrix} = \begin{pmatrix} a & b \\ c & d \end{pmatrix}.$$

Thus the Jacobian is everywhere equal to zero if and only if the determinant of the linear transformation F is equal to zero. We have seen in Sect. 14 that a linear transformation with zero determinant maps the whole plane into a straight line through the origin. Thus, for linear transformations, the property that the Jacobian is everywhere zero is equivalent to the geometrical property that the whole plane is mapped into some straight line.

Returning to the general transformation (18.1), we ask what it means for F to map the whole domain D into a line through the origin. The equation of such a line is $\lambda u + \mu v = 0$, where λ and μ are constants, and the condition that F maps D into that line is that $\lambda f(x, y) + \mu g(x, y) \equiv 0$ in D. The following terminology is used.

Definition 18.1   Two functions $f(x, y)$ and $g(x, y)$ defined in a domain D are *linearly dependent* if there exist numbers λ, μ, not both zero, such that

(18.2) $$\lambda f(x, y) + \mu g(x, y) \equiv 0.$$

(See Ex. 18.3c for an alternative characterization of linear dependence.)

Lemma 18.1   *If F is a differentiable transformation defined by Eq.* (18.1), *and if f and g are linearly dependent, then the Jacobian of F is identically zero.*

PROOF.   Differentiating Eq. (18.2) yields

$$\lambda f_x + \mu g_x = 0$$
$$\lambda f_y + \mu g_y = 0$$

at every point of D. This is a pair of simultaneous linear homogeneous equations for the unknowns λ and μ, whose matrix is the Jacobian matrix of F at a given point of D. If the Jacobian of F were different from zero at some point, then the determinant of this system would be different from zero, and there would be a unique solution $\lambda = 0$, $\mu = 0$. But by assumption, λ and μ are not both zero. Hence, the Jacobian must vanish at each point.   ◆

Now the notion of linearly dependent functions is a special case of the general concept of functional dependence. Intuitively, functional dependence means that there is some relation between the functions f and g. For our purposes the following definition is most appropriate.

Definition 18.2   Two functions $f(x, y)$, $g(x, y)$ defined in a domain D are *functionally dependent* if there exists a continuously differentiable function $\varphi(u, v)$ such that

1. $\varphi(f(x, y), g(x, y)) \equiv 0$.
2. $\nabla \varphi \neq \mathbf{0}$ at each point $(u, v) = (f(x, y), g(x, y))$ for (x, y) in D.

Remark   Geometrically, functional dependence has the following interpretation. Condition 1 means that the transformation F defined by Eq. (18.1) maps the domain D into the set of points satisfying $\varphi(u, v) = 0$, and Condition 2 tells us (by the implicit function theorem) that this set of points lies on a regular curve (Fig. 18.1). (See the note in Ex. 18.5b for further comments on the definition.)

Example 18.1

$$\varphi(u, v) = \lambda u + \mu v.$$

Here Condition 1 is just Eq. (18.2), and Condition 2 says that not both λ and μ are zero, so that f and g are linearly dependent.

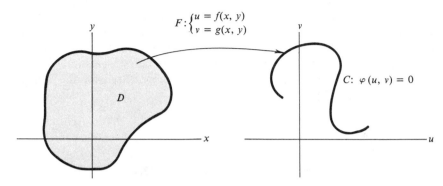

FIGURE 18.1 Geometrical interpretation of functional dependence

Example 18.2

Let

$$f(x, y) = xy, \qquad g(x, y) = 1 + x^2 y^2.$$

Then

$$g(x, y) \equiv 1 + [f(x, y)]^2.$$

This means that Condition 1 is fulfilled if we set

$$\varphi(u, v) = 1 + u^2 - v.$$

Condition 2 is also fulfilled, since

$$\nabla\varphi = \langle 2u, -1 \rangle \neq \mathbf{0}.$$

Thus xy and $1 + x^2 y^2$ are functionally dependent.

Theorem 18.1 *Let F be a differentiable transformation defined by Eq. (18.1). If $f(x, y)$ and $g(x, y)$ are functionally dependent, then the Jacobian of F is identically zero.*

PROOF. If we differentiate the equation of Condition 1, using the chain rule, we find

$$\varphi_u f_x + \varphi_v g_x = 0$$
$$\varphi_u f_y + \varphi_v g_y = 0$$

at every point. Using Condition 2, the reasoning of Lemma 18.1 may be applied without change. ◆

It is natural to ask whether the converse of Th. 18.1 is true. If the Jacobian of a mapping vanishes identically, does that imply functional dependence? The answer is that this is more or less true, but not without qualifications.

Rather than go into detail, we may summarize the situation as follows. Given a pair of functions $f(x, y)$, $g(x, y)$, which we should like to test for functional dependence, the first step is to compute the Jacobian. If the Jacobian is *not* identically zero, then f and g are *not* functionally dependent by Th. 18.1. On the other hand, if the Jacobian *is* identically zero, then this is strong evidence of functional dependence, and the best procedure is to try to find an explicit relationship between the functions. (See also the Remark preceding Ex. 18.27.)

Example 18.3

Let

$$f(x, y) = e^x e^y, \qquad g(x, y) = x^2 + y^2 + 2(x + y + xy).$$

Are these functionally dependent?

We start by finding the Jacobian,

$$f_x = e^x e^y, \qquad\qquad f_y = e^x e^y,$$
$$g_x = 2x + 2(1 + y), \qquad g_y = 2y + 2(1 + x),$$

$$\frac{\partial(f, g)}{\partial(x, y)} = f_x g_y - f_y g_x \equiv 0.$$

Thus we are led to seek a functional relationship. By elementary manipulations we can rewrite f and g in the form,

$$f(x, y) = e^{x+y}, \qquad g(x, y) = (x + y)^2 + 2(x + y).$$

Thus

$$g(x, y) = [\log f(x, y)]^2 + 2 \log f(x, y),$$

so that f and g are functionally dependent.

Diffeomorphisms and Conformal Mappings

The transformations whose Jacobians are identically zero are in a sense the "most singular" of all differentiable transformations. Every point is a singular point, and in general, a whole domain is mapped into a curve. At the other extreme, we have the class of differentiable transformations that are the "most regular." The following terminology is now standard.

Definition 18.3   A differentiable transformation F, which maps a domain D one-to-one onto a domain E and whose Jacobian is never equal to zero, is called a *diffeomorphism*.

Remark   Since the map F of D onto E is one-to-one, there is an inverse map G of E onto D. By Th. 17.3 the map G is also differentiable. One sometimes uses the property that F and F^{-1} are differentiable as the definition of a diffeomorphism.

We conclude our discussion of differentiable transformations by studying a special class of diffeomorphisms that arises in several important connections—the "angle-preserving" transformations. We first make precise the notion of angle between two curves.

Definition 18.4 If two regular curves pass through a point (x_0, y_0), then the angle between the curves at (x_0, y_0) is defined to be the angle between their tangent vectors at the point.

Example 18.4

Find the angle between the curves

$$C_1 : u = t^2 - 1, \quad v = 2t, \qquad C_2 : u = 4t - 4, \quad v = 4t - 2t^2,$$

at the point $(0, 2)$.

We first find on each curve the value of the parameter t that corresponds to the point $(0, 2)$. In both cases it is $t = 1$. For C_1 we have

$$u'(t) = 2t, \qquad v'(t) = 2$$

and for $t = 1$ the tangent vector is

$$\langle u'(1), v'(1) \rangle = \langle 2, 2 \rangle.$$

For C_2 we have

$$u'(t) = 4, \qquad v'(t) = 4 - 4t,$$

and at $t = 1$ the tangent vector is $\langle 4, 0 \rangle$. If θ is the angle between these tangent vectors, then

$$\cos \theta = \frac{\langle 2, 2 \rangle \cdot \langle 4, 0 \rangle}{(|\langle 2, 2 \rangle|)(|\langle 4, 0 \rangle|)} = \frac{8}{(2\sqrt{2})(4)} = \frac{1}{\sqrt{2}}; \qquad \theta = \tfrac{1}{4}\pi.$$

Definition 18.5 A *conformal mapping* is a diffeomorphism that preserves all angles between curves.[2]

Remark To be precise, if F is a diffeomorphism that maps a domain D onto a domain E, then every regular curve in D is mapped onto a regular curve in E (see Ex. 16.16). To be a conformal mapping the following must

---

[2] The term "conformal mapping" is sometimes used to include more general mappings whose Jacobian is different from zero and which preserve angles, even if the mappings are not one-to-one in the whole domain. An example is the mapping $u = x^2 - y^2$, $v = 2xy$, where the domain D is the whole plane minus the origin. Note that Th. 18.2 is valid for such mappings.

hold. Let (x_0, y_0) be any point in D, and (u_0, v_0) its image point in E. For every pair of regular curves through (x_0, y_0), the angle between these curves at (x_0, y_0) must equal the angle between the image curves at (u_0, v_0).

Theorem 18.2 *A diffeomorphism* $F:(x, y) \rightarrow (u, v)$ *is a conformal mapping if and only if one of the following pairs of equations holds throughout* D,

$$(18.3) \qquad u_x = v_y \qquad and \qquad u_y = -v_x$$

or

$$(18.4) \qquad u_x = -v_y \qquad and \qquad u_y = v_x.$$

PROOF.   Let (x_0, y_0) be any point of D. The differential $dF_{(x_0, y_0)}$ defines the correspondence between tangent vectors to curves at (x_0, y_0) and their images. Thus angles between curves at (x_0, y_0) are preserved under F if and only if $dF_{(x_0, y_0)}$ preserves angles. But $dF_{(x_0, y_0)}$ is a linear transformation with matrix

$$\begin{pmatrix} a & b \\ c & d \end{pmatrix} = \begin{pmatrix} u_x(x_0, y_0) & u_y(x_0, y_0) \\ v_x(x_0, y_0) & v_y(x_0, y_0) \end{pmatrix}.$$

By Th. 15.3, $dF_{(x_0, y_0)}$ preserves angles if and only if it is a similarity transformation, and by the Corollary to Th. 15.2 the algebraic condition for this is precisely that either $a = d$ and $b = -c$, or else $a = -d$, $b = c$. Thus F is conformal if and only if at each point (x_0, y_0) either (18.3) or (18.4) holds. Note that if (18.3) holds, the Jacobian is positive, $u_x v_y - u_y v_x = u_x^2 + v_x^2$, while for (18.4) it is negative. Since the Jacobian of a diffeomorphism is never zero, and since the Jacobian is a continuous function in D, it must be either everywhere positive or everywhere negative (see Ex. 5.14). Thus either (18.3) or (18.4) must hold throughout D.   ◆

Remark   The term *conformal* is sometimes reserved for transformations satisfying (18.3) that preserve orientation (since the Jacobian is positive), while a mapping satisfying (18.4), which necessarily reverses orientation, is called *anticonformal*. This is especially true in the study of functions of a complex variable, where one is led to Eqs. (18.3). In that context these equations are called the *Cauchy-Riemann equations*. It is clear from our derivation that these equations are no more nor less than the algebraic condition on the Jacobian matrix at each point that the differential dF be a similarity transformation preserving orientation.

Example 18.5

Let F be the transformation

$$u = x^2 - y^2, \qquad v = 2xy,$$

which we examined in detail at the end of Sect. 13.[3] From the description we gave there it is clear that F maps the first quadrant one-to-one onto the upper half-plane $v > 0$. The Jacobian matrix is

$$J_F = \begin{pmatrix} 2x & -2y \\ 2y & 2x \end{pmatrix}$$

so that Eqs. (18.3) are satisfied, and the Jacobian is equal to $4(x^2 + y^2)$, which is different from zero in the first quadrant. Thus F is a conformal mapping of the first quadrant onto the upper half-plane. (The same would be true for any quadrant, or more generally, any domain D in the x, y plane in which the map F is one-to-one.) Note in particular that the curves $x^2 - y^2 = c$ and $2xy = d$ indicated in Fig. 13.12 must intersect at right angles, since they map onto the perpendicular lines $u = c, v = d$. This gives an example of "orthogonal families of curves." Namely, the set of curves $x^2 - y^2 = c$, for different values of c, forms one family of curves, and the curves $2xy = d$, as d varies, form a second family (see Fig. 18.2). What we have seen is that each curve of the first family intersects each curve of the second family at right angles. This could also be verified by noting that these families are level curves of the functions $x^2 - y^2$ and $2xy$, and the gradients of these functions are the vectors $\langle 2x, -2y \rangle$ and $\langle 2y, 2x \rangle$, respectively, which are perpendicular vectors at each point.

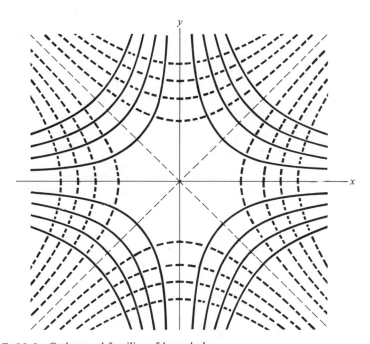

FIGURE 18.2 Orthogonal families of hyperbolas

---

[3] See also Fig. 16.5 and the related discussion of the behavior of this transformation near the point $(1, 1)$.

Exercises

18.1 Test each of the following pairs of functions for functional dependence. (The domain D is the whole plane except in parts a, e, and g, where D is taken to be the first quadrant.) In case of functional dependence, give an explicit relationship of the form $\varphi(u, v) = 0$.

a. $\begin{cases} u = (x + y)/x \\ v = (x + y)/y \end{cases}$
f. $\begin{cases} u = \sin(x^2 + y^2) \\ v = \sin(x + y) \end{cases}$

b. $\begin{cases} u = 6x + 3y \\ v = y + 2x \end{cases}$
g. $\begin{cases} u = \log(x^2 + 2xy + y^2) \\ v = \log(x + y) \end{cases}$

c. $\begin{cases} u = x^2 - y^2 \\ v = x + y \end{cases}$
h. $\begin{cases} u = \cosh(x^2 - y^2) \\ v = \sinh(x^2 - y^2) \end{cases}$

d. $\begin{cases} u = (e^x + 2e^y)/e^y \\ v = x^2 + y^2 - 2xy - 1 \end{cases}$
i. $\begin{cases} u = \cos(x + y) \\ v = \cos(x - y) \end{cases}$

e. $\begin{cases} u = \log(x^2 + y) - \log y \\ v = 1 - (y/x^2) \end{cases}$
*j. $\begin{cases} u = \arctan x + \arctan y \\ v = (x + y)/(1 - xy) \end{cases}$

18.2 Which pairs of functions in Ex. 18.1 are linearly dependent?

18.3 *a.* Show that if one function of a pair is identically zero, then the pair of functions is linearly dependent.

b. Show that if a pair of functions is linearly dependent, and if neither function is identically zero, then each function equals a constant times the other function.

c. Using parts a and b, show that $f(x, y)$ and $g(x, y)$ are linearly dependent if and only if either $f(x, y) \equiv cg(x, y)$ or $g(x, y) \equiv cf(x, y)$ for some constant c (which may be zero).

18.4 *a.* Let $h(t)$ be a continuously differentiable function of one variable. Show that if $g(x, y) \equiv h(f(x, y))$, then $f(x, y)$ and $g(x, y)$ are functionally dependent.

*b. Let $f(x, y)$ and $g(x, y)$ be functionally dependent in a domain D. Let (x_0, y_0) be any point of D. Show that there exists some $r > 0$ and there exists a continuously differentiable function $h(t)$ such that for

$$(x - x_0)^2 + (y - y_0)^2 < r^2,$$

either

$$g(x, y) \equiv h(f(x, y)) \quad \text{or} \quad f(x, y) \equiv h(g(x, y)).$$

18.5 Let

$$f(x, y) = (x^2 + y^2)^2, \qquad g(x, y) = (x^2 + y^2)^3.$$

a. Show that $\partial(f, g)/\partial(x, y) \equiv 0$.

b. Show that although $[f(x, y)]^3 \equiv [g(x, y)]^2$, this does not imply that $f(x, y)$ and $g(x, y)$ are functionally dependent in the whole plane according to Def. 18.2, since the function $\varphi(u, v) = u^3 - v^2$ has zero gradient at the origin.

(*Note*: the definition of functional dependence is sometimes enlarged to allow dependence of this sort, but then a different proof is needed for Th. 18.1. Sometimes functional dependence is defined by the vanishing of the Jacobian, but that does not imply a functional relationship in the sense we envisage intuitively. See, however, Ex. 18.27, and the remark immediately preceding it.)

18.6 a. Let $F(t)$, $G(t)$ be continuously differentiable functions of one variable, and let $f(x, y) = F(h(x, y))$, $g(x, y) = G(h(x, y))$, where $h(x, y) \in \mathscr{C}^1$ in some domain D. Show that $\partial(f, g)/\partial(x, y) \equiv 0$ in D.

b. Give a geometric description of the mapping $u = f(x, y)$, $v = g(x, y)$ in part a.

18.7 Find the angle between each of the following pairs of curves at the point indicated.

a. $C_1: x = t,\quad y = \sqrt{3}\,t;\qquad C_2: x = t,\quad y = \sqrt{3}/t;\qquad$ at $(1, \sqrt{3})$
b. $C_1: x = e^t,\quad y = 2 + 4\sin t;\qquad C_2: x = t^4,\quad y = t + 3;\qquad$ at $(1, 2)$
c. $C_1: x = \log(1 + t^2),\quad y = e^t + e^{-t};$
   $C_2: x = t^3 - t,\quad y = t^2 + t;\qquad$ at $(0, 2)$

18.8 Decide whether each of the following transformations is a diffeomorphism or not in the domain indicated. For those that are diffeomorphisms, use Th. 18.2 to test for conformality.

a. $\begin{cases} u = x + y \\ v = x - y \end{cases};\quad D: \text{all } x, y$

b. $\begin{cases} u = x + y \\ v = y \end{cases};\quad D: \text{all } x, y$

c. $\begin{cases} u = x \\ v = y^3 \end{cases};\quad D: y > 0$

d. $\begin{cases} u = x - 1 \\ v = (y - 1)^3 \end{cases};\quad D: y > 0$

e. $\begin{cases} u = e^x \cos y \\ v = e^x \sin y \end{cases};\quad D: y > 0$

f. $\begin{cases} u = e^x \cos y \\ v = e^x \sin y \end{cases};\quad D: 0 < y < \pi$

g. $\begin{cases} u = x^2 - y^2 - x + y + 2 \\ v = 2xy - x - y \end{cases};\quad D: y > 0$

*h. $\begin{cases} u = x^2 - y^2 - x + y + 2 \\ v = 2xy - x - y \end{cases};\quad D: x > 1, y > 1$

*i. $\begin{cases} u = \dfrac{x}{x^2 + y^2} \\ v = \dfrac{y}{x^2 + y^2} \end{cases};\quad D: (x, y) \neq (0, 0)$

18.9 Consider the transformations

$$F_1 : \begin{cases} u = \tan\left(\tfrac{1}{2}\pi x\right) \\ v = \tan\left(\tfrac{1}{2}\pi y\right) \end{cases}; \quad |x| < 1, \quad |y| < 1,$$

$$F_2 : \begin{cases} z = u/\sqrt{1 + u^2 + v^2} \\ w = v/\sqrt{1 + u^2 + v^2} \end{cases}; \quad \text{all } u, v.$$

a. Show that F_1 is a diffeomorphism of the square $|x| < 1, |y| < 1$ onto the whole plane.

b. Show that F_2 is a diffeomorphism of the whole plane onto the disk $u^2 + v^2 < 1$.

c. Show that $F_2 \circ F_1$ is a diffeomorphism of the square onto the disk.

***d.** Is $F_2 \circ F_1$ conformal?

18.10 Let F_1 be a diffeomorphism of a domain D onto a domain E. Let F_2 be a diffeomorphism defined on E.

a. Show that $F_2 \circ F_1$ is a diffeomorphism.

b. Show that if F_1 and F_2 are both conformal, then so is $F_2 \circ F_1$.

***c.** Do part b using Th. 18.2.

18.11 Let F be a diffeomorphism.

a. Show that F^{-1} is a diffeomorphism.

b. Show that if F is conformal then F^{-1} is conformal.

c. Do part b using Th. 18.2.

18.12 Show that a linear transformation is a conformal mapping if and only if it is a similarity transformation.

18.13 Show that if a conformal mapping of a domain D is given by $u = f(x, y)$, $v = g(x, y)$, then $f(x, y)$ and $g(x, y)$ are harmonic functions in D.

18.14 Show that a diffeomorphism given by $u = f(x, y)$, $v = g(x, y)$, where $f(x, y)$ and $g(x, y)$ are harmonic functions, need not be conformal. (*Hint*: consider linear transformations.)

18.15 **a.** Show that if a diffeomorphism F of a domain D is a conformal mapping, then at each point (x_0, y_0) of D, the vectors

$$\mathbf{w}_1 = \langle u_x(x_0, y_0), v_x(x_0, y_0) \rangle = dF_{(x_0, y_0)}(\langle 1, 0 \rangle)$$

and

$$\mathbf{w}_2 = \langle u_y(x_0, y_0), v_y(x_0, y_0) \rangle = dF_{(x_0, y_0)}(\langle 0, 1 \rangle)$$

satisfy

$$|\mathbf{w}_1| = |\mathbf{w}_2| \neq 0, \qquad \mathbf{w}_1 \perp \mathbf{w}_2.$$

b. Interpret the condition of part a geometrically in terms of the action of F on horizontal and vertical lines.

c. Show that if a diffeomorphism F satisfies the condition of part a at every point of a domain D, then F is a conformal mapping. (*Hint*: use the Corollary to Th. 15.2.)

18.16 Let F be a diffeomorphism defined by $u = f(x, y)$, $v = g(x, y)$ in a domain D.

a. Show that if F is conformal, then at each point of D,

$$\nabla_{\alpha + \pi/2} g = \pm \nabla_\alpha f.$$

b. Show that if F is conformal, then

$$\frac{\partial(f, g)}{\partial(x, y)} = \pm |\nabla f|^2$$

at each point of D.

c. Show that if F is conformal, then f and g satisfy the conditions

$$|\nabla f| = |\nabla g| \neq 0, \qquad \nabla f \perp \nabla g$$

at every point of D.

*d. Show that if f and g satisfy the conditions of part c at each point of D, then F is conformal. (*Hint*: adapt the reasoning used in the proof of Th. 15.2 to get from Eqs. (15.16) to (15.17).)

18.17 Show that if F is a diffeomorphism of a domain D and if the Jacobian of F is positive at each point of D, then F is a conformal mapping if and only if at each point (x_0, y_0) of D the differential $dF_{(x_0, y_0)}$ consists of a dilation by some constant $\lambda > 0$ followed by a rotation. (Note that the value of λ and the amount of rotation will, in general, vary from point to point.)

18.18 Show that if F is a diffeomorphism of a domain D and if the Jacobian of F is negative throughout D, then the transformation $H = G \circ F$, where G is a reflection in the horizontal axis, is a diffeomorphism with positive Jacobian.

Note: it follows that every conformal mapping is either of the form described in Ex. 18.17, or else consists of a transformation of that form followed by a reflection. This characterization of conformal mappings, in terms of the action of dF at each point, provides probably the deepest insight into their nature. Thus, Fig. 16.5, which illustrates a special case, gives a valid picture of the general situation.

*18.19 Let F be a diffeomorphism defined by $u = f(x, y)$, $v = g(x, y)$. Let $(u_0, v_0) = F(x_0, y_0)$. Let C be a regular curve passing through (x_0, y_0) and let Γ be the image of C under F. Let s be the parameter of arc length on C and let σ be the parameter of arc length on Γ. If C is written in the form $x(s)$, $y(s)$, using the arc length s as parameter, then Γ takes the form:

$$u = f(x(s), y(s)), \qquad v = g(x(s), y(s)).$$

Let

$$\langle X, Y \rangle = \langle x'(s_0), y'(s_0) \rangle, \qquad \langle U, V \rangle = \langle u'(s_0), v'(s_0) \rangle,$$

where $(x(s_0), y(s_0)) = (x_0, y_0)$.

a. Show that $|\langle X, Y \rangle| = 1$, $|\langle U, V \rangle| = d\sigma/ds$, evaluated at s_0.

b. Show that $\langle U, V \rangle = dF_{(x_0, y_0)}(\langle X, Y \rangle)$.

c. Using parts a and b, show that $d\sigma/ds$ depends only on the tangent direction of the curve C at (x_0, y_0).

(*Note*: since $d\sigma/ds$ is equal to the rate of change of arc length along Γ with respect to arc length along C, and since this "ratio of arc lengths" depends only on the tangent vector $\langle X, Y \rangle$ to C, we may call it the *dilation of the mapping F at the point (x_0, y_0) in the direction of $\langle X, Y \rangle$*.)

    d. Show that $U^2 + V^2$ is equal to a quadratic form in X and Y, and find the coefficients.

    e. Let λ_1 and λ_2 be the maximum and minimum of the dilation of F at the point (x_0, y_0) as $\langle X, Y \rangle$ ranges over all unit vectors (that is, λ_1 and λ_2 are the maximum and minimum of $d\sigma/ds$ as C ranges over all regular curves through (x_0, y_0)). Show that λ_1^2 and λ_2^2 are the maximum and minimum of the quadratic form in part d.

    f. Show that the quantities λ_1 and λ_2 of part e satisfy

$$\lambda_1 \lambda_2 = \left| \frac{\partial(u, v)}{\partial(x, y)} \right|_{(x_0, y_0)}.$$

    g. Show that $\lambda_1 = \lambda_2$ if and only if $dF_{(x_0, y_0)}$ is a similarity transformation.

    h. Show that F is a conformal mapping if and only if the dilation of F at each point (x_0, y_0) is independent of the direction (that is, the quantity $d\sigma/ds$ is the same for all curves C passing through (x_0, y_0)).

18.20   Note that the mapping $u = x^2 - y^2$, $v = 2xy$ satisfies the Cauchy-Riemann equations (18.3) at every point of the plane. However, according to our description of this mapping in Sect. 13, two lines through the origin map onto two lines through the origin in such a way that the angle between the lines is doubled. Explain why this does not contradict Th. 18.2.

18.21   Let D be the quarter disk $x > 0$, $y > 0$, $x^2 + y^2 < 1$. Let E be the half disk $v < 0$, $u^2 + v^2 < 1$.

    a. Show that the equations $u = x^2 - y^2$, $v = 2xy$ define a conformal mapping of D onto E.

    b. Note that the right angle at the origin in D maps onto a straight angle at the origin in E. Explain why this does not contradict part a.

18.22   The equation $v^2 = 4c^2(c^2 - u)$ represents, for an arbitrary constant $c \neq 0$, a parabola with focus at the origin and opening to the left. As c runs through all positive real numbers we obtain a family of confocal parabolas. Similarly, the equation $v^2 = 4d^2(d^2 + u)$ describes, for arbitrary $d \neq 0$, a parabola with focus at the origin opening to the right. Show that every such parabola meets every parabola of the first family at right angles. Illustrate with a sketch. (*Hint*: consider the images of horizontal and vertical lines under the transformation $u = x^2 - y^2$, $v = 2xy$.)

18.23   *a*. Show that a diffeomorphism $F: (x, y) \rightarrow (u, v)$ has the property that the images of each pair of lines $x = c$, $y = d$ intersect at right angles if and only if $u_x u_y + v_x v_y = 0$.

    b. Give examples of transformations that satisfy the condition in part a and are not conformal.

c. Let $F = (x, y) \rightarrow (u, v)$ be a transformation satisfying the condition in part a, and let $C : (u, v) \rightarrow (z, w)$ be conformal. Show that $G \circ F$ satisfies the condition in part a.

18.24 Show that a diffeomorphism is a conformal mapping if and only if every pair of perpendicular lines map onto curves that intersect at right angles. (*Hint*: apply Th. 15.3.)

18.25 Show that the curves $x^3 - 3xy^2 = c$, where $c \neq 0$, and $3x^2y - y^3 = d$, $d \neq 0$, intersect at right angles.

18.26 Show that the two families of curves

$$x^2 + y^2 - cx = 0, \qquad c \neq 0$$

and

$$x^2 + y^2 - dy = 0, \qquad d \neq 0$$

are orthogonal families; describe the curves of each family and sketch several of them. (*Hint*: consider the mapping: $u = x/(x^2 + y^2)$, $v = y/(x^2 + y^2)$.)

Remark The following result may be considered as a partial converse of Th. 18.1: *Suppose* $f(x, y)$, $g(x, y) \in \mathscr{C}^1$ *in a domain* D *and* $\partial(f, g)/\partial(x, y) \equiv 0$ *in* D. *If* $f_y(x_0, y_0) \neq 0$, *then there is a function* $F(t)$ *of one variable such that for all* (x, y) *in some disk* $(x - x_0)^2 + (y - y_0) < r^2$, $g(x, y) = F(f(x, y))$. For a proof, see [36], Sect. 12.3. Exercise 18.27 is intended to give some insight into why such a result might be expected to hold.

*18.27 Let F be defined in a domain D by $u = f(x, y)$, $v = g(x, y)$, where $\partial(u, v)/\partial(x, y) \equiv 0$.

a. Show that at any point where $u_y \neq 0$, there exists a number λ such that $\nabla g = \lambda \nabla f$.

b. Show that the image under F of a curve $C : x(t)$, $y(t)$, is the curve $\Gamma : u = f_C(t)$, $v = g_C(t)$, and that at a point where $u_y \neq 0$, $g_C'(t) = \lambda f_C'(t)$, where λ is the same as in part a.

c. Show that if C is a level curve $f(x, y) = u_0$, and if $u_y \neq 0$ at each point of C, then $g_C'(t) \equiv 0$ and $g(x, y)$ is constant along C.

d. Show that if C is a segment of a vertical line $x = x_0$ and if $u_y \neq 0$ along C, then the curve $\Gamma : u = f(x_0, y)$, $v = g(x_0, y)$ can be expressed in non-parametric form as $v = F(u)$. (*Hint*: the equation $u = f(x_0, y)$ can be solved for y in terms of u.)

Remark (continued) The remainder of the argument may be pictured intuitively as follows. If $u_y(x_0, y_0) \neq 0$, then by continuity, $u_y \neq 0$ in some disk $(x - x_0)^2 + (y - y_0)^2 < r^2$. Let $u_0 = f(x_0, y_0)$, $v_0 = g(x_0, y_0)$. By part d, the segment $x = x_0$, $y_0 - r \leq y \leq y_0 + r$ maps onto a curve $v = F(u)$ through (u_0, v_0). Each point (u_1, v_1) of this curve is the image of a point (x_1, y_1) in the disk, and since $u_y(x_1, y_1) \neq 0$ we may apply the implicit function theorem to deduce that the level curve $f(x, y) = u_1$ through (x_1, y_1) is a regular curve C. By part c, $g(x, y) \equiv v_1$ on C. Thus the curve C is the inverse image of the single point

(u_1, v_1), and the inverse image of each point on $v = F(u)$ is a whole curve. This gives a family of curves each of which maps onto a point of $v = F(u)$. (Note the analogy with the behavior of linear transformations whose determinant is zero, as described in Ex. 14.5c.) All that remains to complete the proof is to show that this family of curves fills out some disk about the point (x_0, y_0), and that can be done by slightly more refined arguments.

19 Vector fields

At the beginning of this chapter we noted that one way in which pairs of functions arise is as the components of a vector field. We now investigate some of the basic concepts and problems that occur in this connection.

Definition 19.1 If $u(x, y) \in \mathscr{C}^k$ and $v(x, y) \in \mathscr{C}^k$ in a domain D, then the mapping that assigns to each point (x, y) in D the vector $\langle u(x, y), v(x, y) \rangle$ is called a $\mathscr{C}^k$ *vector field in D*.

A *differentiable vector field* is a vector field that is $\mathscr{C}^k$ for some $k \geq 1$.

We gave several examples in Sect. 13, including the vector fields obtained by forming the gradient of a given function.

Definition 19.2 A vector field $\langle u(x, y), v(x, y) \rangle$ is called *conservative*, if there exists a function $f(x, y)$ such that $\langle u(x, y), v(x, y) \rangle = \nabla f(x, y)$; the function $f(x, y)$ is called a *potential function* for the field.

Remark Much of the terminology in the theory of vector fields is of physical origin. We do not try to explain each expression as it is introduced, but the physical interpretation will be discussed at an appropriate point.

By definition, if $f(x, y)$ is a potential function, we have

(19.1) $$u = f_x, \qquad v = f_y.$$

In particular, for $\langle u, v \rangle$ to be a differentiable vector field, the functions $u = f_x$ and $v = f_y$ must be continuously differentiable, and hence $f(x, y)$ must be in $\mathscr{C}^2$.

A natural question is the following: "how does one tell whether a given vector field is conservative; that is, equal to the gradient of some function?" The complete answer to this question turns out to have many fascinating ramifications, which we shall explore in the following chapter. There is, however, a partial answer that is easily given.

Lemma 19.1   *If a differentiable vector field $\langle u(x, y), v(x, y)\rangle$ has a potential function in a domain D, then*

(19.2) $$v_x - u_y = 0$$

throughout D.

PROOF.   By assumption, there exists a function $f(x, y) \in \mathscr{C}^2$ in D satisfying Eq. (19.1). By the equality of mixed derivatives (Th. 11.1) we have $(f_x)_y = (f_y)_x$, which in view of Eq. (19.1) yields Eq. (19.2).   ♦

We should be clear as to the precise value of this lemma. It tells us that if a given vector field does *not* satisfy Eq. (19.2), then it *cannot* have a potential function. However, if Eq. (19.2) is satisfied, Lemma 19.1 does not allow us to assert that a potential function exists; furthermore if a potential function does exist, there remains the problem of how to find it. These are matters that we shall investigate in the next chapter.

There is one more elementary observation to be made here.

Lemma 19.2   *If a vector field in a domain D has a potential function, then this function is determined up to an additive constant.*

PROOF.   If f and g are potential functions of the same vector field, then they both have the same gradient, and by the Corollary to Th. 7.2, they differ by a constant.   ♦

Example 19.1

Let

$$u = y, \qquad v = x.$$

To see if the vector field $\langle y, x\rangle$ is conservative, we check Eq. 19.2. We have $u_y = 1$, $v_x = 1$. Hence, Eq. 19.2 holds, and a potential function *may* exist. If it does, we have $f_x = y$, $f_y = x$. But by inspection we can exhibit such a function: $f(x, y) = xy$.

Example 19.2

$$u = -y, \qquad v = x.$$

In this case $u_y = -1$, $v_x = 1$. Equation (19.2) does not hold, and hence a potential function cannot exist.

Example 19.3

$$u = -y/(x^2 + y^2), \qquad v = x/(x^2 + y^2).$$

Here $\langle u, v\rangle$ is a differentiable vector field in the domain D consisting of the whole plane except for the origin. We find that

$$u_y = v_x = (y^2 - x^2)/(x^2 + y^2)^2.$$

Thus a potential function may exist. In this case it is not as easy as in Example 19.1 to construct a potential function explicitly. We shall return to this example in Sect. 22, where we shall show (in the discussion preceding Th. 22.2) that in fact there does not exist a potential function in D for this vector field.

It is a useful observation that when the vanishing of an expression, as for example in Eq. (19.2), is of fundamental importance, then a study of that expression itself is often fruitful. We shall in fact again meet the quantity

$$(19.3) \qquad\qquad v_x - u_y,$$

in connection with Green's theorem (Sects. 25–27). At this point, we introduce a similar expression, which is also of importance.

Definition 19.3 If $\langle u(x, y), v(x, y) \rangle$ is a differentiable vector field, then the quantity

$$(19.4) \qquad\qquad u_x + v_y$$

is called the *divergence* of the vector field.

Remark It would be natural to consider expression (19.4) if for no other reason than that it represents the *trace* of the Jacobian matrix

$$\begin{pmatrix} u_x & u_y \\ v_x & v_y \end{pmatrix}$$

of the pair of functions $u(x, y)$, $v(x, y)$. We have seen that the trace of a matrix plays a basic role in the theory of quadratic forms, and in fact the trace and the determinant are the two most fundamental quantities associated with a given matrix.

We now introduce two important classes of vector fields.

Definition 19.4 A differentiable vector field $\langle u(x, y), v(x, y) \rangle$ in a domain D is called *solenoidal* or *divergence-free* if it satisfies

$$(19.5) \qquad\qquad u_x + v_y = 0$$

throughout D.

Definition 19.5 A differentiable vector field that satisfies both Eqs. (19.2) and (19.5) is called a *harmonic vector field*.

The reason for this terminology is the following.

Lemma 19.3 *If $f(x, y)$ is the potential function of a differentiable vector field $\langle u(x, y), v(x, y) \rangle$, then $\langle u(x, y), v(x, y) \rangle$ is a harmonic vector field if and only if $f(x, y)$ is a harmonic function.*

PROOF. Since $u = f_x$ and $v = f_y$, Eq. (19.2) automatically holds, and since $u_x + v_y = f_{xx} + f_{yy}$, Eq. (19.5) is equivalent to f harmonic. ◆

In case it may have passed unobserved, we call attention to the fact that the pair of equations (19.2) and (19.5) are precisely the same as the Eqs. (18.4), which arose in connection with conformal mapping. This is a striking example of what is often called the "unity of mathematics." The investigation of two apparently totally unconnected subjects, such as a geometric question about angle-preserving transformations, and a class of vector fields that arise in physics, may lead to precisely the same mathematical formulation—in this case, the same pair of equations that must be satisfied.

We conclude this section with a brief description of some of the physical problems that led to the consideration of the above types of vector fields. Historically, in the theory of vector fields the physical interpretation and the mathematical treatment have gone hand in hand, each aiding in the development of the other. Most of the early theorems in the subject were inspired by physical considerations. The most famous example of the reverse development is Maxwell's theory of electromagnetism, in which mathematical reasoning led Maxwell to predict the existence of electromagnetic waves, including the many familiar types since discovered—radio, radar, cosmic rays, etc. We should emphasize, however, that no knowledge of physics is required to understand the mathematical material presented in this book. To the extent that one or another physical interpretation aids the intuition, it should certainly be utilized; wherever it does not, it can be ignored.

Heat Flow

A simple example of a function of two variables is provided by the temperature at each point on the bottom of a frying pan, which is being heated by a gas burner. A somewhat idealized version of the same situation is the following. Suppose that constant heat sources are applied at different points along the circumference of a metal disk, and enough time is allowed to elapse so that the temperature distribution attains an equilibrium. The temperature then becomes a function of position, and the gradient of the temperature is a vector field which at each point indicates the direction and magnitude of heat flow. This vector field is therefore conservative. From the fact that there are no heat sources in the interior of the disk, it can be shown that the vector field is also divergence-free, and hence is a harmonic vector field. By Lemma 19.3, the temperature, which is the potential function of the field, is a harmonic function. Conversely, given a harmonic function in a disk, one may picture it as defining a temperature distribution, and its gradient defines the vector field of heat flow.

We may mention that this physical interpretation accounts very nicely for the property of harmonic functions known as the "maximum principle" (see Example 11.6). Namely, if a harmonic function, pictured as the potential function of a steady heat flow, had a local maximum at a point, then the heat would be flowing away from that point, and the temperature would not have reached an equilibrium.

Fluid Flow

Consider the flow of a liquid or gas enclosed in a tube. The tube may have variable diameter, and the velocity of flow may vary from point to point, but we assume that the velocity at each point does not vary with time. Let us further assume that the flow can be represented by a lateral section of the tube (Fig. 19.1). The physical situation may then be described by a *velocity field*; that is, a vector field such that the magnitude and direction of the vector at each point represents the speed and direction of flow at that point. We shall return frequently to this physical model of a vector field, and we shall see that various quantities, such as the divergence, are most easily visualized in terms of such a flow.

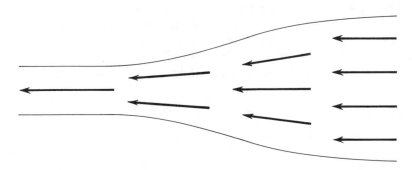

FIGURE 19.1   Velocity field of flow in a tube

Force Fields

Assume, as an approximation, that the sun, planets, and other bodies in the solar system lie in a plane. If a particle of unit mass is placed at some point in this plane, it is attracted to each of these objects in varying degree, and the total effect is that of a force exerted on the particle, imparting to it a given acceleration in a given direction. The magnitude and direction of this force determines a vector, and the set of all vectors obtained in this way (by placing the particle at arbitrary points) defines a vector field called a *gravitational force field*. Theoretically, the force field determines the path any particle follows if it is placed at an arbitrary point either at rest or with an arbitrary initial velocity. It is a consequence of Newton's laws that the gravitational force field (at all points outside the solar bodies) is a harmonic vector field.

Similarly a distribution of electrostatic charges determines a force field, which in turn determines the motion of a charged particle placed in the field. Perhaps the simplest force field to visualize is a magnetic field, whose direction at each point is found by placing a magnetic compass at the point and observing the direction of the needle. The term "solenoidal" for a vector field originated in the properties of a magnetic field inside a solenoid.

For further discussion of these matters from the physical point of view, we refer to references [9], [33].

Exercises

19.1   For each of the following vector fields $\langle u(x, y), v(x, y) \rangle$, compute the divergence, and decide which vector fields are solenoidal.

    a. $\langle a, b \rangle$,   *a* and *b* constants

    b. $\langle ax, by \rangle$,   *a* and *b* constants

    c. $\langle \varphi(x), \psi(y) \rangle$, $\varphi(x)$ and $\psi(y)$ continuously differentiable

    d. $\langle \varphi(y), \psi(x) \rangle$, $\varphi(y)$ and $\psi(x)$ continuously differentiable

    e. $\langle 2xy, x^2 \rangle$

    f. $\langle -x^2 + y^2, 2xy \rangle$

    g. $\langle 2yx^2, 3x^2y \rangle$

    h. $\langle e^{x+y}, e^{x-y} \rangle$

    i. $\langle \sin x \cos y, \cos x \sin y \rangle$

    j. $\langle y/x, \log x \rangle$

    k. $\langle e^x \cos y, e^x \sin y \rangle$

    l. $\langle e^x \sin y, e^x \cos y \rangle$

19.2   For each of the vector fields in Ex. 19.1, decide whether or not it is conservative, and if so, find a potential function.

19.3   Which of the vector fields in Ex. 19.1 are harmonic vector fields?

19.4   Which of the following vector fields are harmonic vector fields?

    a. $\langle x + y, x - y \rangle$

    b. $\langle x^2 + y^2, 2xy \rangle$

    c. $\langle x^3 - 3x^2 + 3x - 3xy^2 + 3y^2 - 1, y^3 - 3yx^2 + 6yx - 3y \rangle$

    d. $\langle -y/(x^2 + y^2), x/(x^2 + y^2) \rangle$

    e. $\langle \log (x^2 + y^2), 2 \arctan y/x \rangle$

19.5   Show that if a vector field $\langle u, v \rangle$ is conservative, then the vector field $\langle \tilde{u}, \tilde{v} \rangle = \langle -v, u \rangle$ is solenoidal.

19.6   Let $\mathbf{w} = \langle u, v \rangle$ be a $\mathscr{C}^2$ harmonic vector field.

    a. Show that *u* and *v* are harmonic functions.

    b. Show that for any constant vector **d**, the projection of **w** in the direction of **d** is a harmonic function.

19.7   Given $f(x, y) \in \mathscr{C}^2$ in a domain D, consider the vector field $\mathbf{w} = \nabla f$. Express the divergence of the vector field $\mathbf{w}$ in terms of f.

19.8   A revealing way to think of the gradient operator ∇ is as a "vector valued operator," which assigns to each function f, the vector field ∇f. Symbolically, we may write

$$\nabla = \left\langle \frac{\partial}{\partial x}, \frac{\partial}{\partial y} \right\rangle$$

meaning

$$\nabla f = \left\langle \frac{\partial}{\partial x}, \frac{\partial}{\partial y} \right\rangle f = \left\langle \frac{\partial f}{\partial x}, \frac{\partial f}{\partial y} \right\rangle.$$

This leads us to write

$$\nabla^2 = \nabla \cdot \nabla = \left\langle \frac{\partial}{\partial x}, \frac{\partial}{\partial y} \right\rangle \cdot \left\langle \frac{\partial}{\partial x}, \frac{\partial}{\partial y} \right\rangle = \frac{\partial^2}{\partial x^2} + \frac{\partial^2}{\partial y^2}$$

meaning

$$\nabla^2 f = f_{xx} + f_{yy}.$$

(The operator $\partial^2/\partial x^2 + \partial^2/\partial y^2$ is called the *Laplacian*, and is also frequently denoted by Δ. Thus, if $f(x, y) \in \mathscr{C}^2$, the Laplacian of f is $\Delta f = \nabla^2 f = f_{xx} + f_{yy}$.) For a vector field $\mathbf{w} = \langle u, v \rangle$, we write

$$\nabla \cdot \mathbf{w} = \left\langle \frac{\partial}{\partial x}, \frac{\partial}{\partial y} \right\rangle \cdot \langle u, v \rangle = \frac{\partial u}{\partial x} + \frac{\partial v}{\partial y},$$

which is the divergence of $\mathbf{w}$.

Show that the following equations involving this notation are valid.

  a. $\nabla^2 f = \nabla \cdot (\nabla f)$
  b. $\nabla \cdot (f\mathbf{w}) = (\nabla f) \cdot \mathbf{w} + f(\nabla \cdot \mathbf{w})$
  c. $\nabla \cdot (f\nabla g - g\nabla f) = f\nabla^2 g - g\nabla^2 f$
  d. $\nabla^2(fg) = f\nabla^2 g + g\nabla^2 f + 2(\nabla f \cdot \nabla g)$

19.9   For each of the following functions $f(x, y)$, find the vector field $\langle u, v \rangle = \nabla f$. Try to visualize the vector field by sketching a number of its vectors, where the vector $\langle u(x, y), v(x, y) \rangle$ is drawn as a displacement vector starting at the point (x, y). (See, for example, Fig. 13.3.)

  a. $f(x, y) = x$ *d.* $f(x, y) = x^2 + 2xy + y^2$
  b. $f(x, y) = 2x + y$ *e.* $f(x, y) = y/x$
  c. $f(x, y) = x^2 + y^2$ *f.* $f(x, y) = e^{-x^2 - y^2}$

19.10   If the plane is rotated about the origin with a constant angular velocity ω, then a point with coordinates (x_0, y_0) at time $t = 0$ has coordinates (x, y) at time t, where

$$x = x_0 \cos \omega t - y_0 \sin \omega t$$
$$y = x_0 \sin \omega t + y_0 \cos \omega t.$$

This motion defines a velocity field that may be described as follows: at any time t_1, the velocity vector at the point $(x(t_1), y(t_1))$ is given by $\langle u, v \rangle = \langle x'(t_1), y'(t_1) \rangle$.

a. Show that this vector field does not depend on the time, but only on the position. In fact, $\langle x'(t_1), y'(t_1)\rangle = \omega\langle -y(t_1), x(t_1)\rangle$ so that $\langle u(x, y), v(x, y)\rangle = \omega\langle -y, x\rangle$.

b. Sketch this vector field, choosing the constant $\omega = \frac{1}{4}$.

Remark: given a vector field $\langle u(x, y), v(x, y)\rangle$ in a domain D, a regular curve C is called an *integral curve* of this vector field if the tangent vector to C at each point has the same direction as the vector field at that point. If the curve is given by $x(t)$, $y(t)$, this means that if $x_0 = x(t_0)$, $y_0 = y(t_0)$, then $\langle x'(t_0), y'(t_0)\rangle$ should have the same direction as $\langle u(x_0, y_0), v(x_0, y_0)\rangle$. The problem is essentially that of solving a pair of simultaneous differential equations:

$$\frac{dx}{dt} = u(x, y)$$

$$\frac{dy}{dt} = v(x, y).$$

However, since it is only the direction of the vectors that must coincide, the problem is really that of solving the single equation

$$\frac{dy}{dx} = F(x, y),$$

where $F(x, y) = v(x, y)/u(x, y)$, which is obtained by dividing the second equation above by the first. This problem is discussed in detail in books on differential equations (see, for example, [1]). As an aid in visualization, it is useful to picture the vector field as the velocity field of a fluid flow. Then if we follow the motion of a particle carried along by the flow, it describes a path whose direction at each point is the direction of the velocity field at that point. These paths are called *stream lines* of the flow, and we see that stream lines are precisely integral curves of the vector field. A function $g(x, y)$ is called a *stream function* of the flow, if its level curves $g(x, y) = c$ are stream lines. Examples and further comments are given in Exs. 19.11–19.15.

19.11   Sketch each of the following vector fields. Show that the given function $g(x, y)$ is a stream function, and sketch some of the stream lines.

  a. $\langle 1, 0\rangle$; $g(x, y) = y$
  b. $\langle y, 0\rangle$; $g(x, y) = y$
  c. $\langle -y, x\rangle$; $g(x, y) = x^2 + y^2$

  d. $\left\langle \dfrac{x}{(x^2 + y^2)^{1/2}}, \dfrac{y}{(x^2 + y^2)^{1/2}} \right\rangle$; $g(x, y) = \dfrac{y}{x}$

19.12   Show that $g(x, y)$ is a stream function of a vector field $\mathbf{w} = \langle u, v\rangle$ if and only if $\nabla g \perp \mathbf{w}$ at each point. Verify that this is the case for each part of Ex. 19.11.

19.13   Let $\langle f(x, y), g(x, y)\rangle$ define a harmonic vector field. Show that $g(x, y)$ is a stream function for the vector field ∇f, and $f(x, y)$ is a stream function for the vector field ∇g.

*19.14 Let

$$f(x, y) = x\left(1 + \frac{1}{x^2 + y^2}\right), \qquad g(x, y) = y\left(1 - \frac{1}{x^2 + y^2}\right).$$

a. Show that the vector field $\mathbf{w} = \nabla f$ has the following properties:
   (1) it is horizontal along the x axis
   (2) it is tangent to the unit circle $x^2 + y^2 = 1$ at every point of that circle
   (3) it vanishes at the points $(1, 0)$ and $(-1, 0)$
   (4) it tends to the constant vector field $\langle 1, 0 \rangle$ as $x^2 + y^2 \to \infty$.

b. Using the properties in part a, make a very rough sketch of the vector field $\mathbf{w}$ in the domain D consisting of the entire exterior of the unit circle. See if these properties coincide with what you might visualize to be the velocity field of the following flow: a fluid is flowing uniformly in the positive x direction, with velocity vector $\langle 1, 0 \rangle$ throughout the plane; then an obstacle in the form of a circular disk is placed at the origin and the fluid is constrained to flow about the obstacle.

c. Verify that $\langle f(x, y), g(x, y) \rangle$ is a harmonic vector field, and hence, by Ex. 19.13, $g(x, y)$ is a stream function of the vector field in part a.

d. Sketch several level curves $g(x, y) = c$ (or rather, the part of the level curves lying outside $x^2 + y^2 = 1$), including the case $c = 0$, and compare the form of these curves as stream lines with the properties of the flow described in part a. (*Hint:* write the curve in the form $x^2 = h(y)$. Show that $h(y) \geq 0$ in an interval, at one end of which $x = 0$, and at the other end $x \to \infty$. Thus the curve lies in a horizontal strip and has a horizontal asymptote.)

*19.15 Let F be a diffeomorphism of a domain D onto a domain E.

a. Show that the differential dF assigns in a natural way to each vector field $\mathbf{v}$ in D a vector field $\mathbf{w}$ in E.

b. Show that if $\mathbf{v}$ and $\mathbf{w}$ are as in part a, then F maps integral curves of $\mathbf{v}$ onto integral curves of $\mathbf{w}$.

19.16 The force field set up by a charged particle placed at the origin is given by

$$\left\langle -\frac{x}{x^2 + y^2}, -\frac{y}{x^2 + y^2} \right\rangle.$$

a. Show that this field is conservative and that $-\frac{1}{2} \log (x^2 + y^2)$ is a potential function.

b. Show that this field is solenoidal.

20 Change of coordinates

Having discussed pairs of functions first from the point of view of plane mappings, and then as vector fields, we come now to our third and final interpretation, that of coordinate transformations. We begin with a few general remarks on coordinate systems.

We note first that a function is usually defined by assigning a number to each point of some set. Thus, we may assign to each position on a mountain range its altitude above sea level, or to each point on a frying pan its temperature at a given time. Only later, after each point of the set has been assigned coordinates, can the function be expressed in an explicit form, such as x^2, or $x + \cos y$. If the function represents the curvature of a railroad track, then it can be represented as a function of one variable, since we can identify each point along the track by a single coordinate, such as the distance along the track from a fixed point. To represent temperature in the atmosphere would require a function of three variables, since each point is identified by three coordinates. The two-dimensionality of the plane, translated into our terms, means that each point can be represented by a pair of coordinates. This may be done in many different ways; in order to study a particular function we may choose coordinates that are most convenient for that purpose. Thus, when we were presented with a quadratic form $ax^2 + 2bxy + cy^2$, we found it useful to make a rotation of coordinates,

$$(20.1) \qquad \begin{aligned} X &= x \cos \alpha + y \sin \alpha \\ Y &= -x \sin \alpha + y \cos \alpha, \end{aligned}$$

so that in the new coordinates the same function would be represented as $AX^2 + CY^2$.

If (x, y) are rectangular coordinates, then the new coordinates (X, Y) given by Eqs. (20.1) are rectangular also, meaning that the "coordinate curves," $X = c$, $Y = d$, are straight lines intersecting at right angles. It is often useful to consider more general systems of coordinates. All that is needed is a method of assigning a unique pair of numbers X, Y to each point, as a means of quantitatively distinguishing one point from another. If we had previously chosen some other coordinate system, which assigned a pair of numbers x, y to each point, then X and Y would each be a function of the two variables x, y; that is,

$$(20.2) \qquad \begin{aligned} X &= f(x, y) \\ Y &= g(x, y), \end{aligned}$$

where f and g are specific functions. One sometimes uses the expression "curvilinear coordinates" for the general case in which the coordinate curves $X = c$, $Y = d$ are not straight lines. The crucial property is that the curves $X = c$, $Y = d$ should intersect at a unique point, so that this point may be identified in the new coordinate system by the pair of values (c, d) (Fig. 20.1).

Conversely, given any pair of functions $f(x, y)$, $g(x, y)$ such that Eqs. (20.2) set up a one-to-one correspondence between (x, y) and (X, Y), we may use these equations to define new coordinates.

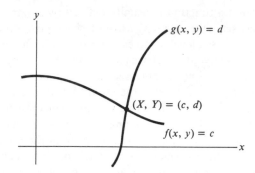

FIGURE 20.1 Curvilinear coordinates

Example 20.1

Let (x, y) be Cartesian coordinates, and let

(20.3)
$$X = \sqrt{x^2 + y^2}$$
$$Y = \tan^{-1} y/x.$$

Here the coordinate curves $X = c$, $Y = d$, are circles about the origin and rays from the origin, respectively (Fig. 20.2). The coordinates X, Y are called *polar coordinates*, and are more commonly designated as r, θ.

There are two points that should be made in connection with general coordinate systems. The first is that a particular coordinate system may not be defined in the whole plane, but only in some domain D. For example, the functions $f(x, y)$, $g(x, y)$ used in Eq. (20.2) may be defined only in a part of the plane. Thus, in the case of polar coordinates, the function $\tan^{-1}(y/x)$

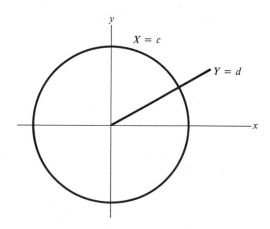

FIGURE 20.2 Polar coordinates

is not defined at the origin, and it is not defined in a single-valued way in the rest of the plane. To avoid all difficulties we must restrict polar coordinates to some domain D (such as the first quadrant, or the upper half-plane) in which it is possible to select a single-valued branch of $\tan^{-1}(y/x)$.

Secondly, in order to be able to apply calculus, the relation between coordinate systems should be sufficiently differentiable so that the derivatives of a function with respect to one coordinate system can be computed in terms of its derivatives with respect to the other.

Suppose then that $u(x, y)$ is continuously differentiable, and let new coordinates X, Y be introduced by Eq. (20.2). Then u becomes a function of X, Y, and using the chain rule, we have

(20.4)
$$u_x = u_X f_x + u_Y g_x$$
$$u_y = u_X f_y + u_Y g_y.$$

These equations may be solved to express u_X, u_Y in terms of u_x, u_y providing the determinant is not zero. But the determinant is $f_x g_y - g_x f_y$, the Jacobian of f and g with respect to x and y. Our basic requirements for change of coordinates—that there should be a one-to-one correspondence between the pairs x, y and X, Y, such that each pair is given by continuously differentiable functions of the other—are precisely the conditions stated in Th. 17.2, in the context of transformations, guaranteeing that the Jacobian cannot vanish at any point.

Example 20.2

A rotation of coordinates, given by (20.1), is of the form (20.2) with

$$f(x, y) = x \cos \alpha + y \sin \alpha, \qquad g(x, y) = -x \sin \alpha + y \cos \alpha.$$

Substituting into (20.4), we have

(20.5)
$$u_x = u_X \cos \alpha - u_Y \sin \alpha$$
$$u_y = u_X \sin \alpha + u_Y \cos \alpha.$$

Solving for u_X, u_Y, we obtain

(20.6)
$$u_X = u_x \cos \alpha + u_y \sin \alpha$$
$$u_Y = -u_x \sin \alpha + u_y \cos \alpha.$$

Similarly, we can compute derivatives of arbitrary order. For example, applying the chain rule to the function u_X, we find

$$(u_X)_x = (u_X)_x f_x + (u_X)_Y g_x = u_{XX} f_x + u_{XY} g_x,$$

and similarly for u_Y. Thus, differentiating the first equation of (20.4) with respect to x, we have

(20.7) $$u_{xx} = u_{XX} f_x^2 + 2u_{XY} f_x g_x + u_{YY} g_x^2 + u_X f_{xx} + u_Y g_{xx}.$$

In a similar fashion we can express all partial derivatives of u with respect to x, y in terms of its derivatives with respect to X, Y and the derivatives of the change-of-coordinate functions f, g. We note here the expression for the Laplacian,

$$(20.8) \quad u_{xx} + u_{yy} = u_{XX}(f_x^2 + f_y^2) + 2u_{XY}(f_x g_x + f_y g_y) + u_{YY}(g_x^2 + g_y^2)$$
$$+ u_X(f_{xx} + f_{yy}) + u_Y(g_{xx} + g_{yy}).$$

It is quite clear that if we are too capricious in our choice of new coordinates, then the expressions for most basic quantities become unmanageable. There are, however, many possible choices that greatly simplify these expressions (see, for example, Exs. 20.1, 2). If we substitute $f(x, y) = (x^2 + y^2)^{1/2}$, $g(x, y) = \tan^{-1}(y/x)$ into (20.8), then we find the expression for the Laplacian in polar coordinates

$$u_{xx} + u_{yy} = u_{XX} + \frac{1}{X^2} u_{YY} + \frac{1}{X} u_X$$

or, in the more usual notation, with $X = r$, $Y = \theta$,

$$(20.9) \quad u_{xx} + u_{yy} = u_{rr} + \frac{1}{r} u_r + \frac{1}{r^2} u_{\theta\theta}.$$

Since $(ru_r)_r = ru_{rr} + u_r = r(u_{rr} + u_r/r)$, we may rewrite Eq. (20.9) in the form

$$(20.10) \quad \Delta u = \frac{1}{r} \frac{\partial}{\partial r}\left(r \frac{\partial u}{\partial r}\right) + \frac{1}{r^2} \frac{\partial^2 u}{\partial \theta^2}.$$

Let us note just one consequence of this formula.

Lemma 20.1 *If $u(x, y)$ is a harmonic function that depends only on the distance r to the origin, then*

$$(20.11) \quad u = c \log r + d,$$

for suitable constants c and d.

PROOF.   If we express u in polar coordinates, then $u_\theta = 0$, and since u is harmonic, $\Delta u = 0$. From Eq. (20.10), we have

$$\frac{\partial}{\partial r}\left(r \frac{\partial u}{\partial r}\right) = 0$$

or

$$r \frac{\partial u}{\partial r} = c$$

$$\frac{\partial u}{\partial r} = \frac{c}{r}$$

from which Eq. (20.11) follows immediately.   ◆

We consider next what happens to the components of a vector under a change of coordinates. We restrict ourselves to the case of a rotation of coordinates, given by Eqs. (20.1). We recall that the components with respect to a given coordinate system were originally defined in terms of a displacement vector as the difference in coordinates between the beginning and the endpoint. Thus, in the original coordinate system the components would be

$$\langle u, v \rangle = \langle x_2 - x_1, y_2 - y_1 \rangle,$$

while the new components would be

$$\langle U, V \rangle = \langle X_2 - X_1, Y_2 - Y_1 \rangle$$

(Fig. 20.3). Substituting the coordinates of the beginning and endpoints in Eq. (20.1), and subtracting, we obtain

(20.12)
$$U = \quad u \cos \alpha + v \sin \alpha$$
$$V = -u \sin \alpha + v \cos \alpha.$$

Using this fact about displacement vectors as a model, we use Eqs. (20.12) to define the new components $\langle U, V \rangle$ of an arbitrary vector, after the rotation of coordinates (20.1), if the vector had components $\langle u, v \rangle$ with respect to the original coordinate system. Thus the relation (20.12) between the old and new components of a vector is precisely the same as that between the old and new coordinates of a point.

Let us consider, for example, a function f defined in the plane. It has one expression with respect to the coordinates x, y, and another with respect to

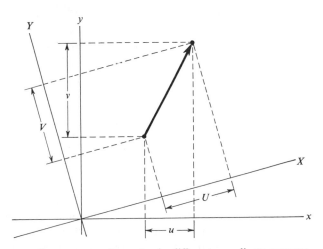

FIGURE 20.3   Components of a vector in different coordinate systems

X, Y. At each point we can consider the vector

$$(20.13) \qquad \langle u, v \rangle = \langle f_x, f_y \rangle,$$

which is the gradient of f with respect to the coordinates x, y. By Eq. (20.12), the components of this vector with respect to the new coordinates X, Y are given by

$$(20.14) \qquad \langle U, V \rangle = \langle f_x \cos \alpha + f_y \sin \alpha, -f_x \sin \alpha + f_y \cos \alpha \rangle.$$

But using Eqs. (20.6), we find that Eq. (20.14) reduces to

$$(20.15) \qquad \langle U, V \rangle = \langle f_X, f_Y \rangle.$$

This computation shows that the pairs of partial derivatives f_x, f_y and f_X, f_Y represent the components of the *same* vector with respect to two different coordinate systems. If we carry out an analogous computation with other pairs, such as $f_x, -f_y$ and $f_X, -f_Y$, or f_y, f_x and f_Y, f_X, we find that these do *not* represent the same vector. Of course, this is not fortuitous, but is a consequence of the fact that the gradient of a function may be described in an intrinsic fashion in terms of the direction and magnitude of the maximal directional derivative. Since this characterization is independent of coordinates, it follows that whatever expression we obtain for the gradient vector in one (Cartesian) coordinate system must be valid in any other.

We examine next the case of an arbitrary differentiable vector field in some domain D. Again, we assume that we have two coordinate systems x, y and X, Y, related by (20.1). Let the components of this vector field be $\langle u, v \rangle$ and $\langle U, V \rangle$ with respect to the two systems, respectively. Each of the four quantities u, v, U, V may be considered as functions of either x, y or X, Y. Differentiation of the first equation in (20.12) yields

$$U_X = u_X \cos \alpha + v_X \sin \alpha.$$

Inserting the first of Eqs. (20.6) (and the analogous equation for the function v), we obtain

$$(20.16) \qquad U_X = u_x \cos^2 \alpha + (u_y + v_x) \sin \alpha \cos \alpha + v_y \sin^2 \alpha.$$

In a similar way, we may express the other partial derivatives U_Y, V_X, and V_Y in terms of u_x, u_y, v_x, v_y. We note the result for V_Y:

$$(20.17) \qquad V_Y = u_x \sin^2 \alpha - (u_y + v_x) \sin \alpha \cos \alpha + v_y \cos^2 \alpha.$$

Adding (20.16) and (20.17), we find

$$(20.18) \qquad U_X + V_Y = u_x + v_y.$$

Equation (20.18) states that the divergence of a vector field assigns to each point a number in an "invariant" manner; that is, independent of the choice of (Cartesian) coordinates. As in the case of the gradient, it is the particular combination of derivatives that has this property, and had we considered other expressions, such as $u_x - v_y$, or $u_y + v_x$, the analog of (20.18) would not hold. Unlike the gradient, the divergence has not been described up to now in any intrinsic fashion. However, its invariance under coordinate change, proved above by a purely formal computation, leads us to suspect that there should be a coordinate-free description, from which Eq. (20.18) would follow immediately. We shall, in fact, be able to give just such a description of the divergence of a vector field in Sect. 27, but only after we have developed the basic properties of integration in several variables.

Exercises

20.1   For each of the following change-of-coordinate functions $X = f(x, y)$, $Y = g(x, y)$, use Eq. (20.8) to express $u_{xx} + u_{yy}$ in terms of the new coordinates X, Y.

$$\textbf{\textit{a.}} \quad \begin{cases} X = ax + by \\ Y = cx + dy \end{cases} \quad \textbf{\textit{b.}} \quad \begin{cases} X = x^2 - y^2 \\ Y = 2xy \end{cases} \quad \textbf{\textit{c.}} \quad \begin{cases} X = e^x \cos y \\ Y = e^x \sin y \end{cases}$$

20.2   Show that if $f(x, y)$, $g(x, y)$ define a conformal mapping and if we introduce new coordinates $X = f(x, y)$, $Y = g(x, y)$, then

$$u_{xx} + u_{yy} = \left| \frac{\partial(f, g)}{\partial(x, y)} \right| (u_{XX} + u_{YY}),$$

so that a function u is harmonic with respect to x, y if and only if it is harmonic with respect to X, Y.

20.3   Show that if the coordinates (X, Y) are obtained from (x, y) by a rotation of axes, then $u_{xx} + u_{yy} = u_{XX} + u_{YY}$.

20.4   Carry out the derivation of Eq. (20.9) from Eq. (20.8).

20.5   Show that for any integer n the function $u(x, y)$ that takes the form $r^n \cos n\theta$ in polar coordinates is a harmonic function.

20.6   **a.** Write down the form that Eqs. (20.4) take when X and Y are polar coordinates r, θ, and compute $u_x^2 + u_y^2$.
   b. Solve the equations in part a for u_r, u_θ in terms of u_x, u_y.
   c. Let $u(x, y)$ and $v(x, y)$ satisfy the Cauchy-Riemann equations. Find corresponding equations, which relate the partial derivatives of u and v with respect to r and θ.
   d. Show that the functions $u(x, y)$, $v(x, y)$, which are given in polar coordinates by $r^n \cos n\theta$, $r^n \sin n\theta$, satisfy the Cauchy-Riemann equations.
   e. Let n be a positive integer. Study the transformation $F: u(x, y), v(x, y)$, which takes the form $R = r^n$, $\varphi = n\theta$, if polar coordinates (r, θ) and (R, φ) are introduced in the x, y plane and u, v plane, respectively. Describe this transformation geometrically, and show that it is a diffeomorphism of the angular sector

$0 < \theta < \pi/n$ onto the upper half-plane. Using part d, show that it is a conformal mapping.

20.7 Let $u(x, y) \in \mathscr{C}^1$ in the upper half-plane and $xu_y \equiv yu_x$. Show that there exists a function $g(t)$ of one variable such that $u(x, y) = g(x^2 + y^2)$. (*Hint:* see Exs. 7.26 and 7.27, and use Ex. 20.6a.)

20.8 Find the function $u(x, y)$ defined for all $(x, y) \neq (0, 0)$, which satisfies the equation

$$xu_y - yu_x = \frac{x}{\sqrt{x^2 + y^2}},$$

with $u(x, 0) = 0$ for all $x \neq 0$. (*Hint:* see Ex. 20.7.)

20.9 Find the most general form of a harmonic function in the upper half-plane that is constant along each ray through the origin.

20.10 Show that if $u(x, y) \in \mathscr{C}^2$ in the whole plane, and $u_{xy} \equiv 0$, then there exist functions $G(t)$, $H(t)$ of one variable such that $u(x, y) = G(x) + H(y)$. (*Hint:* use Ex. 7.26 to show that $u_x = g(x)$, $u_y = h(y)$. Set

$$G(x) = \int_0^x g(t)\, dt, \qquad H(y) = \int_0^y h(t)\, dt,$$

and use the Corollary to Th. 7.2.)

20.11 The equation

$$u_{tt} = c^2 u_{xx},$$

where c is a nonzero constant, is called the *one-dimensional wave equation.*

    a. Show that under the change of variables $X = x - ct$, $Y = x + ct$, we have

$$c^2 u_{xx} - u_{tt} = 4c^2 u_{XY}.$$

    b. Show that if $u(x, t) \in \mathscr{C}^2$ in the whole plane and $u(x, t)$ satisfies the wave equation, then $u(x, t)$ is of the form

$$u(x, t) = G(x - ct) + H(x + ct).$$

(*Hint:* use part a and Ex. 20.10.)

20.12 Derive the expressions for u_{xy} and u_{yy} analogous to Eq. (20.7) for u_{xx}.

20.13   *a.* Show that under the change of coordinates $x = e^X$, $y = e^Y$, we have

$$x^2 u_{xx} + y^2 u_{yy} + xu_x + yu_y = u_{XX} + u_{YY}.$$

    b. Use part a to write down some solutions of the equation

$$x^2 u_{xx} + y^2 u_{yy} + xu_x + yu_y = 0,$$

and then verify directly that this equation is satisfied.

20.14 *a.* Let a, b, c be arbitrary constants. Show that under a rotation of coordinates defined by Eqs. (20.1), we have

$$au_{xx} + 2bu_{xy} + cu_{yy} = Au_{XX} + 2Bu_{XY} + Cu_{YY},$$

and find the explicit expressions for A, B, C in terms of a, b, c and α. (*Hint*: use the expressions in Ex. 20.12.)

b. Show that as a special case of part a,

$$u_{xx} - u_{yy} = (u_{XX} - u_{YY}) \cos 2\alpha - u_{XY} \sin 2\alpha.$$

(In particular, the expression $u_{xx} - u_{yy}$ is not invariant under a rotation of coordinates, as is $u_{xx} + u_{yy}$ by Ex. 20.3.)

c. Let $q(x, y) = ax^2 + 2bxy + cy^2$. Show that under the rotation of coordinates defined by (20.1),

$$q(x, y) = Q(X, Y) = AX^2 + 2BXY + CY^2,$$

where A, B, C are given by the same expressions as in part a.

d. Let λ_1, λ_2 be the maximum and minimum of the quadratic form $q(x, y)$ of part c subject to the condition $x^2 + y^2 = 1$. Show that under a suitable rotation of coordinates,

$$au_{xx} + 2bu_{xy} + cu_{yy} = \lambda_1 u_{XX} + \lambda_2 u_{YY}.$$

***20.15** Find the most general function $u(x, y)$ that satisfies the equation $3u_{xx} + 10u_{xy} + 3u_{yy} = 0$, throughout the plane. (*Hint*: use Exs. 20.14d and 20.11b.)

20.16 *a.* Show that under the change of coordinates $X = kx$, $Y = ly$, we have

$$\lambda_1 u_{XX} + \lambda_2 u_{YY} = \frac{\lambda_1}{k^2} u_{xx} + \frac{\lambda_2}{l^2} u_{yy};$$

thus, choosing $k = \sqrt{|\lambda_1|}$ and $l = \sqrt{|\lambda_2|}$, both coefficients can be reduced to ± 1.

b. Show that under a suitable linear change of coordinates in each case, every sum of second-order derivatives with constant coefficients can be reduced to one of the following three forms:

$$au_{xx} + 2bu_{xy} + cu_{yy} = \begin{cases} \pm(u_{XX} + u_{YY}) & \text{if } ac - b^2 > 0 \\ u_{XX} - u_{YY} & \text{if } ac - b^2 < 0 \\ u_{XX} & \text{if } ac - b^2 = 0, \ a + c \neq 0. \end{cases}$$

(*Hint*: use the composition of two linear transformations, first applying Ex. 20.14d, and then Ex. 20.16a.)

c. Find the most general function $u(x, y)$ that satisfies $u_{xx} \equiv 0$ throughout the plane. Deduce that all solutions of the equation $au_{xx} + 2bu_{xy} + cu_{yy} = 0$ in the whole plane can be written down explicitly in case $ac - b^2 < 0$ or $ac - b^2 = 0$, whereas for $ac - b^2 > 0$ they can be expressed in terms of solutions to Laplace's equation $\Delta u = 0$.

*20.17 **a.** Let $X = f(x, y)$, $Y = g(x, y)$ define a change of coordinates. Let (x_0, y_0) be a point such that either

(1) $u_x(x_0, y_0) = 0$ and $u_y(x_0, y_0) = 0$, or
(2) the second derivatives of f and g are all zero.

Show that at the point (x_0, y_0), the expressions derived in Ex. 20.12, together with Eq. (20.7), can be summarized in a single matrix equation:

$$\begin{pmatrix} u_{xx} & u_{xy} \\ u_{xy} & u_{yy} \end{pmatrix} = \begin{pmatrix} f_x & g_x \\ f_y & g_y \end{pmatrix} \begin{pmatrix} u_{XX} & u_{XY} \\ u_{XY} & u_{YY} \end{pmatrix} \begin{pmatrix} f_x & f_y \\ g_x & g_y \end{pmatrix}.$$

(*Note:* matrix multiplication is discussed in the Remark following Ex. 15.21.)
b. Show that under the hypotheses of part a,

$$u_{xx}u_{yy} - u_{xy}^2 = \left[\frac{\partial(f, g)}{\partial(x, y)}\right]^2 (u_{XX}u_{YY} - u_{XY}^2).$$

c. Show that under the hypotheses of part a,

$$au_{xx} + 2bu_{xy} + cu_{yy} = Au_{XX} + 2Bu_{XY} + Cu_{YY},$$

where

$$\begin{pmatrix} A & B \\ B & C \end{pmatrix} = \begin{pmatrix} f_x & f_y \\ g_x & g_y \end{pmatrix} \begin{pmatrix} a & b \\ b & c \end{pmatrix} \begin{pmatrix} f_x & g_x \\ f_y & g_y \end{pmatrix};$$

show that this equation contains as a special case the answer to Ex. 20.14a.
d. Let the functions $f(x, y)$, $g(x, y)$ of part a define a diffeomorphism F. For any numbers r, s, let

$$\langle R, S \rangle = dF_{(x_0, y_0)}(\langle r, s \rangle).$$

Show that under the hypotheses of part a,

$$u_{xx}r^2 + 2u_{xy}rs + u_{yy}s^2 = u_{XX}R^2 + 2u_{XY}RS + u_{YY}S^2,$$

where the derivatives on the left are evaluated at the point (x_0, y_0) and those on the right are evaluated at the corresponding point $(X_0, Y_0) = (f(x_0, y_0), g(x_0, y_0))$.
e. Show that under the hypotheses of part a, the quadratic form for the second directional derivatives $\nabla_\alpha^2 u$ has the same nature whether computed with respect to x, y or X, Y; in other words,

$$u_{xx}(x_0, y_0) \cos^2 \alpha + 2u_{xy}(x_0, y_0) \cos \alpha \sin \alpha + u_{yy}(x_0, y_0) \sin^2 \alpha$$

and

$$u_{XX}(X_0, Y_0) \cos^2 \alpha + 2u_{XY}(X_0, Y_0) \cos \alpha \sin \alpha + u_{YY}(X_0, Y_0) \sin^2 \alpha$$

are simultaneously positive definite or negative definite, or positive semidefinite, etc.
f. Let F be the diffeomorphism $X = f(x, y)$, $Y = g(x, y)$; given a surface $z = h(X, Y)$, consider the corresponding surface $z = \tilde{h}(x, y)$, where $\tilde{h} = h \circ F$ (see Ex. 17.4). Give a geometric interpretation of part e in terms of these two surfaces. (Note that if F is linear, then hypothesis 2 of part a holds at

every point, and therefore so do all subsequent parts of this exercise, for an arbitrary function $u(x, y)$. On the other hand, for an *arbitrary* diffeomorphism F, these results are valid at any point where $\nabla u = 0$.)

20.18   Verify Eq. (20.17) and derive similar expressions for U_Y and V_X.

20.19   Using Eqs. (20.16) and (20.17) and the answer to Ex. 20.18, transform each of the following expressions into expressions involving u_x, u_y, v_x, v_y.

    a. $U_X - V_Y$
    b. $U_Y + V_X$
    c. $U_Y - V_X$

20.20   Show that if $\langle u, v \rangle$ is a harmonic vector field, then the corresponding vector field $\langle U, V \rangle$ obtained by rotating coordinates is also harmonic.

*20.21   *a.* Show that Eqs. (20.16) and (20.17) and the answer to Ex. 20.18 can be written in the form of a single matrix equation.

$$\begin{pmatrix} U_X & U_Y \\ V_X & V_Y \end{pmatrix} = \begin{pmatrix} \cos\alpha & \sin\alpha \\ -\sin\alpha & \cos\alpha \end{pmatrix} \begin{pmatrix} u_x & u_y \\ v_x & v_y \end{pmatrix} \begin{pmatrix} \cos\alpha & -\sin\alpha \\ \sin\alpha & \cos\alpha \end{pmatrix}.$$

    b. Show that $U_X V_Y - U_Y V_X = u_x v_y - u_y v_x$.

Remark It may have become apparent in the course of these exercises that the standard notation that we have been using may lead to some real difficulties. These become even more serious later on, when dealing with functions of more than two variables. To cite one example, if w is given as a function of the three variables x, y, z, and if z is in turn a function of x and y, then what is meant by $\partial w/\partial x$? It can have two interpretations, depending on whether we hold y and z fixed and let x vary, or we substitute z as a function of x and y and consider w just as a function of x and y. In the latter case, for fixed y, as x varies z also varies. Thus the notation $\partial w/\partial x$ should not be used in that situation, but be replaced by a more precise notation. Also in the present section, any possible confusion concerning the notation u_x and u_X could be avoided by writing $u = h(X, Y)$ and $u = \tilde{h}(x, y)$, where $\tilde{h}(x, y) = h(f(x, y), g(x, y))$. Then h_X, h_Y should be related to $\tilde{h}_x, \tilde{h}_y$.

It may be appropriate in this connection to touch on a basic problem in functional notation. Consider, for example, the question "what happens to the function $f(x, y)$ under a rotation of coordinates?" This question, taken at face value, is meaningless. The function $f(x, y)$ assigns to every pair of numbers a third number. If $f(x, y) = x^2 + y$, then $f(\pi, 1) = \pi^2 + 1$, $f(X, Y) = X^2 + Y$, $f(r, \theta) = r^2 + \theta$. What is implied in the above question is that we consider the pair of numbers x, y to define a point p in the plane; we use the function of two variables $f(x, y)$ to define a function, say g, which assigns to each point p of the plane a number $g(p)$ by the rule $g(p) = f(x, y)$ if (x, y) are the coordinates of the point p in the given coordinate system. Then if the point p is represented by (X, Y) in terms of a new coordinate system, we may define a new function of two variables, say $\tilde{f}(X, Y)$, by setting $\tilde{f}(X, Y) = g(p)$. The question then, is how this new function $\tilde{f}(X, Y)$ is related to the original function $f(x, y)$.

The problem, in brief, is whether in speaking of a function f, we mean a point function (such as $g(p)$ in the above discussion), which assigns a number to each point, and which takes on different forms in different coordinate systems, or a function of two variables, which assigns a fixed number to each pair of numbers according to a given rule, and which has nothing to do with the choice of coordinates. The fact is that both meanings are in common use, and in any given case the intended meaning must be deduced from the context. The important thing is to understand precisely what is meant in a given situation, and then the notation should not pose any problems. In some cases a more cumbersome notation is desirable in order to eliminate any possible confusion, but in others it may defeat one of the basic purposes of mathematical symbolism, which is to compress a large amount of information into compact form. It is well to remember that mathematical notation can be an invaluable aid to thought, but it cannot be a substitute for thought.

CHAPTER FOUR

Integration

21 Line integrals

In our study of functions of two variables, we used the technique of restricting a function to an arbitrary curve, thus obtaining a function of a single variable. Given a function $f(x, y)$ defined at every point of a curve

$$(21.1) \qquad C: x(t), y(t), \qquad a \le t \le b,$$

we used the notation

$$(21.2) \qquad f_C(t) = f(x(t), y(t)), \qquad a \le t \le b,$$

for the function of one variable obtained in this way.

We adopt a parallel method in our study of integration of functions of two variables and introduce so-called *line integrals*. There are a number of possibilities concerning the variable with respect to which we integrate. We start with the following definition.

Definition 21.1 Let $f(x, y)$ be a function that is defined and continuous at every point of a differentiable curve (21.1). We define $\int_C f \, dx$ and $\int_C f \, dy$ by

$$(21.3) \qquad \int_C f \, dx = \int_a^b f_C(t) x'(t) \, dt,$$

$$(21.4) \qquad \int_C f \, dy = \int_a^b f_C(t) y'(t) \, dt.$$

Example 21.1

If $f(x, y) = xy^2$, find $\int_C f \, dx$ for

$$C: x = \cos t, \ y = \sin t, \qquad 0 \le t \le \pi/2.$$

We have

$$x'(t) = -\sin t, \qquad f_C(t) = \cos t \sin^2 t;$$

$$\int_C xy^2 \, dx = -\int_0^{\pi/2} \cos t \sin^3 t \, dt = -\frac{1}{4}\sin^4 t \Big|_0^{\pi/2} = -\frac{1}{4}.$$

Example 21.2

If $f(x, y) = xy^2$, find $\int_C f \, dx$ for

$$C: x = \frac{1 - t^2}{1 + t^2}, \qquad y = \frac{2t}{1 + t^2}, \qquad 0 \le t \le 1.$$

Here

$$f_C(t) = \frac{4t^2(1 - t^2)}{(1 + t^2)^3}, \qquad x'(t) = -\frac{4t}{(1 + t^2)^2};$$

$$\int_C xy^2 \, dx = -\int_0^1 \frac{16t^3(1 - t^2)}{(1 + t^2)^2} \, dt.$$

The answer is thus expressed as an ordinary integral of a rational function. We do not carry out the computation, since we shall find the value of this integral by a different reasoning a little later on.

Example 21.3

If $f(x, y) = xy^2$, find $\int_C f \, dx$ for

$$C: x = 1 - t, \, y = t, \qquad 0 \le t \le 1.$$

In this case

$$x'(t) = -1, \qquad f_C(t) = t^2(1 - t),$$

and

$$\int_C xy^2 \, dx = -\int_0^1 t^2(1 - t) \, dt = \left[\frac{t^4}{4} - \frac{t^3}{3}\right]_0^1 = -\frac{1}{12}.$$

As in Examples 21.1–21.3 we are interested in the way a line integral for a given function f depends on the choice of the curve C. The first important observation may be stated roughly as follows: the values of the line integrals (21.3) and (21.4) do not depend on the choice of the parameter t used to describe the curve C. More precisely, we have the following statement.[1]

---

[1] Actually the statement in Lemma 21.1 is more general than what is usually referred to as "change of parameter," since we do not assume that the function $h(\tau)$ is monotone; that is, that the correspondence between τ and t is one-to-one.

Lemma 21.1 *Let C be a differentiable curve defined by (21.1). Let $h(\tau) \in \mathscr{C}^1$ for $\alpha \le \tau \le \beta$, where $a \le h(\tau) \le b$, and $h(\alpha) = a$, $h(\beta) = b$. Consider the curve obtained from C by setting $t = h(\tau)$;*

$$(21.5) \qquad \tilde{C}: x(h(\tau)), \, y(h(\tau)), \qquad \alpha \le \tau \le \beta$$

(Fig. 21.1). Then for any continuous function $f(x, y)$,

$$(21.6) \qquad \int_C f \, dx = \int_{\tilde{C}} f \, dx; \qquad \int_C f \, dy = \int_{\tilde{C}} f \, dy.$$

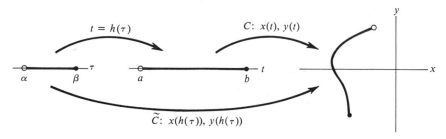

FIGURE 21.1 Change of parameter defining a curve

PROOF. We have

$$\int_{\tilde{C}} f \, dx = \int_\alpha^\beta f_{\tilde{C}}(\tau) \frac{dx}{d\tau} \, d\tau = \int_\alpha^\beta f_{\tilde{C}}(\tau) \frac{dx}{dt} \frac{dt}{d\tau} \, d\tau = \int_a^b f_C(t) \frac{dx}{dt} \, dt = \int_C f \, dx;$$

the first and last in this line of equalities are statements of the definition (21.3) of the line integral. The second equality is an application of the chain rule to the function $x(h(\tau))$, and the third follows from the rule for change of variables in integration,

$$\int_a^b g(t) \, dt = \int_\alpha^\beta g(t(\tau)) \frac{dt}{d\tau} \, d\tau, \qquad t(\alpha) = a, \, t(\beta) = b,$$

this rule being applied to the function $g(t) = f_C(t) \, dx/dt$. Thus the first equation in (21.6) is verified. The second follows in the same manner. ◆

Corollary *When expressing a line integral, Eqs. (21.3) or (21.4), we may assume that the parameter for the curve C varies over the interval from 0 to 1.*

PROOF. Given an arbitrary curve C of the form (21.1), we may set $t = a + (b - a)\tau$, $0 \le \tau \le 1$, and we obtain the same values for the line integrals (21.3) and (21.4) using the parameter τ. ◆

Let us now reconsider Examples 21.1 and 21.2. In Example 21.1, the curve C consists of the arc of the unit circle lying in the first quadrant, starting at $(1, 0)$ and ending at $(0, 1)$ (Fig. 21.2). In Example 21.2, we may confirm by a direct computation that $x(t)^2 + y(t)^2 \equiv 1$, so that the curve C lies on the unit circle. Furthermore, we see that $y \geq 0$ throughout, while x decreases from 1 to 0 as t goes from 0 to 1. Thus the curve C describes the same arc of the unit circle in the same direction as in Example 21.1. Since the parameter in Example 21.1 represents the polar angle at the origin, if we denote that angle by τ, it is clear that the parameter t in Example 21.2 should be expressible as a function of τ. In fact, setting $t = \tan(\tau/2)$ in the curve C of Example 21.2 yields $x = \cos \tau$, $y = \sin \tau$, and by Lemma 21.1 the value of the line integral is the same as in Example 21.1.

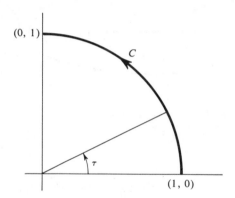

FIGURE 21.2 Circular arc: $x = \cos \tau$, $y = \sin \tau$, $0 \leq \tau \leq \frac{1}{2}\pi$

It is important to observe that a curve defined parametrically in the form (21.1) has a natural direction or orientation, namely the direction associated with increasing values of the parameter t. Thus the initial point on the curve is the point $x(a)$, $y(a)$, and the endpoint is $x(b)$, $y(b)$. The changes of parameter allowed in Lemma 21.1 preserved the orientation, since we required that $h(\alpha) = a$, $h(\beta) = b$. If we choose a parameter that reverses the sense in which the curve is described, then the line integral changes sign. To see this, we may assume by the Corollary to Lemma 21.1 that the parameter describes the interval from 0 to 1. We may then state the result as follows.

Lemma 21.2 *Let C be a differentiable curve of the form*

$$(21.7) \qquad\qquad C: x(t), y(t), \qquad\qquad 0 \leq t \leq 1.$$

Let $\tilde{C}$ be the curve obtained from C by setting

$$(21.8) \qquad\qquad t = 1 - \tau, \qquad\qquad 0 \leq \tau \leq 1.$$

Then for any continuous function f(x, y),

(21.9) $$\int_{\tilde{C}} f\, dx = -\int_C f\, dx, \qquad \int_{\tilde{C}} f\, dy = -\int_C f\, dy.$$

PROOF.   Just as in the proof of Lemma 21.1, we have

$$\int_{\tilde{C}} f\, dx = \int_0^1 f_{\tilde{C}}(\tau)\frac{dx}{d\tau}\, d\tau = \int_0^1 f_{\tilde{C}}(\tau)\frac{dx}{dt}\frac{dt}{d\tau}\, d\tau$$

$$= \int_1^0 f_C(t)\frac{dx}{dt}\, dt = -\int_0^1 f_C(t)\frac{dx}{dt}\, dt = -\int_C f\, dx.$$

The only difference is that when applying the rule for changing the variable of integration, we must be careful to insert the correct values for the limits of integration; that is, $\tau = 0$ corresponds to $t = 1$, and $\tau = 1$ to $t = 0$. Then an extra step is required to restore the normal order of integration, and it is in this step that the minus sign is introduced. Similar reasoning yields the second equation in (21.9).   ◆

We next note that it is often convenient to combine a pair of line integrals of the form (21.3) and (21.4). Thus, given a curve C and a pair of continuous functions $p(x, y), q(x, y)$ on C, we define $\int_C p\, dx + q\, dy$ by

(21.10) $$\int_C p\, dx + q\, dy = \int_C p\, dx + \int_C q\, dy.$$

The left-hand side of Eq. (21.10) is a frequently used notation for the sum of the two integrals on the right.

We note first that the combined form $\int_C p\, dx + q\, dy$ includes both of the line integrals (21.3) and (21.4) as special cases, by choosing either $p = f$ and $q = 0$ or $p = 0$ and $q = f$. There is a more important reason for considering the sum (21.10) rather than each integral separately. Recalling the definition in Eqs. (21.3) and (21.4), we find

(21.11) $$\int_C p\, dx + q\, dy = \int_a^b p_C(t)x'(t)\, dt + \int_a^b q_C(t)y'(t)\, dt$$

$$= \int_a^b [p_C(t)x'(t) + q_C(t)y'(t)]\, dt.$$

The integrand on the right-hand side has the form of a scalar product. From our discussion in Sect. 3, we recognize the vector

$$\langle x'(t), y'(t)\rangle = s'(t)\mathbf{T}(t),$$

where s is the parameter of arc length along C, and $\mathbf{T}(t)$ is the unit tangent to C at the point $x(t), y(t)$. If we introduce at each point of the curve the vector

$$\mathbf{v}(x, y) = \langle p(x, y), q(x, y)\rangle,$$

then Eq. (21.11) takes the form

(21.12) $$\int_C p\,dx + q\,dy = \int_a^b (\mathbf{v}\cdot\mathbf{T})s'(t)\,dt = \int_0^L (\mathbf{v}\cdot\mathbf{T})\,ds,$$

where L is the length of the curve C. The integrand $\mathbf{v}\cdot\mathbf{T}$ is equal to the component of the vector $\mathbf{v}$ in the direction of the unit vector $\mathbf{T}$, or the *tangential component* of $\mathbf{v}$; it is frequently denoted by v_T. Equation (21.12) then takes the form

(21.13) $$\int_C p\,dx + q\,dy = \int_0^L v_T\,ds.$$

In order to guarantee the existence of the unit tangent $\mathbf{T}$ it is necessary to assume that C is a regular curve. We may then describe Eq. (21.13) as follows (Fig. 21.3). Let C be a regular curve of length L, and let s denote the parameter of arc length along C. To each point of C is assigned a vector $\mathbf{v}$ whose x and y components, p and q, vary continuously along C. Then the tangential component v_T is a continuous function of s, and the ordinary integral of this function from 0 to L is equal to the line integral (21.10).

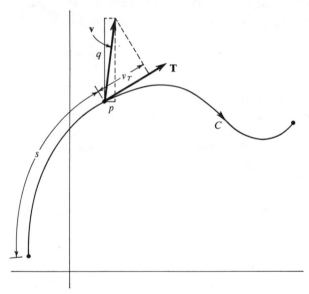

FIGURE 21.3 Geometric interpretation of $\int_C p\,dx + q\,dy = \int_0^L v_T\,ds$

Remark In analogy with Eqs. (21.3) and (21.4), we define $\int_C f\,ds$, the *integral of f with respect to arc length*, by

$$\int_C f\,ds = \int_a^b f_C(t)\frac{ds}{dt}\,dt.$$

By the chain rule, we have

$$\int_a^b f_C(t) \frac{ds}{dt}\, dt = \int_0^L f_C(t(s))\, ds,$$

where the integrand on the right is simply the function f referred to the parameter of arc length s. In this notation Eq. (21.13) takes the form

(21.13a) $$\int_C p\, dx + q\, dy = \int_C v_T\, ds.$$

Example 21.4

Consider the case where the components of $\mathbf{v}$ are constant. Let us set $\mathbf{v} = \langle c, d \rangle$, where $|\mathbf{v}| = 1$. Then for the curve C defined by (21.1), we have

(21.14)
$$\int_C c\, dx + d\, dy = c \int_a^b x'(t)\, dt + d \int_a^b y'(t)\, dt$$
$$= c[x(b) - x(a)] + d[y(b) - y(a)]$$
$$= \mathbf{v} \cdot \langle x(b) - x(a), y(b) - y(a) \rangle.$$

The right-hand side of Eq. (21.13) may be described in this case as the integral of the component of the unit tangent $\mathbf{T}$ in the direction of the fixed unit vector $\mathbf{v}$, and the result yields, by Eq. (21.14), the projection in the direction of $\mathbf{v}$ of the displacement vector from the initial point to the endpoint of C.

An interesting feature of Example 21.4 is that if we consider the constant vector field $\mathbf{v}(x, y) = \langle c, d \rangle$ defined in the whole plane, and compute the line integral (21.14) for different choices of the curve C, then the resulting value depends only on the initial point and the endpoint of C, and is completely independent of the path taken by C in between. We shall encounter independence of path in a more general setting in the next section.

We conclude the present section with a brief discussion of the interpretation of the line integral in the case where the vector $\mathbf{v} = \langle p, q \rangle$ has one of the physical interpretations discussed at the end of Sect. 19.

First, consider the case where $\mathbf{v}$ represents a force, and C is the trajectory of a particle that is subjected at each point to the force $\mathbf{v}$. Then v_T represents the component of the force in the direction of motion, and the integral $\int_0^L v_T\, ds$ is defined in physics to be the *work* done by the force in moving the particle along the path C. By Eq. (21.13), this physical quantity is represented by a line integral in terms of the components p and q of the force at each point. From Eq. (21.14) we see that in the special case where the force is constant and the partial moves along a straight line in the direction of the force, the work done is equal to the magnitude of the force times the distance traversed. It follows from the same equation that if the force is constant, but the path is not straight, the work done depends not on the distance traversed, but on the displacement between beginning and endpoint (which is not at all evident on physical grounds).

The second physical interpretation is that of a fluid flow, where **v** is the velocity vector of the flow at each point. Then v_T represents the component of the flow velocity in the direction of the curve C, and $\int_0^L v_T \, ds$ may be described as the "total rate of flow along C."

Exercises

21.1 Let $f(x, y) = x^2 - y^2$. Compute the line integrals $\int_C f \, dx$ for the following curves.

a. $C: x = (1 - t)^2,$ $\quad y = t,$ $\quad 0 \le t \le 1$
b. $C: x = 1 - t^2,$ $\quad y = t^2,$ $\quad 0 \le t \le 1$
c. $C: x = \cos t,$ $\quad y = \sin t,$ $\quad 0 \le t \le \frac{1}{2}\pi$.
d. $C: x = \cosh t,$ $\quad y = \sinh t,$ $\quad 0 \le t \le \log 2$

21.2 Verify Lemma 21.1 for the case $f(x, y) = xy$, where

$$C: x = t^2, \qquad y = t^3, \qquad\qquad 0 \le t \le 1$$

and

$$h(\tau) = \sin \tau, \qquad\qquad 0 \le \tau \le \tfrac{1}{2}\pi,$$

so that

$$\tilde{C}: x = \sin^2 \tau, \qquad y = \sin^3 \tau, \qquad\qquad 0 \le \tau \le \tfrac{1}{2}\pi.$$

21.3 Verify Lemma 21.2 for the case $f(x, y) = 1/x$, where

$$C: x = 1 + t^2, \qquad y = 2t, \qquad\qquad 0 \le t \le 1.$$

21.4 Evaluate $\int_C y \, dx - x \, dy$ for each of the following curves.

a. $C: x = t^3,$ $\quad y = t^2,$ $\quad 0 \le t \le 1$
b. $C: x = t,$ $\quad y = t^n,$ $\quad 0 \le t \le 1$ $\quad$ (n, a positive integer)
c. $C: x = \cos t,$ $\quad y = 1 + \sin t,$ $\quad \frac{3}{2}\pi \le t \le 2\pi$
d. C is the straight-line segment from $(0, 0)$ to $(1, 1)$.
e. C is the straight-line segment from $(1, 1)$ to $(0, 0)$.

21.5 Evaluate $\int_C y \, dx + x \, dy$ for each of the curves of Ex. 21.4.

21.6 Evaluate

$$\int_C -\frac{y}{x^2 + y^2} \, dx + \frac{x}{x^2 + y^2} \, dy,$$

where

a. C is the upper arc of the circle $x^2 + y^2 = 2$ going from $(1, 1)$ to $(-1, 1)$.
b. C is the straight-line segment from $(1, 1)$ to $(-1, 1)$.
c. C is the lower arc of the circle $x^2 + y^2 = 2$ going from $(1, 1)$ to $(-1, 1)$.

21.7 For each of the following vector fields $v(x, y)$, evaluate $\int_C v_T \, ds$, where C is the ellipse,

$$C: x = a \cos t, \qquad y = b \sin t, \qquad 0 \le t \le 2\pi.$$

a. $v = \langle x, y \rangle$ e. $v = \langle y, 0 \rangle$
b. $v = \langle y, x \rangle$ f. $v = \langle 0, y \rangle$
c. $v = \langle -y, x \rangle$ g. $v = \langle c, d \rangle$, c and d are constants
d. $v = \langle -x, y \rangle$

21.8 Explain how the answer to Ex. 21.7g could be predicted by Eq. (21.14).

21.9 Evaluate $\int_{\tilde{C}} v_T \, ds$, where v is the vector field of Ex. 21.7c, and $\tilde{C}$ is the ellipse

$$\tilde{C}: x = a \cos t, \qquad y = -b \sin t, \qquad 0 \le t \le 2\pi.$$

Compare your answer with the answer to Ex. 21.7c and explain how it could have been predicted.

21.10 Show that if a curve C lies on a vertical line, then for any function $f(x, y)$, $\int_C f \, dx = 0$. If C lies on a horizontal line, then $\int_C f \, dy = 0$ for any f.

21.11 Let C be defined by $x = \varphi(t)$, $y = \psi(t)$, $a \le t \le b$. Choose values $t_0, t_1, \ldots, t_n$ such that

$$a = t_0 < t_1 < t_2 < \cdots < t_{n-1} < t_n = b.$$

Let C_k be the curve defined by $x = \varphi(t)$, $y = \psi(t)$, $t_{k-1} \le t \le t_k$, where $k = 1, 2, \ldots, n$.

a. Draw a sketch indicating the curves $C_1, \ldots, C_n$ and their relation to C.

b. Show that for any function $f(x, y)$,

$$\int_C f \, dx = \int_{C_1} f \, dx + \cdots + \int_{C_n} f \, dx$$

$$\int_C f \, dy = \int_{C_1} f \, dy + \cdots + \int_{C_n} f \, dy$$

$$\int_C f \, ds = \int_{C_1} f \, ds + \cdots + \int_{C_n} f \, ds.$$

*21.12 Let C be a curve defined by $x = \varphi(t)$, $y = \psi(t)$, $a \le t \le b$.

a. If $\varphi'(t) > 0$ for $a \le t \le b$, show that the equation $x = \varphi(t)$, $a \le t \le b$ can be inverted in the form $t = h(x)$, $\alpha \le x \le \beta$.

b. In the notation of part a, if $g(x) = \psi(h(x))$, show that

$$\int_C f \, dx = \int_\alpha^\beta f(x, g(x)) \, dx.$$

c. Define $F(x)$ by $F(x) = f(x, g(x))$, $\alpha \leq x \leq \beta$. Note that the equation $y = g(x)$ is simply the nonparametric form of the curve C. Draw a sketch illustrating the function $F(x)$ as follows. Draw the curve C in the form $y = g(x)$. For each value of x, $\alpha \leq x \leq \beta$, erect a line perpendicular to the x axis and find the point (x, y) where this line intersects the curve C. Then $F(x)$ is the value of f at this point (x, y). By part b, the ordinary integral of the function $F(x)$ with respect to x is equal to the line integral $\int_C f\, dx$. (This could also be described in terms of Lemma 21.1 as making a change of parameter such that $\tilde{C}$ is the curve $x = \tau$, $y = g(\tau)$, $\alpha \leq \tau \leq \beta$.)

d. If $\varphi'(t) < 0$ for $a \leq t \leq b$, then setting $t = a + b(1 - \tau)$, $0 \leq \tau \leq 1$, we obtain a new curve $\tilde{C}$ defined by $x = \tilde{\varphi}(\tau) = \varphi(t(\tau))$, $y = \tilde{\psi}(\tau) = \psi(t(\tau))$, which consists of the curve C described in the opposite direction. Using parts a and b and Lemmas 21.1 and 21.2, show that

$$\int_C f\, dx = -\int_{\tilde{C}} f\, dx = -\int_\alpha^\beta f(x, g(x))\, dx,$$

where $y = g(x)$, $\alpha \leq x \leq \beta$ is the nonparametric form of the curve $\tilde{C}$.

e. Combining parts a–d we may form the following picture of a line integral $\int_C f\, dx$. If C_1 is a part of the curve C along which x is an increasing function of t, then $\int_{C_1} f\, dx$ is the ordinary integral with respect to x of f considered as a function of its x coordinate alone along the curve C_1. If C_2 is a part of the curve along which x is a decreasing function of t, then let $\tilde{C}_2$ be the curve C_2 described in the opposite direction, $\int_{C_2} f\, dx = -\int_{\tilde{C}_2} f\, dx$, and the latter integral may be described as above. Finally, if C_3 is a part of the curve on which x is constant, then $\int_{C_3} f\, dx = 0$. Describe in analogous fashion $\int_C f\, dy$.

21.13 Evaluate $\int_C f\, ds$ for the following functions $f(x, y)$ and curves C.

a. $f(x, y) = y$, C is the straight-line segment from $(0, 1)$ to $(1, 0)$.
b. $f(x, y) = x^2 + y^2$, C is the circle $x = 2 \cos t$, $y = 2 \sin t$, $0 \leq t \leq 2\pi$.
c. $f(x, y) = (1 - x^2 - y^2)^{1/2}$, C the same as in part a.
d. $f(x, y) = y$, C the straight-line segment from $(1, 0)$ to $(0, 1)$.
e. $f(x, y) = xy$, C is the part of the ellipse $x^2/a^2 + y^2/b^2 = 1$, lying in the first quadrant.

21.14 Prove that $\int_C f\, ds$ does not depend on the orientation of the curve C; that is, given a curve

$$C: x(t), y(t), \qquad\qquad 0 \leq t \leq 1,$$

let $\tilde{C}$ be the curve obtained from C by setting

$$t = 1 - \tau, \qquad\qquad 0 \leq \tau \leq 1.$$

Show that

$$\int_{\tilde{C}} f\, ds = \int_C f\, ds.$$

(*Hint:* follow the proof of Lemma 21.2, but note carefully the form of the chain rule for ds/dt; the equation

$$\frac{ds}{d\tau} = \frac{ds}{dt}\frac{dt}{d\tau}$$

is *not* always true. What is the correct form of this equation?)

Remark The ordinary integral of a function of a single variable, $\int_a^b F(x)\,dx$, has a well-known geometric interpretation in terms of area. The line integral $\int_C f\,ds$ may also be pictured in terms of areas, but in this case the interpretation is in terms of areas of curved surfaces. Since the surfaces involved are of special kinds, rather than general curved surfaces, it is possible to obtain an intuitive picture without having studied the general theory. In Exs. 21.5–21.7 we discuss some of the relationships between the line integral and the area of certain surfaces. The facts stated concerning the area of these surfaces follow easily from the general theory of surface area. (See, for example, the discussion of surface area in [11], Chapter IV, section 6.)

21.15 Let C be a curve $x(t)$, $y(t)$, $a \le t \le b$, and let $f(x, y)$ be a function that is defined and continuous along C. Assume further that $f(x, y) \ge 0$ on C. Using rectangular coordinates x, y, z in three-dimensional space, consider the curve C to lie in the x, y plane, and erect above each point (x, y) of C, the vertical line segment of height $f(x, y)$. As the point (x, y) traces out the curve C, from $(x(a), y(a))$ to $(x(b), y(b))$, these vertical line segments sweep out a "cylindrical surface" S. The area of this surface S is equal to $\int_C f\,ds$.

a. Draw a sketch illustrating the curve C and the surface S just described.

b. If the function $f(x, y)$ is defined in a domain D in which the curve C lies, the surface S may be described as "the surface lying above the x, y plane and below the space curve defined as the intersection of the surface $z = f(x, y)$ with the cylindrical surface generated by all vertical lines through C." Illustrate this with a sketch.

c. Explain Ex. 21.14 (at least for positive functions f) in terms of this geometric interpretation of the line integral.

*d. Using this interpretation, show that the answers to Ex. 21.13a, b, and c consist of the areas of a triangle, cylinder, and semicircle, respectively. Use elementary geometry to compute these areas, and verify the results obtained previously.

21.16 Using the geometric interpretation of the line integral given in Ex. 21.15, find the area of the cylindrical surfaces S described as follows, and illustrate each surface with a sketch.

a. S lies over the circle $x^2 + y^2 = 1$ and under the surface $z = 1 + x^2$.

b. S lies over the circular arc $x = a \cos t$, $y = a \sin t$, $-\frac{1}{4}\pi \le t \le \frac{1}{4}\pi$, and under the surface $z = x^2 - y^2$.

c. S lies over the elliptic arc $y = 2(1 - x^2)^{1/2}$, $0 \le x \le 1$, and under the surface $z = xy$.

d. S is the part of a circular cylinder of radius a cut out by another cylinder of equal radius, the axes of the two cylinders intersecting at right angles.

e. Consider a vertical cylinder of radius $\frac{1}{2}$ whose axis intersects the y axis at the point $(0, \frac{1}{2})$. S is that part of the cylinder which lies in the first octant and is cut out by a sphere of radius 1 with center at the origin.

21.17 Let C be a curve of the form $y = g(x)$, $a \le x \le b$, where $g(x) \ge 0$ for $a \le x \le b$. Let S be the surface obtained by revolving the curve C about the x axis. Then S is a surface of revolution, and its area A is given by the formula

$$A = 2\pi \int_C y \, ds.$$

Use this formula to compute the area of the following surfaces, and sketch each surface.

a. A cone, where $C: x = ht, y = rt, 0 \le t \le 1$.

b. A frustrum of a cone, $C: y = ((r_2 - r_1)/h)x + r_1$, $0 \le x \le h$; show that the area of the cone of part a, and that of a circular cylinder may both be obtained as special cases.

c. A paraboloid of revolution; $C: y = \sqrt{x}, 0 \le x \le 2$.

d. A zone of a sphere; $C: y = \sqrt{a^2 - x^2}, h_1 \le x \le h_2$.

(Note the curious fact that the area depends only on the width of the zone.)

e. A *catenoid*; $C: y = \cosh x$ (a *catenary*), $-h \le x \le h$.

f. A *pseudosphere*; C is defined for $x > 0$ to be a curve $y = g(x)$, where at each point

$$\frac{dy}{dx} = -\frac{y}{\sqrt{a^2 - y^2}},$$

and $g(0) = a$ (the *tractrix*). It has the properties $g(x) > 0$ for $x > 0$ and $\lim_{x \to \infty} g(x) = 0$. If this curve is reflected in the y axis by setting $g(-x) = g(x)$, and the entire curve thus obtained is revolved about the x axis, the resulting surface is called the *pseudosphere of pseudoradius a*. Show that for any x_1, x_2 such that $0 < x_1 < x_2$, the zone of the pseudosphere satisfying $x_1 \le x \le x_2$ has surface area $2\pi a[g(x_1) - g(x_2)]$.

g. By taking the limit as x_1 tends to zero and x_2 tends to infinity in part f, find the area of the entire pseudosphere, and compare with the area of the entire sphere obtained by setting $h_1 = -a, h_2 = a$ in part d.

21.18 A physical interpretation of the line integral $\int_C f \, ds$ is the following. Picture a wire of variable density bent into the shape of the curve C. If the density at each point (in mass per unit length) is equal to f, then the *total mass* of the wire is equal to $\int_C f \, ds$. Compute the total mass in each of the following cases.

a. $f(x, y) = x\sqrt{1 + 4y}$; $C: y = x^2$, $0 \le x \le 1$

b. $f(x, y) = |x| + |y|$; $C: x = \cos t, y = \sin t$, $0 \le t \le \pi$

c. $f(x, y) = 1/\sqrt{x^2 + y^2}$; $C: x = e^t \cos t, y = e^t \sin t, \quad -\frac{1}{2}\pi \leq t \leq \frac{1}{2}\pi$

d. $f(x, y) = \sqrt{y}$; $C: x = t - \sin t, y = 1 - \cos t, \quad 0 \leq t \leq 2\pi$

21.19 In the notation of Ex. 21.18, where $f(x, y)$ represents the density in mass per unit length, the following quantities have physical importance. (For the corresponding quantities in the case of plane figures, see Ex. 24.12,

Moment about the x axis:	$\int_C yf\, ds$
Moment about the y axis:	$\int_C xf\, ds$
Center of gravity:	the point $(\bar{x}, \bar{y})$, where

$$\bar{x} = \int_C xf\, ds / \int_C f\, ds \qquad \bar{y} = \int_C yf\, ds / \int_C f\, ds$$

For each part of Ex. 21.18, compute both moments, and use them to find the center of gravity.

21.20 If in Ex. 21.19, the density f is constant, then this constant cancels out in the definition of $\bar{x}$ and $\bar{y}$, and hence the center of gravity $(\bar{x}, \bar{y})$ depends only on the curve C; it is called the *centroid* of the curve C, and is given by $(\bar{x}, \bar{y})$, with

$$\bar{x} = \frac{1}{L} \int_C x\, ds, \qquad \bar{y} = \frac{1}{L} \int_C y\, ds,$$

where L is the length of C. Find the centroid of each of the following curves C.

a. The circular arc: $x = \cos t, y = \sin t, 0 \leq t \leq \alpha$.
b. The straight-line segment from (x_1, y_1) to (x_2, y_2).
*c. The first arch of the cycloid: $x = t - \sin t, y = 1 - \cos t, 0 \leq t \leq 2\pi$.
d. The first quadrant of the *hypocycloid* or *astroid*: $x = \cos^3 t, y = \sin^3 t$,
$0 \leq t \leq \frac{1}{2}\pi$.

21.21 Let S be a surface of revolution formed by revolving a plane curve C about a fixed line that does not cross the curve. Let R be the distance of the centroid of C from the fixed line, and let L be the length of the curve C. The *Theorem of Pappus* states that the area A of the surface S is given by the formula

$$A = 2\pi RL.$$

a. Prove the theorem of Pappus. (Note that it may be assumed that the curve C lies in the upper half-plane and that the fixed line coincides with the x axis.)
b. Verify the theorem of Pappus for the case of a frustrum of a cone, by comparing the answers to Exs. 21.17b and 21.20b.
c. Use the theorem of Pappus and the centroid of the circle,

$$C: x^2 + (y - b)^2 = a^2 \qquad (b > a),$$

to find the surface area of the torus obtained by revolving the circle C about the x axis.
d. Use the theorem of Pappus and the known area of a sphere (Ex. 21.17d) to find the centroid of a semicircle, $x^2 + y^2 = a^2, y \geq 0$.

21.22 In each of the following cases interpret the vector field $\mathbf{v} = \langle p, q \rangle$ as a force field, and compute the work done in moving a particle along the given curve C.

$\quad$ *a.* $\mathbf{v} = \langle x - y, x + y \rangle$; $\quad$ $C: x = \cos t, y = \sin t,$ $\quad$ $0 \le t \le \pi$
$\quad$ *b.* $\mathbf{v} = \langle x^2 + y^2, x^2 - y^2 \rangle$; $\quad$ $C: x = \cos t, y = \sin t,$ $\quad$ $0 \le t \le \pi$
$\quad$ *c.* $\mathbf{v} = \langle 6xy^2, 5xy - 2 \rangle$; $\quad$ $C: y = x^2,$ $\quad$ $0 \le x \le 1$
$\quad$ *d.* $\mathbf{v} = \langle 1, x \rangle$; $\quad$ $C: y = \log x,$ $\quad$ $1 \le x \le 2$

21.23 According to the "inverse square law," the force exerted is inversely proportional to the square of the distance from the body exerting the force. If that body is placed at the origin, the field takes the form

$$\mathbf{v} = -k \frac{\langle x, y \rangle}{(x^2 + y^2)^{3/2}}.$$

$\quad$ *a.* Find the work done by this force in moving a particle along a straight-line segment from the point $(r_2 \cos \alpha, r_2 \sin \alpha)$ to $(r_1 \cos \alpha, r_1 \sin \alpha)$, where α is fixed and $r_1 < r_2$.
$\quad$ *b.* Find the work done in moving a particle along a horizontal line segment from $(r_2 \cos \alpha, r_2 \sin \alpha)$ to $(r_1 \cos \alpha, r_2 \sin \alpha)$.
$\quad$ *c.* Find the work done in moving a particle along a vertical line segment from $(r_1 \cos \alpha, r_2 \sin \alpha)$ to $(r_1 \cos \alpha, r_1 \sin \alpha)$.
$\quad$ *d.* Show that the total work done in parts b and c is the same as the work done in part a.
$\quad$ *e.* Let C be a path consisting of an arbitrary succession of horizontal and vertical line segments, starting at a point whose distance from the origin is R_2 and ending at a point whose distance from the origin is R_1. Show that the total work done in moving a particle along the path C is equal to $k(1/R_1 - 1/R_2)$. (*Hint:* show that this is true for each of the line segments making up the path, and deduce that it must be true for the whole path.)

*21.24 The definition we have given of line integrals is best suited for the actual computation of line integrals, but it has two theoretical disadvantages. First, it was expressed in terms of the parameter t, and we had to show later that the value of the line integral was actually independent of the choice of parameter. Second, the definition can only be applied to differentiable curves. There is an alternative definition, which avoids both of these difficulties. Given a curve C with a fixed orientation and a function $f(x, y)$ continuous on C, choose an arbitrary sequence of points (x_n, y_n) along C (ordered according to the given orientation). Let (ξ_n, η_n) be an arbitrary point on C between (x_{n-1}, y_{n-1}) and (x_n, y_n). Use the notation

$$\Delta x_n = x_n - x_{n-1}, \quad \Delta y_n = y_n - y_{n-1}, \quad \Delta s_n = \sqrt{|\Delta x_n|^2 + |\Delta y_n|^2}.$$

Form the sums

$$\sum f(\xi_n, \eta_n) \Delta x_n, \quad \sum f(\xi_n, \eta_n) \Delta y_n, \quad \sum f(\xi_n, \eta_n) \Delta s_n.$$

If these sums tend to a limit as the maximum of Δs_n tends to zero, this limit being independent of the subdivision and of the choice of intermediate points (ξ_n, η_n), then the limits are defined to be

$$\int_C f\,dx, \qquad \int_C f\,dy, \qquad \int_C f\,ds,$$

respectively.

   a. Draw a sketch indicating the quantities Δx_n, Δy_n, Δs_n and the point (ξ_n, η_n).
   b. For C a differentiable curve, so that x and y are differentiable functions of a parameter t, write down the above sums in terms of t by applying the mean value theorem for functions of one variable to the terms Δx_n and Δy_n. Show that it is reasonable to expect these sums to tend in the limit to the expressions we used to define the corresponding line integrals.

   21.25   The line integrals

$$\int_C f\,dx, \qquad \int_C f\,dy, \qquad \int_C f\,ds,$$

which we have defined in this section, are special cases of a more general situation. Suppose that $f(x, y)$ and $g(x, y)$ are defined along C. We wish to define $\int_C f\,dg$. This may be done in the manner described in Ex. 21.24, or else, in case $g_C'(t)$ exists and is continuous, we may follow the procedure used in Eqs. (21.3) and (21.4), and use as our definition

$$\int_C f\,dg = \int_a^b f_C(t)g_C'(t)\,dt.$$

In particular, if C is a differentiable curve lying in a domain D, if $g(x, y) \in \mathscr{C}^1$ in D, and if $f(x, y)$ is continuous along C, then the integral on the right exists, and we use it to define $\int_C f\,dg$. An important case is that in which the curve C does not pass through the origin, and we choose for $g(x, y)$ one of the functions

$$r(x, y) = \sqrt{x^2 + y^2} \qquad \text{or} \qquad \theta(x, y) = \arctan y/x.$$

Note that any two branches of $\theta(x, y)$ differ by a constant, and therefore the expression

$$\theta_C'(t) = \theta_x x'(t) + \theta_y y'(t) = -\frac{y}{x^2 + y^2} x'(t) + \frac{x}{x^2 + y^2} y'(t)$$

is well-defined for each t. Thus the line integrals

$$\int_C f\,dr \qquad \text{and} \qquad \int_C f\,d\theta$$

are uniquely defined, independently of how θ itself is chosen along any part of the curve C.

In case the function $f(x, y) \equiv 1$, we use the notation

$$\int_C dg \qquad \text{for} \qquad \int_C 1 \, dg.$$

a. Suppose that C does not pass through the origin. Show that

$$\int_C r^2 \, d\theta = \int_C -y \, dx + x \, dy.$$

b. Under the same hypotheses as part a, show that

$$\int_C d\theta = \int_C \frac{-y}{x^2 + y^2} \, dx + \frac{x}{x^2 + y^2} \, dy.$$

c. Show that

$$\int_C dg = \int_C g_x \, dx + g_y \, dy.$$

22 Potential functions; independence of path

In the previous section we computed line integrals by first expressing them as ordinary (definite) integrals, and then evaluating these integrals in the usual fashion by finding indefinite integrals; that is, by using the fundamental theorem of calculus. In the present section we prove a theorem that may be considered to be a generalization of the fundamental theorem of calculus to functions of two variables. We replace definite integrals by line integrals, and indefinite integrals or antiderivatives by potential functions. One consequence is a rapid method of evaluating many line integrals.

We begin by stating the two standard forms of the fundamental theorem of calculus.[2]

Let $g(x)$ be continuous on an interval $a \le x \le b$. Then

$$(22.1) \qquad f(X) = \int_a^X g(x) \, dx \Rightarrow f'(X) = g(X);$$

$$(22.2) \qquad f'(x) = g(x), \quad a \le x \le b \Rightarrow \int_a^b g(x) \, dx = f(b) - f(a).$$

We can summarize these two statements by saying "differentiation and integration are inverse processes." We shall show that in a certain sense "forming gradients and line integrals are inverse processes." To be made precise, this rough statement must be separated into several parts. We start with a result generalizing Eq. (22.2).

---

[2] For further discussion and a proof of the fundamental theorem, see [14], Sections 3.4, 3.5, and 9.5.

Lemma 22.1 *Let $f(x, y) \in \mathscr{C}^1$ in a domain D. Consider the gradient vector field*

(22.3)
$$\langle p(x, y), q(x, y) \rangle = \nabla f(x, y)$$

in D. Let (x_1, y_1), (x_2, y_2) be any two points in D, and consider a differentiable curve

(22.4)
$$C : x(t), y(t), \qquad a \leq t \leq b,$$

lying in D, such that

(22.5) $(x(a), y(a)) = (x_1, y_1), \qquad (x(b), y(b)) = (x_2, y_2).$

Then

(22.6)
$$\int_C p \, dx + q \, dy = f(x_2, y_2) - f(x_1, y_1).$$

PROOF.   By definition of the line integral,

$$\int_C p \, dx + q \, dy = \int_a^b [p_C(t)x'(t) + q_C(t)y'(t)] \, dt.$$

But by (22.3) and Th. 7.1, we have at each point of the curve C

$$px'(t) + qy'(t) = f_x x'(t) + f_y y'(t) = f_C'(t),$$

whence, by (22.2) and (22.5),

$$\int_C p \, dx + q \, dy = \int_a^b f_C'(t) \, dt = f_C(b) - f_C(a)$$

$$= f(x_2, y_2) - f(x_1, y_1). \qquad \blacklozenge$$

Remark   Recalling that a function f satisfying (22.3) in a domain is called a potential function for the vector field $\langle p, q \rangle$, we may state the content of Lemma 22.1 as follows: *if a vector field $\langle p, q \rangle$ has a potential function f in a domain D, then for any curve C lying in D, the value of the line integral $\int_C p \, dx + q \, dy$ is obtained by simply forming the difference between the values of f at the endpoint and at the beginning point of C.*

Example 22.1

Consider Eq. (21.14). There the vector field $\mathbf{v}$ was constant, $\mathbf{v} = \langle c, d \rangle$. We may set $\mathbf{v} = \nabla f$, where $f = cx + dy$. Then by Eq. (22.6),

$$\int_C c \, dx + d \, dy = (cx_2 + dy_2) - (cx_1 + dy_1) = \langle c, d \rangle \cdot \langle x_2 - x_1, y_2 - y_1 \rangle,$$

which, by virtue of (22.5), is the same as the result obtained previously.

Example 22.2

Evaluate $\int_C y\, dx + x\, dy$, where

$$C: x \cos t,\, y = \sin t, \qquad\qquad 0 \le t \le \tfrac{1}{2}\pi.$$

Method 1. Using the definition of the line integral, since $x'(t) = -\sin t$, $y'(t) = \cos t$,

$$\int_C y\, dx + x\, dy = \int_0^{\pi/2} (-\sin^2 t + \cos^2 t)\, dt = \int_0^{\pi/2} \cos 2t\, dt = \tfrac{1}{2} \sin 2t \Big|_0^{\pi/2} = 0.$$

Method 2. Using Lemma 22.1, with $p = y$, $q = x$, clearly the function $f(x, y) = xy$ satisfies (22.3). Since the curve C starts at $(1, 0)$ and ends at $(0, 1)$, we have by (22.6)

$$\int_C y\, dx + x\, dy = 0 \cdot 1 - 1 \cdot 0 = 0.$$

For various applications it is desirable to enlarge the class of curves C over which we form line integrals, and to observe that Lemma 22.1 continues to hold for this wider class of curves.

Definition 22.1 A *piecewise smooth curve* is a curve $C: x(t), y(t), a \le t \le b$, where $x(t), y(t)$ are continuous for $a \le t \le b$, and where there exist a finite number of values $t_1 < t_2 < \cdots < t_n$ between a and b such that $x(t), y(t)$ define a regular curve on each of the intervals

$$a \le t \le t_1, \quad t_1 \le t \le t_2, \quad \ldots, \quad t_n \le t \le b.$$

Remark When referring to a differentiable function on a closed interval, it is understood that only the right- or left-hand derivative is defined at the endpoints. Thus, for a piecewise smooth curve, the functions $x(t)$ and $y(t)$ in general have distinct right- and left-hand derivatives at the points t_k. Geometrically speaking, a piecewise smooth curve is allowed to have a finite number of corners.

Definition 22.2 If C is a piecewise smooth curve as defined above, and if $p(x, y), q(x, y)$ are continuous along C, then $\int_C p\, dx + q\, dy$ is defined by

$$\int_C p\, dx + q\, dy = \int_{C_1} p\, dx + q\, dy + \cdots + \int_{C_{n+1}} p\, dx + q\, dy,$$

where

$$C_1: \quad x(t), y(t), \qquad a \le t \le t_1$$

$$C_2: \quad x(t), y(t), \qquad t_1 \le t \le t_2$$

$$\vdots$$

$$C_{n+1}: x(t), y(t), \qquad t_n \le t \le b.$$

Lemma 22.2 *Lemma 22.1 remains true if we allow C to be any piecewise smooth curve lying in D.*

PROOF. Applying Lemma 22.1 to each of the curves $C_1, \ldots, C_{n+1}$, in the notation of Def. 22.2 and adding, we find

$$\int_C p \, dx + q \, dy = [f(x(t_1), y(t_1)) - f(x(a), y(a))]$$

$$+ [f(x(t_2), y(t_2)) - f(x(t_1), y(t_1))]$$

$$+ \cdots$$

$$+ [f(x(b), y(b)) - f(x(t_n), y(t_n))]$$

$$= f(x(b), y(b)) - f(x(a), y(a)). \qquad \blacklozenge$$

This is the desired generalization of the fundamental theorem of calculus in its second form. Thus, we may state Eq. (22.2) in the following form: *the integral of the derivative of a function over an interval yields the difference in values of the function at the endpoints of the interval.* The corresponding statement for a function of two variables takes the form: *the line integral of the gradient of a function over a curve yields the difference in values of the function at the endpoints of the curve.* Let us recall also that the proof of the second statement consists essentially in reducing it to the first one by our familiar device of forming the function $f_C(t)$ of a single variable out of a function $f(x, y)$ of two variables restricted to the curve C.

We turn next to the problem of generalizing statement (22.1). In essence, this statement may be paraphrased as follows: *the definite integral of a function over an interval with fixed initial point and variable endpoint yields a function of the variable endpoint whose derivative is the original function.* In two variables, the corresponding problem would be to find a function having given partial derivatives, and if possible, to do so by representing this function as an integral with variable endpoint. We shall divide this problem into two parts—one for the x derivative and one for the y derivative.

Lemma 22.3 *Let $p(x, y)$, $q(x, y)$ be arbitrary continuous functions in a disk D: $(x - x_1)^2 + (y - y_1)^2 < r^2$. For each point (X, Y) in D, define $g(X, Y)$ as follows: let C be the piecewise smooth curve consisting of the vertical* *line segment from (x_1, y_1) to (x_1, Y) followed by the horizontal line segment from (x_1, Y) to (X, Y) (see Fig. 22.1). Set*

(22.7) $$g(X, Y) = \int_C p \, dx + q \, dy.$$

Then at each point of D,

(22.8) $$g_X(X, Y) = p(X, Y).$$

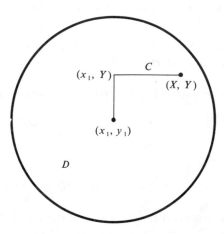

FIGURE 22.1 Notation for Lemma 22.3

PROOF. If we denote the vertical line segment by C_1, and the horizontal one by C_2, then along C_1, $x \equiv x_1$, and $dx/dt \equiv 0$, whatever the choice of the parameter t. Similarly, along C_2, $dy/dt \equiv 0$. If we then make a change of parameter so that along C_1 we use simply y as the parameter, and along C_2 we use x, we have

$$g(X, Y) = \int_{C_1} p \, dx + q \, dy + \int_{C_2} p \, dx + q \, dy$$

$$= \int_{y_1}^{Y} q(x_1, y) \, dy + \int_{x_1}^{X} p(x, Y) \, dx.$$

To form $\partial g/\partial X$, we hold Y fixed and take the ordinary derivative with respect to X. But with Y fixed, the first term on the right is a constant, whereas the second term on the right becomes the integral of a function of the single variable x, over an interval with fixed initial point and variable endpoint. Application of (22.1) therefore gives

$$\frac{\partial}{\partial X} g(X, Y) = \frac{\partial}{\partial X} \int_{y_1}^{Y} q(x_1, y) \, dy + \frac{\partial}{\partial X} \int_{x_1}^{X} p(x, Y) \, dx = 0 + p(X, Y),$$

which is the desired result, Eq. (22.8). ◆

Remark The fact that D was a disk, rather than an arbitrary domain, was needed to guarantee that the curve C lay entirely in D for any choice of (X, Y). It may appear that by choosing C in a different manner the proof could be made to work in any domain D. However, given a continuous function $p(x, y)$ in an arbitrary domain D, there may not exist a continuous function $g(x, y)$ in D satisfying (22.8). An analogous statement holds for Eq. (22.10). For further details, see Ex. 22.18.

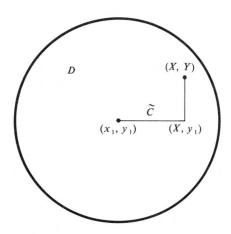

FIGURE 22.2 Notation for Lemma 22.4

Lemma 22.4 *Using the same notation as in Lemma 22.3, let $\tilde{C}$ be the curve consisting of the horizontal segment from (x_1, y_1) to (X, y_1) followed by the vertical segment from (X, y_1) to (X, Y) (Fig. 22.2). Set*

(22.9)
$$h(X, Y) = \int_{\tilde{C}} p \, dx + q \, dy;$$

then

(22.10)
$$h_Y(X, Y) = q(X, Y).$$

PROOF. As in Lemma 22.3, we find

$$h(X, Y) = \int_{x_1}^{X} p(x, y_1) \, dx + \int_{y_1}^{Y} q(X, y) \, dy.$$

Fixing X and letting Y vary, the first term becomes constant and we may apply (22.1) to the second term, yielding (22.10). ♦

Combining the statements of these two lemmas, we may summarize the situation as follows: *given an arbitrary continuous function in a neighborhood of a point* (x_1, y_1) *we can find a function* $g(X, Y)$ *in this neighborhood having the given function as its* X *derivative, and another function* $h(X, Y)$ *having the given function as its* Y *derivative. Furthermore, each of the functions* $g(X, Y)$, $h(X, Y)$ *is represented as a line integral from the fixed point* (x_1, y_1) *to the variable point* (X, Y).

The next question that arises is whether we can construct a single function, having prescribed both partial derivatives. We know already from Lemma 19.1 that the answer is, in general, "no." There does not exist, for example, a function $f(x, y)$ satisfying simultaneously

$$f_x = y^2 \qquad f_y = x^2,$$

since it would follow that on the one hand $f_{xy} = 2y$ and on the other hand $f_{xy} = 2x$. In general, if $p(x, y)$, $q(x, y) \in \mathscr{C}^1$, then it is impossible to have simultaneously

(22.11) $\qquad f_x(x, y) = p(x, y), \qquad f_y(x, y) = q(x, y)$

unless $p_y = q_x$.

It is important to observe that if, for given p and q, there does exist a function $f(x, y)$ satisfying (22.11), then this function may be represented by a line integral. Namely, if $p(x, y)$, $q(x, y)$ are continuous in a domain D, and if there exists a function $f(x, y)$ in D satisfying (22.11), then for any fixed point (x_1, y_1) in D, and for a variable point (X, Y), if we let C be a curve in D starting at (x_1, y_1) and ending at (X, Y), Lemma 22.1 shows that

$$f(X, Y) = \int_C p \, dx + q \, dy + f(x_1, y_1).$$

One consequence of Lemma 22.1 is that for vector fields possessing a potential function, the value of the line integral over a curve C depends only on the beginning and endpoint of C. This property of a vector field has a special designation.

Definition 22.3 Let $p(x, y)$, $q(x, y)$ be continuous in a domain D. The line integral $\int_C p \, dx + q \, dy$ is called *independent of path* if for any two piecewise smooth curves C_1, C_2, lying in D and having the same beginning and endpoints,

(22.12) $$\int_{C_1} p \, dx + q \, dy = \int_{C_2} p \, dx + q \, dy.$$

Combining Lemmas 22.2, 22.3, and 22.4, we arrive at the following fundamental result.

Theorem 22.1   *Let $p(x, y)$, $q(x, y)$ be continuous functions in a domain D. Then the following are equivalent:*

1. *there exists a function $f(x, y)$ in D satisfying (22.11);*
2. *$\int_C p\,dx + q\,dy$ is independent of path.*

PROOF.   It follows immediately from Lemma 22.2 that the first statement implies the second. To prove the converse, let (x_0, y_0) be any fixed point of D. For an arbitrary point (X, Y) in D, let C be a piecewise smooth curve starting at (x_0, y_0) and ending at (X, Y). By assumption, the value of $\int_C p\,dx + q\,dy$ does not depend on the choice of C, but only on the endpoint (X, Y). We may therefore write

$$f(X, Y) = \int_C p\,dx + q\,dy.$$

Consider next an arbitrary point (x_1, y_1) in D and a neighborhood N of (x_1, y_1) such that N lies in D. For any point (X, Y) in N, we may choose the curve C to consist of a fixed curve C_1 from (x_0, y_0) to (x_1, y_1) followed by a vertical line segment C_2 followed by a horizontal segment C_3. Then

$$f(X, Y) = \int_C p\,dx + q\,dy$$

$$= \int_{C_1} p\,dx + q\,dy + \int_{C_2} p\,dx + q\,dy + \int_{C_3} p\,dx + q\,dy$$

$$= f(x_1, y_1) + g(X, Y),$$

where $g(X, Y)$ is the function defined in Lemma 22.3. It follows from Lemma 22.3 that $f_X(X, Y) = g_X(X, Y) = p(X, Y)$ at all points of N, and in particular, at (x_1, y_1). Similarly, we may choose C to consist of C_1 followed by a horizontal segment $\tilde{C}_2$ followed by a vertical segment $\tilde{C}_3$. Then, in the notation of Lemma 22.4,

$$f(X, Y) = f(x_1, y_1) + h(X, Y) \qquad \text{and} \qquad f_Y(X, Y) = h_Y(X, Y) = q(X, Y)$$

throughout N. Since (x_1, y_1) was an arbitrary point of D, it follows that (22.11) is satisfied throughout D, and the theorem is proved.   ◆

It is often convenient to treat the notion of independence of path in a slightly modified form.

Definition 22.4   A curve $C: x(t)$, $y(t)$, $a \leq t \leq b$, is called *closed* if $(x(b), y(b)) = (x(a), y(a))$; that is, if its endpoint is the same as its beginning point.

Lemma 22.5 *Let $p(x, y)$, $q(x, y)$ be continuous in a domain D. Then the following are equivalent:*

1. $\int_C p \, dx + q \, dy$ *is independent of path*;

2. $\int_C p \, dx + q \, dy = 0$ *for every piecewise smooth closed curve C lying in D.*

PROOF. Assume that the first condition holds, and let $C: x(t)$, $y(t)$, $a \leq t \leq b$, be any closed curve in D, where $x(a) = x(b) = x_0$, $y(a) = y(b) = y_0$. Let C_1 and C_2 be the parts of C corresponding to the parameter values $a \leq t \leq (a + b)/2$ and $(a + b)/2 \leq t \leq b$, respectively. Let $\tilde{C}_2$ be the curve C_2 traversed in the opposite direction. Then C_1 and $\tilde{C}_2$ have the same beginning and endpoints, so that

$$\int_C p \, dx + q \, dy = \int_{C_1} p \, dx + q \, dy + \int_{\tilde{C}_2} p \, dx + q \, dy$$

$$= \int_{C_1} p \, dx + q \, dy - \int_{\tilde{C}_2} p \, dx + q \, dy$$

$$= 0.$$

Conversely, suppose that the second condition holds. Let C_1 and C_2 be any two curves in D starting at the same point (x_1, y_1), and ending at the same point (x_2, y_2). Let $\tilde{C}_2$ consist of the curve C_2 traversed in the opposite direction. Then the curve C consisting of C_1 followed by $\tilde{C}_2$ is a closed curve starting and ending at (x_1, y_1), By virtue of Lemma 21.2 we have

$$\int_{C_1} p \, dx + q \, dy - \int_{C_2} p \, dx + q \, dy = \int_{C_1} p \, dx + q \, dy + \int_{\tilde{C}_2} p \, dx + q \, dy$$

$$= \int_C p \, dx + q \, dy = 0.$$

But this is just the condition (22.12) for independence of path, and the lemma is proved. ◆

Example 22.3

Consider the vector field

$$(22.13) \qquad \mathbf{v}(x, y) = \langle p(x, y), q(x, y) \rangle = \left\langle -\frac{y}{x^2 + y^2}, \frac{x}{x^2 + y^2} \right\rangle,$$

defined in the domain D consisting of the whole plane with the origin deleted. Consider the curve

$$C: x = \cos t, \, y = \sin t, \qquad\qquad 0 \leq t \leq 2\pi.$$

Then C is a closed curve in D, starting and ending at $(1, 0)$. On C we have

$$x^2 + y^2 = 1, \qquad x'(t) = -\sin t, \qquad y'(t) = \cos t,$$

and therefore

$$\int_C -\frac{y}{x^2 + y^2}\, dx + \frac{x}{x^2 + y^2}\, dy = \int_0^{2\pi} (\sin^2 t + \cos^2 t)\, dt = 2\pi.$$

Thus the second condition of Lemma 22.5 fails to hold for the vector field (22.13), and consequently the first condition also fails. It follows from Th. 22.1 that there cannot exist a potential function for this vector field. On the other hand, by direct verification, the condition $p_y = q_x$ does hold throughout D. We thus have an example that shows that the converse of Lemma 19.1 is not true.

If we now combine Lemma 22.5 with Th. 22.1 and Lemma 19.1, we arrive at the following result.

Theorem 22.2 *Let $p(x, y)$, $q(x, y) \in \mathscr{C}^1$ in a domain D, then*

1. $\langle p, q \rangle = \nabla f$ *for some $f(x, y)$ in D*

   $\Leftrightarrow$

2. $\int_C p\, dx + q\, dy$ *is independent of path* $\left.\begin{array}{c}\\ \\ \\ \end{array}\right\}$ $\Rightarrow$ 4. $p_y = q_x$ *in D.*

   $\Leftrightarrow$

3. $\int_C p\, dx + q\, dy = 0$ *for all closed C in D*

Thus the first three conditions are equivalent, and each of them implies the fourth. Condition 4 does not imply the first three, as the example just given shows.

It is illuminating to examine the content of this theorem in the light of various physical interpretations of the vector field $\mathbf{v} = \langle p, q \rangle$.

Consider, first, the case where $\mathbf{v}$ represents a force field. Then, as we have indicated in Sect. 21, the line integral $\int_C p\, dx + q\, dy$ represents work. The line integral's independence of path is interpreted to mean physically that the work done in moving a particle from one point to another is independent of the choice of path. In this case we can assign a "potential energy" to each point of the field, such that the difference in potential energy at any two points represents the work done by the field in moving a unit particle from the point of higher potential to one of lower potential, or equivalently, the work that has to be done against the field to move the particle from the point of lower potential to that of higher potential. Thus, lifting a weight through a certain height against the force of gravity is often interpreted as increasing the weight's potential energy, which can be converted back into work by letting the weight drop. In the proof of Th. 22.1, we have seen that potential energy defined in this way is precisely a potential function for the vector force field, that is to say, a function $f(x, y)$ satisfying (22.3). It is from this application that the terminology derives. Condition 3 of Th. 22.2 is also easily interpreted in these terms. It states that it is impossible to increase the potential energy by following some trajectory and returning to the starting point. This is one form of the so-called "law of conservation of energy," and it is in this connection that the term "conservative vector field" arose, indicating a vector field satisfying the equivalent conditions 1, 2, 3 of Th. 22.2.

We turn next to the interpretation of the vector field **v** as the velocity field of a fluid flow. In that case we noted that the line integral $\int_C p \, dx + q \, dy$ may be interpreted as the total rate of flow along the curve C. If C is a closed curve, the line integral would represent the total rate of flow around the curve C. For this reason, a fluid flow satisfying condition 3 of Th. 22.2 (and hence the equivalent conditions 1 and 2) is called *irrotational*, and the same word is used to denote an arbitrary vector field having this property. Thus the designations "conservative" and "irrotational" indicate identical properties of a vector field, but arise from different applications.

In conclusion we note that the terms "conservative" and "irrotational" are frequently used for vector fields satisfying condition 4 of Th. 22.2. As we have seen, this condition is not strictly equivalent to the other three, and consequently there may be some confusion in using the same terms for different properties. The reason that distinct words were not chosen is that under certain restrictions all four properties in Th. 22.2 are equivalent. (See Th. 25.2 and the discussion following it.) In order to show this it is necessary to introduce double integrals, which we shall define and study in the following sections.

Exercises

22.1   Evaluate each of the following line integrals by finding a potential function $f(x, y)$ and applying Lemma 22.1.

    a. $\int_C y \, dx + x \, dy$; $C: x = 3 \cos t, y = 2 \sin t,$ $0 \le t \le \frac{1}{2}\pi$

    b. $\int_C x \, dx + y \, dy$; $C: x = 3 \cos t, y = 2 \sin t,$ $0 \le t \le \frac{1}{2}\pi$

    c. $\int_C 2xy \, dx + x^2 \, dy$; $C: x = t^2 + 1, y = 3 - 2t,$ $0 \le t \le 1$

    d. $\int_C 2 \cos x \sin y \, dx + 2 \sin x \cos y \, dy$; $C: x = \frac{1}{6}\pi t, y = \frac{1}{4}\pi t,$

                                                      $1 \le t \le 2$

    e. $\int_C y/x \, dx + \log x \, dy$; $C: x = e^t, y = t^e,$ $1 \le t \le 2$

    f. $\int_C e^x \sin y \, dx + e^x \cos y \, dy$; $C: x = e^t, y = \pi[t^2 - \frac{1}{2}(t - 1)],$

                                                       $0 \le t \le 1$

22.2   Do Ex. 21.5, using Lemma 22.1.

22.3   Illustrate the definition of a piecewise smooth curve by a sketch.

22.4   Let C be the curve from $(1, 1)$ to $(3, 2)$ consisting of a horizontal segment C_1 followed by a vertical segment C_2.

    a. Write C explicitly as a piecewise smooth curve.

    b. Evaluate $\int_C x^2 y\, dx - 3x\, dy$.

    c. Evaluate $\int_C y^2\, dx + 2xy\, dy$, first by a direct computation, and then by using Lemma 22.2.

22.5   Let C be the curve from $(3, 2)$ to $(1, 1)$ consisting of a horizontal line segment C_1 followed by a vertical line segment C_2. Answer parts a, b, and c of Ex. 22.4 for this curve.

22.6   Let C be the rectangle consisting of the four successive line segments:

    $C_1:(0, 0)$ to $(a, 0)$; $C_2:(a, 0)$ to (a, b);

    $C_3:(a, b)$ to $(0, b)$; $C_4:(0, b)$ to $(0, 0)$.

    a. Write C explicitly as a piecewise smooth curve.

    b. Evaluate $\int_C x^2\, dx + y^2\, dy$.

    c. Evaluate $\int_C -y\, dx + x\, dy$.

22.7   Let $\langle p, q \rangle$ be a conservative vector field in a domain D. Then by the observation following Eq. (22.11), a potential function can be written down explicitly in the form of a line integral. If the domain D is of a simple type, such as a disk, or a rectangle, or the whole plane, then the path of integration can be chosen in a consistent simple way. For example, choosing the path consisting of a horizontal line segment followed by a vertical line segment, which joins $(0, 0)$ to (X, Y), show that the function f defined by

$$f(X, Y) = \int_0^X p(x, 0)\, dx + \int_0^Y q(X, y)\, dy$$

is a potential function.

22.8   Use the method of Ex. 22.7 to find potential functions for each of the following vector fields. (The domain D is the whole plane.)

    a. $\langle 3x^2 y^2 + 2xy^3, 3x^2 y^2 + 2x^3 y \rangle$

    b. $\langle 2x + 2xy - y, x^2 - 3y^2 - x \rangle$

    c. $\left\langle \dfrac{x(1 + x^2 + y^2)}{(1 + x^2)^2}, -\dfrac{y}{1 + x^2} \right\rangle$

    d. $\langle e^{xy} + xye^{xy}, x^2 e^{xy} \rangle$

22.9   Use the answers to Ex. 22.8 to evaluate the following line integrals.

    a. $\int_C (3x^2 y^2 + 2xy^3)\, dx + (3x^2 y^2 + 2x^3 y)\, dy$;

              $C: x = 1 + \sin^3 t, y = \cos^2 t - 1$, $0 \le t \le \frac{1}{2}\pi$

    b. $\int_C (2x + 2xy - y)\, dx + (x^2 - 3y^2 - x)\, dy$; C is the curve of Ex. 22.4

    c. $\int_C \dfrac{x(1 + x^2 + y^2)}{(1 + x^2)^2}\, dx - \dfrac{y}{1 + x^2}\, dy$; $C: y = 2x^2$, $0 \le x \le 1$.

    d. $\int_C (e^{xy} + xye^{xy})\, dx + x^2 e^{xy}\, dy$; $C: x = (y^2 - 3)/2, -2 \le y \le 2$.

22.10 Show that two vector fields may have the same direction at every point and one of the vector fields may be conservative, while the other one is not. (*Hint:* consider the effect on a vector field of multiplying through by a nonzero function.)

22.11 In each of the following cases we give a vector field $\langle p, q \rangle$ and a function h. Show that the vector field $\langle p, q \rangle$ is not conservative, but that the vector field $\langle \tilde{p}, \tilde{q} \rangle = h\langle p, q \rangle = \langle hp, hq \rangle$ is. Find a potential function f for $\langle \tilde{p}, \tilde{q} \rangle$.

a. $\langle p, q \rangle = \langle 3y, 6x - 3y \rangle$; $h = y$
b. $\langle p, q \rangle = \langle y^2, xy \rangle$; $h = 2x$
c. $\langle p, q \rangle = \langle y^2, xy \rangle$; $h = x/(1 + x^2 y^2)$

22.12 Let $f(x, y)$ and $g(x, y)$ be functionally dependent in a domain D. Show that the vector fields ∇f and ∇g have the same direction at every point where neither of them vanishes.

22.13 Let $u(x, y)$ be a harmonic function in a domain D. A function $v(x, y)$ in D is called a *conjugate harmonic function* to $u(x, y)$ if u and v satisfy the Cauchy-Riemann equations $u_x = v_y$, $u_y = -v_x$. Given $u(x, y)$ the problem of finding a conjugate harmonic function is equivalent to that of finding a potential function $v(x, y)$ for the vector field $\langle p, q \rangle = \langle -u_y, u_x \rangle$. Thus, a conjugate harmonic function can often be found by the method of Ex. 22.7. Carry this out for the following functions.

a. $u = x - y$ *d.* $u = \cos x \cosh y$
b. $u = x^2 - y^2$ *e.* $u = e^x \cos y$
c. $u = \sin x \sinh y$ *f.* $u = x^3 y - xy^3$

22.14 Consider the vector field arising from a force that obeys the inverse square law

$$\langle p, q \rangle = -k \frac{1}{(x^2 + y^2)^{3/2}} \langle x, y \rangle.$$

According to Ex. 21.23e, if (x_1, y_1) is a fixed point and (X, Y) a variable point, then

$$(22.14) \qquad \int_C p \, dx + q \, dy = k \left(\frac{1}{\sqrt{X^2 + Y^2}} - \frac{1}{\sqrt{x_1^2 + y_1^2}} \right),$$

where C is a curve starting at (x_1, y_1), ending at (X, Y), and consisting of horizontal and vertical line segments.

    a. Deduce from the above facts that if this vector field is conservative, then its potential function must equal $k/(x^2 + y^2)^{1/2}$ up to an additive constant.
    b. Show that the function $f(x, y) = k/(x^2 + y^2)^{1/2}$ is in fact a potential function for this field, and deduce that Eq. (22.14) holds for *any* path C joining the given points.

22.15 Interpret the following vector fields as force fields. Show that each is conservative and find the work done in moving a particle from the point $(-2, 1)$ to the point $(3, -2)$.

a. $\langle -x, y \rangle$ *c.* $\langle y^2, 2xy \rangle$

b. $\langle 2x, 3y^2 \rangle$ *d.* $2\langle ax + by, bx + cy \rangle$

22.16 Let a particle move along a path $C: x(t), y(t), a \le t \le b$, under the action of a force field $\mathbf{F} = \langle p, q \rangle$. Consider at each point

| the velocity vector | $\mathbf{v} = \langle dx/dt, dy/dt \rangle$, |
| the acceleration vector | $\mathbf{a} = d\mathbf{v}/dt$. |

If the particle has mass m, then the *kinetic energy* of the particle at each point of the path C is defined by

$$\text{kinetic energy} = \tfrac{1}{2}m|\mathbf{v}|^2.$$

Newton's second law of motion states that at each point,

$$\mathbf{F} = m\mathbf{a}.$$

Using this notation, show the following.

    a. The work done by the force $\mathbf{F}$ in moving the particle along the path C is equal to

$$\int_a^b (\mathbf{F} \cdot \mathbf{v}) \, dt.$$

    b. At each point of the curve C, the rate of change of kinetic energy is given by

$$\frac{d}{dt} (\tfrac{1}{2}m|\mathbf{v}|^2) = \mathbf{F} \cdot \mathbf{v}.$$

    c. The kinetic energy of the particle at the end of the path C is equal to its kinetic energy at the beginning plus the total work done by the force in moving the particle along the path. (This is the *work-energy principle*.)

22.17 Let $\mathbf{F} = \langle p, q \rangle$ be a conservative force field, and let $f(x, y)$ be a potential function. The *potential energy* of a particle is defined at each point by

$$\text{potential energy} = -f(x, y) + c,$$

where c is a constant that may be chosen in any convenient manner. (The value of c is not important, since it is always the *difference* in potential energy between two points that plays a role.)

    a. Show that if the force $\mathbf{F}$ moves a particle along a path C, then the potential energy of the particle is *decreased* by the amount of work done by the force $\mathbf{F}$.

(Note that the force required at each point to move the particle against the field would be $-\mathbf{F}$. Thus the potential energy is *increased* by the amount of work done *against* the field. For example, lifting a weight against the force of gravity increases its potential energy. If the weight is allowed to drop, its potential energy decreases by the amount of work done by the gravitational field during the motion. These considerations explain why potential energy in physics is chosen to be the *negative* of the potential function in mathematics.)

b. Prove the *law of conservation of energy*: if a particle moves along a path C under the action of a conservative force field, then the sum of its potential energy and kinetic energy remains constant.

*22.18 Let $f(x, y)$ be continuous in a domain D. A natural question is whether there exists a solution $g(x, y)$ of the equation

$$g_x = f$$

in D. According to Lemma 22.3, a solution always exists if D is a disk. Similarly, Lemma 22.4 guarantees that the equation

$$h_y = f$$

has a solution $h(x, y)$, if D is a disk. However, these equations do not necessarily have solutions in an arbitrary domain D. Suppose for example, that we choose

$$D: 1 < x^2 + y^2 < 4, y > 0, \qquad f(x, y) = \frac{1}{x^2 + (y - 1)^2}.$$

Then $f(x, y)$ is continuous in D (and in fact in the whole plane except for the point $(0, 1)$, which is not in D, although it is a boundary point of D). Show that there cannot exist a continuous function $g(x, y)$ in D satisfying $g_x = f$. (*Hint:* proceed by contradiction. Suppose there were such a function $g(x, y)$. Then

$$g(1, 1) - g(-1, 1) = \lim_{\epsilon \to 0} [g(1, 1 + \epsilon) - g(-1, 1 + \epsilon)]$$

$$= \lim_{\epsilon \to 0} \int_{-1}^{1} g_x(x, 1 + \epsilon) \, dx$$

$$= \lim_{\epsilon \to 0} \int_{-1}^{1} \frac{1}{x^2 + \epsilon^2} \, dx.$$

However, for fixed $\epsilon > 0$ the latter integral can be evaluated explicitly, and it may be verified that

$$\lim_{\epsilon \to 0} \int_{-1}^{1} \frac{1}{x^2 + \epsilon^2} \, dx = \infty.$$

This contradiction proves the result.)

Remark The problem of finding a function $f(x, y)$ that satisfies

$$f_x = p, \qquad f_y = q$$

for given p and q, arises in a variety of contexts and often under different names. In Ex. 22.17 it is the problem of finding the potential energy of a given force field. More generally, we have described it as the problem of finding a potential function of a given vector field. We note two more versions of the same problem.

First, using the notation of differentials (see the Remark following Ex. 16.16) the differential of a function $f(x, y)$ is written as

$$df = f_x \, dx + f_y \, dy.$$

Thus, if $f_x = p$ and $f_y = q$, this may be written as

$$df = p \, dx + q \, dy.$$

Starting with arbitrary functions $p(x, y)$, $q(x, y)$, the expression

$$p \, dx + q \, dy$$

is called a *differential form* (since it has the form of a differential of a function without necessarily being one). If there exists a function $f(x, y)$ satisfying $f_x = p$, $f_y = q$, then $p \, dx + q \, dy$ is called an *exact differential form* or an *exact differential*. Thus the problem of deciding whether a differential form $p \, dx + q \, dy$ is exact, and if it is exact, of representing it as the differential of a given function, is precisely the same as deciding whether the vector field $\langle p, q \rangle$ is conservative, and if it is conservative, finding a potential function. Using the terminology of differentials, Th. 22.1 may be stated in the form

$$p \, dx + q \, dy \text{ is exact} \Leftrightarrow \int_C p \, dx + q \, dy \text{ is independent of path.}$$

If C joins (x_1, y_1) to (x_2, y_2), then the notation

$$\int_C p \, dx + q \, dy = \int_{(x_1, y_1)}^{(x_2, y_2)} p \, dx + q \, dy$$

is permissible whenever the integral is independent of path. Lemma 22.1 may then be stated as follows: *if $p \, dx + q \, dy$ is exact, and equal to df, then*

$$\int_{(x_1, y_1)}^{(x_2, y_2)} p \, dx + q \, dy = \int_{(x_1, y_1)}^{(x_2, y_2)} df = f(x_2, y_2) - f(x_1, y_1).$$

Thus, the differential notation is highly suggestive. For the actual computation of the function $f(x, y)$ from a given p and q, the method of Ex. 22.7 may be used.

As a final example of the same problem in a different context, consider the ordinary differential equation

$$\frac{dy}{dx} = \varphi(x, y).$$

A solution consists of a function $y = g(x)$ such that

$$g'(x) \equiv \varphi(x, g(x)).$$

Such equations often arise in the form

$$(22.15) \qquad\qquad p(x, y) + q(x, y) \frac{dy}{dx} = 0,$$

in which case $\varphi(x, y) = -p(x, y)/q(x, y)$. If the vector field $\langle p, q \rangle$ has a potential function $f(x, y)$, then applying the implicit function theorem to a level curve

$f(x, y) = c$ near a point where $f_y \neq 0$, we find that this level curve may be written as $y = g(x)$, where

$$\frac{dy}{dx} = -\frac{f_x}{f_y} = -\frac{p}{q} = \varphi.$$

The level curves of $f(x, y)$ are therefore implicit forms of solutions to the given equation. If such a function f exists, Eq. (22.15) is called an *exact differential equation*. Equation (22.15) is also often written in the form

$$p(x, y) \, dx + q(x, y) \, dy = 0,$$

using the notation of differentials. In fact, the following three statements are equivalent:

1. $p(x, y) + q(x, y) \, dy/dx = 0$ is an exact differential equation.
2. $p \, dx + q \, dy$ is an exact differential form.
3. $\langle p, q \rangle$ is a conservative vector field.

Thus, solutions of an exact differential equation may be represented as level curves of a function $f(x, y)$, where $f(x, y)$ may be found explicitly by the method of Ex. 22.7.

We note in conclusion that multiplying through Eq. (22.15) by a nonzero function $h(x, y)$ gives a new equation

$$(22.16) \qquad \tilde{p}(x, y) + \tilde{q}(x, y) \frac{dy}{dx} = 0,$$

where $\tilde{p} = hp$, $\tilde{q} = hq$. The solutions of Eq. (22.16) coincide with those of Eq. (22.15). (Equations (22.15) and (22.16) are in fact two forms of the same equation.) In particular, if Eq. (22.15) is not exact, it is of interest to look for a function $h(x, y)$ that makes Eq. (22.16) exact, in which case solutions may be found by the method indicated above. Such a function $h(x, y)$ is called an *integrating factor* for Eq. (22.15). Finding an integrating factor is equivalent in our terminology to making a given vector field conservative by changing its magnitude at each point (see Exs. 22.10 and 22.11). There is no general solution to the problem of finding an integrating factor, but there are methods for treating a variety of important special cases. These are discussed in detail in books on differential equations.

23 Area

The object of this and the following section is to develop the notion of double integration. Rather than present the theory in detail, we prefer to stress certain points and merely sketch others. There are two reasons for this mode of presentation. The first is that the theory of multiple integration is essentially identical for any number of variables, and one may as well present it in general form at the appropriate place. The second is that there is a large gap between theory and practice in multiple integration. In practice, multiple integrals are almost invariably reduced to a succession of ordinary integrals.

Thus, for our present purposes, as well as for most elementary applications, the principal properties of double integrals are those stated in Th. 24.3, and in Th. 24.4 and its Corollary. Our aim, therefore, is to present first some of the fundamental ideas and then to proceed to the properties needed in applications.[3]

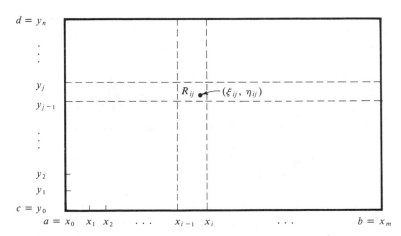

FIGURE 23.1   Subdivision of a rectangle for defining the double integral

To begin, consider a function $f(x, y)$ defined on a rectangle

$$R: a \leq x \leq b, \; c \leq y \leq d.$$

To define the double integral of f over R, we proceed exactly as in the case of the single integral. Divide the rectangle R into a number of smaller rectangles

$$R_{ij}: x_{i-1} \leq x \leq x_i, \; y_{j-1} \leq y \leq y_j,$$

where

$$a = x_0 < x_1 < \cdots < x_m = b; \qquad c = y_0 < y_1 < \cdots < y_n = d$$

(see Fig. 23.1). Choose an arbitrary point (ξ_{ij}, η_{ij}) in the rectangle R_{ij}, and

---

[3] The development followed here is in broad outline an adaptation to two dimensions of the treatment of multiple integration given by Maak [25]. Some changes in approach and organization make it impossible to give specific references in Maak's book for results stated here without proof. However, with the exception of one point in the proof of Th. 23.3 (see Footnote 6, p. 335) all results stated in this and the following section are consequences of the theorems proved (in the n-dimensional case) in the first three sections of Chapter 11 of Maak. For a different treatment of multiple integration, carried out specifically in the case of two dimensions, see [6], Sections 3.1 and 3.2.

form the sum

$$(23.1) \qquad \sum_{\substack{i=1,\ldots,m \\ j=1,\ldots,n}} f(\xi_{ij}, \eta_{ij})(x_i - x_{i-1})(y_j - y_{j-1}).$$

If these sums tend to a limit as the size of the subdivisions tends to zero, then $f(x, y)$ is said to be *integrable*, and the value of the limit is called the *double integral of f over R*. Analogs of the basic properties of integration in one variable can be established without difficulty for integration in two variables over a rectangle. In particular, if $f(x, y)$ is continuous on R, then the integral exists.

As we have seen from the beginning in our study of functions of two variables, a crucial difficulty is that it is not possible in general to restrict our attention to functions defined on rectangles. First of all, the set of points where a function is defined may be far more complicated. Furthermore, even for functions defined on a rectangle (or in the whole plane) it may be necessary to perform an integration over a part of the domain of definition. Thus, even for the simple case of a circle, it may not be at all clear how to adapt the above definition.

As a first step toward obtaining a general definition of the integral, let us note that the expression (23.1) may also be written in the form

$$(23.2) \qquad \sum f(\xi_{ij}, \eta_{ij}) A_{ij},$$

where A_{ij} is the area of the rectangle R_{ij}. To define the integral of $f(x, y)$ over a more complicated set of points than a rectangle, one approach would be to divide up the set into a number of small parts, choose a point in each part, multiply the value of the function at the point by the area of the part containing that point, form the sum over all the parts, as in (23.2), and then take the limit as the subdivisions become progressively finer. It is impossible to carry this out unless we have a clear-cut notion of the area of each of the parts in the subdivision. We shall, therefore, devote the remainder of this section to a discussion of area, and then return to the subject of integration proper in the following section.

On an intuitive basis, the meaning of area seems immediately clear, whereas that of the integral appears very difficult to grasp on first encounter. Nevertheless a careful discussion of the two notions involves exactly the same order of difficulty. Let us start with a brief discussion of area as it might be introduced in a course in plane geometry.

To begin with, it is assumed that a unit of length has been chosen. The basic unit of area is a square each of whose sides has unit length. For any positive integers m and n, a rectangle whose length is m units and whose width is n units can be subdivided into m-times-n unit squares. This rectangle is therefore assigned the area mn. Dividing the length of this rectangle into p

equal parts, and the width into q equal parts, leads to a subdivision of the rectangle into p-times-q smaller rectangles, each of the same size; namely, of length m/p and width n/q. Since the total area of the large rectangle is mn, the area of each of the small rectangles must be mn/pq or $(m/p)(n/q)$. Thus if r and s are any two (positive) rational numbers, we are led to assign the number rs for the area of a rectangle of length r and width s. The same value is used if r and s are arbitrary (positive) real numbers. (The bridge between rational and real numbers is often ignored, but it is not hard to justify, once the basic properties of area and of the real numbers have been studied.)

The next step is to note that from a given parallelogram, we obtain a rectangle having the same area by cutting a triangle off one end and translating it over to the other end. This gives the formula "base times height" for the area of the parallelogram. Since an arbitrary triangle can be represented as half of a parallelogram, we obtain the formula "one half base times height" for the area of a triangle. Finally, an arbitrary polygon may be divided up into triangles, and its area thereby computed.

Following a discussion of this nature, it is customary to include the formula $A = \pi r^2$ for the area of a circle of radius r. It is not always made clear that this formula falls into an entirely different category from the previous ones. Assuming that the number π has been adequately described, it still remains to define what is meant by the area enclosed by a curve, and then show that in the case of a circle this definition yields the desired expression. The most common procedure is to find the area of a regular polygon inscribed in the circle, and then to consider the limit of the area as the number of sides tends to infinity. This is obviously a method specially adapted to the case of a circle.

There are many possible approaches to the definition of area, the choice depending in part on the degree of generality of the point sets whose area one wishes to define. We limit ourselves to a relatively simple class, which we denote as "figures."

Definition 23.1   Let D be a bounded plane domain whose boundary consists of a finite number of piecewise smooth curves. Then the set of points in D together with its boundary points is called a *figure*.

Remark   We generally use the letter F to denote a figure. We recall that D bounded means that it lies in some disk $x^2 + y^2 < R^2$. Then F lies in the closed disk $x^2 + y^2 \le R^2$.

Definition 23.2   A *polynomial figure* is a figure defined by a set of inequalities $P_1(x, y) \ge 0, \ldots, P_n(x, y) \ge 0$, where $P_1(x, y), \ldots, P_n(x, y)$ are polynomials.

We refer to the discussion of polynomial figures in Sect. 4 and to Examples 4.5–4.10 given there. The great majority of figures one encounters in computing double integrals are of this type.

We now define the area of an arbitrary figure.

Definition 23.3 Let F be an arbitrary figure. For each integer $k \geq 0$, divide the plane into squares by the lines $x = m/2^k$, $y = n/2^k$, where m and n run through all the integers. Let N_k be the number of squares of the form

$$(23.3) \qquad \frac{m-1}{2^k} \leq x \leq \frac{m}{2^k}, \qquad \frac{n-1}{2^k} \leq y \leq \frac{n}{2^k},$$

which lie entirely within the figure F. Let

$$(23.4) \qquad A_k = N_k \left(\frac{1}{2^k}\right)^2.$$

Then the number

$$(23.5) \qquad A = \lim_{k \to \infty} A_k$$

is called the *area* of F.

Remark In Fig. 23.2 we have indicated for two successive values of k the operation described in Def. 23.3. Let us note that the quantity $(1/2^k)^2$ represents the value we would normally assign for the area of the squares (23.3), and consequently the number A_k would correspond to the total area of those squares which lie inside F. Regardless of the geometric interpretation of A_k, the numbers N_k satisfy the relation $N_{k+1} \geq 4N_k$, since to each square of the form (23.3) that lies in F correspond four squares of sides $1/2^{k+1}$, which lie in F. Therefore,

$$(23.6) \qquad A_{k+1} = N_{k+1}\left(\frac{1}{2^{k+1}}\right)^2 \geq 4N_k\left(\frac{1}{2^{k+1}}\right)^2 = N_k\left(\frac{1}{2^k}\right)^2 = A_k.$$

Thus the numbers A_k form a sequence in which each term is at least equal to or greater than the previous one. On the other hand, since by definition every figure F lies in some circle $x^2 + y^2 \leq R^2$, it also lies in a square $-M \leq x \leq M$, $-M \leq y \leq M$, for some integer M, and it follows that $A_k \leq (2M)^2$ for all k. It is a basic property of real numbers (see, for example, [3] Vol. I, Section 9.3) that the sequence A_k must then tend to a limit, so that the area as given by (23.5) is well defined for an arbitrary figure F.[4]

---

[4] The procedure described in the definition of area may be applied to *any* bounded set of points F. Even if F is not a figure, the numbers A_k will still converge to a number A which is called the *inner area* of F. However, for complicated point sets F, the inner area so defined will not have the basic properties listed in Th. 23.1.

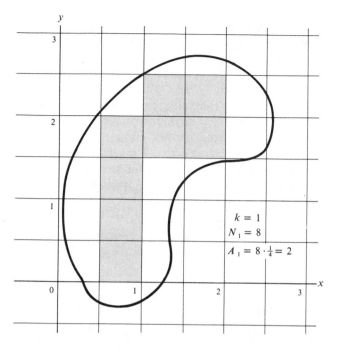

$$k = 1$$
$$N_1 = 8$$
$$A_1 = 8 \cdot \tfrac{1}{4} = 2$$

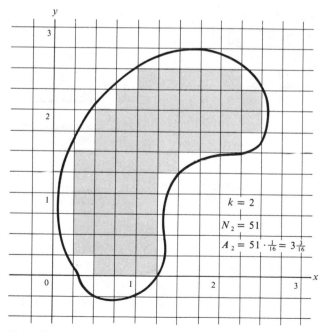

$$k = 2$$
$$N_2 = 51$$
$$A_2 = 51 \cdot \tfrac{1}{16} = 3\tfrac{3}{16}$$

FIGURE 23.2 Illustration of the definition of area

There are two questions that must be asked about our definition (or for that matter, about any definition of area that is offered). The first is, "does it correspond to our intuitive notion of area?" and the second is, "can it be used practically to compute the area of a given figure?" The answer to the second question is "no," although the definition does give an approximate value of area by computing explicitly a value of A_k. (This can be done in practice by placing a transparent square grid over the figure and counting the number of squares that fall inside.) On the other hand, using our definition as a basis, we shall derive a practical method for computing the area of many figures (Th. 23.3). As for the first question, it will be completely answered in the two following theorems. To state them, we introduce some useful terminology.

Definition 23.4 Two figures are said to *overlap* if they have at least one point in common other than boundary points (Fig. 23.3).

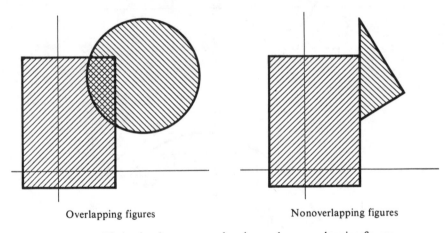

Overlapping figures   Nonoverlapping figures

FIGURE 23.3 Distinction between overlapping and nonoverlapping figures

Definition 23.5 F is called the *union* of $F_1, \ldots, F_n$ if the points of F consist of all points that lie in at least one of $F_1, \ldots, F_n$. We write $F = F_1 \cup \cdots \cup F_n$ (Fig. 23.4).

Definition 2.36 A *partition* of a figure F is a collection of figures $F_1, \ldots, F_n$ such that no two overlap, and $F = F_1 \cup \cdots \cup F_n$ (Fig. 23.5).

Theorem 23.1 *The notion of area introduced above has the following properties.*

1. *For every figure F, the area of F is a nonnegative real number.*

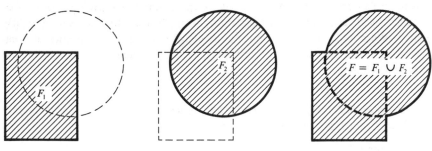

FIGURE 23.4 The union of two figures

2. *If $F_1, \ldots, F_n$ form a partition of F, then the area of F is the sum of the areas of $F_1, \ldots, F_n$.*
3. *If $\tilde{F}$ is obtained from F by a translation, then F and $\tilde{F}$ have the same area.*
4. *If F is the square $0 \leq x \leq 1, 0 \leq y \leq 1$, then the area of F is 1.*

We do not give the details of the proof for this theorem. Let us note merely that properties 1 and 4 follow immediately from the definition, while properties 2 and 3 are considerably more difficult to prove. It is precisely in order to have property 2 that we so strictly limited the notion of a figure. In fact, if we apply our definition to more general point sets, property 2 need not hold at all (see Ex. 23.15).

The important fact to observe concerning Th. 23.1 is that the four properties listed are absolutely basic to our intuitive notion of area. Indeed, a careful examination of the elementary discussion given above, in deriving

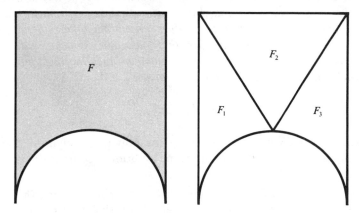

FIGURE 23.5 The partition of a figure

the formula for the area of a triangle, reveals that these four properties of area are used repeatedly in the reasoning, even though they are not explicitly stated. We could, of course, continue to list further basic properties of area, but that turns out to be unnecessary, in consequence of the following theorem.

Theorem 23.2 *Suppose that to each figure F is assigned a real number A(F) such that the following conditions are satisfied:*

1. *For all F, $A(F) \geq 0$.*
2. *If $F_1, \ldots, F_n$ form a partition of F, then $A(F) = A(F_1) + \cdots + A(F_n)$.*
3. *If $\tilde{F}$ is obtained from F by a translation, then $A(\tilde{F}) = A(F)$.*
4. *If F is the square $0 \leq x \leq 1, 0 \leq y \leq 1$, then $A(F) = 1$.*

Then for every figure F, A(F) equals the area of F.

Once again, we omit the proof of this theorem (see, however, Ex. 23.13). The principal thing is to understand its content. We may summarize the situation as follows.

First, there is no unique way to define area. The particular definition we have given is one that seems intuitively reasonable and at the same time can be used to derive the four basic properties listed above.

Second, any other definition having these four properties must assign the same value for the area of an arbitrary figure as that obtained from our definition.

The next result is an illustration of how Th. 23.2 may be used.

Theorem 23.3 *Let F be an arbitrary figure, and let L be any straight line. Let L(t) be the straight line parallel to L, at a distance $|t|$ from L, where one side of the line L is chosen for positive values of t and the other side for negative values of t. Let h(t) denote the total length of those line segments on L(t) which lie in F (Fig. 23.6). Let a and b be any two numbers such that L(t) fails to meet F if $t < a$ or if $t > b$. Then the area of F equals*

(23.7)
$$\int_a^b h(t) \, dt.$$

PROOF. We indicate the proof only for polynomial figures. The same ideas apply in general, but there are some additional complications.[5]

---

[5] The proof for a general figure follows from Th. 17 in Chapter III of [16]. The function *f* in that theorem should be chosen to be equal to 1 at each point of *F*, and equal to 0 everywhere else.

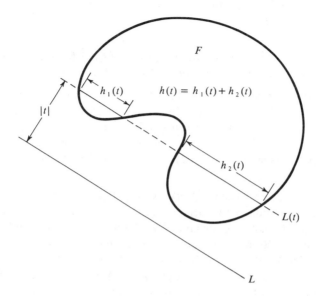

FIGURE 23.6 Notation for Th. 23.3

We consider a fixed line L, and we assign to each figure F the value[6] of the integral (23.7), which we denote by $A(F)$. In order to prove the theorem, it is sufficient to show that the number $A(F)$ so defined verifies the four conditions of Th. 23.2. The first of these is obvious, since $h(t) \geq 0$ for all t. For the second, we denote by $h_1(t), \ldots, h_n(t)$ the value of the function $h(t)$ corresponding to the figures $F_1, \ldots, F_n$, respectively. Then, by the definition of a partition,

$$(23.8) \qquad h(t) = h_1(t) + \cdots + h_n(t)$$

except in the case where the line $L(t)$ contains a line segment that lies on the boundary of two distinct figures in the partition. However, assuming that $F_1, \ldots, F_n$ are all polynomial figures, this can occur for at most a finite number of values of t. Thus Eq. (23.8) holds except for at most a finite number of values of t, and integrating (23.8) from a to b yields condition 2. Condition 3 is clear, since for the figure $\tilde{F}$, the corresponding function $\tilde{h}(t)$ is of the form $\tilde{h}(t) = h(t + c)$ for a fixed constant c, and the integral (23.7) gives the same result for $h(t)$ and $\tilde{h}(t)$. Finally, condition 4 may be verified by finding

---

[6] The fact that the integral (23.7) exists follows in the general case from the reference given in the previous footnote. In the special case of polynomial figures one can show that the function $h(t)$ is bounded and continuous except for at most a finite number of values of t, from which the existence of the integral (23.7) follows immediately.

explicitly the function $h(t)$ in this case, and carrying out the integration (23.7) (see Ex. 23.14). ◆

Corollary 1 *Let F be an arbitrary figure, and let $\hat{F}$ be obtained from F by a rotation or a reflection. Then the area of $\hat{F}$ equals the area of F.*

PROOF. In the case of a rotation, if the line L is subjected to the same rotation to yield a line $\hat{L}$, then the function $\hat{h}(t)$ corresponding to $\hat{L}$ and $\hat{F}$ equals $h(t)$ for all t, and the integral (23.7) yields the same value for both figures. In the case of a reflection, choose L to be the line in which the figure is reflected. Then $\hat{h}(t) = h(-t)$, and again the integral is the same. ◆

Remark Clearly, a basic property we would expect from any reasonable definition of area would be invariance under all Euclidean motions. In the fundamental list of properties given in Th. 23.1 we only mentioned invariance under translation. We now see that it would have been superfluous to add invariance under rotation and reflection, since by Ths. 23.2 and 23.3 these are a consequence of the four stated properties, and indeed, must hold for any definition of area satisfying the conditions of Th. 23.2.

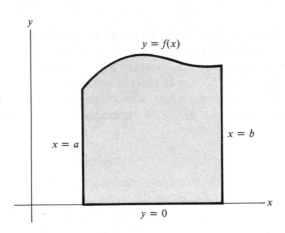

FIGURE 23.7 Area under a curve

Corollary 2 *Let $f(x) \in \mathscr{C}^1$ for $a \le x \le b$, suppose $f(x) > 0$. Then the area of the figure (Fig. 23.7) defined by*

(23.9) $$x \ge a, \quad x \le b, \qquad y \ge 0, \quad y \le f(x)$$

is equal to

(23.10) $$\int_a^b f(x)\, dx.$$

PROOF.   Note first that the point-set defined by (23.9) is indeed a figure in our sense. Its boundary consists of three line segments and a regular curve. If we choose the line L in Th. 23.3 to be the y axis, then $t = x$, $h(t) = f(x)$, and (23.7) reduces to (23.10).   ◆

Remark   In case $f(x)$ is a polynomial, Eqs. (23.9) define a polynomial figure. This was the case for which we carried out the proof of Th. 23.3. Actually one can show that if $f(x)$ is merely continuous, our definition of area gives the same value as the integral (23.10).

We may note that it is often the practice in elementary calculus not to define area, but to assume that it can be defined in a way that satisfies certain elementary properties (for example, if one figure is included in another, then the area of the first is less than or equal to that of the second), and to deduce from these properties that the integral (23.10) must represent the area. Thus, the discussion in the present section indicates one way to fill in the gap, by giving a definition of area having the necessary properties to obtain formula (23.10).

Exercises

23.1   Using the notation of Th. 23.3, let L be the y axis and let $L(t)$ be the line $x = t$. Sketch each of the following figures F and find explicitly the function $h(t)$.

  a. $F: x \geq 0,\ y \geq 0,\ x + y \leq 1.$
  b. $F: 2x - y \geq 0,\ 2y + x \geq 0,\ 2 - x \geq 0$
  c. $F: x - y \geq 0,\ y - x^2 \geq 0$
  d. $F: x + y \geq 1,\ x^2 + y^2 \leq 1$
  e. $F: x^2 + y^2 \leq 1,\ y - 2x^2 + 1 \geq 0$
  f. $F: (1 + x^2)y \leq 1,\ 2y - x^2 \geq 0$
  g. $F: y \leq e^x,\ y \geq e^{-x},\ x \leq 2$
  h. $F: y \leq \cos x,\ y \geq \sin x,\ x \geq 0,\ x \leq 1$
  i. $F: 2x + y \leq 2,\ y - 2x \leq 2,\ y \geq 0$
  j. $F: |x| + |y| \leq 1$
  k. $F: x + 3y \geq 0,\ 3x - y \geq 0,\ 2x + y \leq 5$
  l. $F: 4x - y^2 \geq 0,\ 2x - y \leq 4$
  m. $F: x^2 + y^2 \leq 9,\ x^2 + y^2 \geq 4$
  n. $F: 2x - y^2 \geq 0,\ 4(x - 1) - y^2 \leq 0$
  o. $F: x^2 - 4 \leq 0,\ y^2 - 3y \leq 0,\ 4x^2 + y^2 \geq 4$
  *p. $F: x^2 + y^2 \leq a^2,\ x^2 \leq b^2;\ 0 < b \leq a$

23.2   Find the area of each figure in Ex. 23.1, using the value of $h(t)$ obtained there and applying Th. 23.3. Check your answers, wherever possible, by means of standard expressions for area.

23.3 For each of the figures in Ex. 23.1, if L is the x axis, and $L(t)$ is the line $y = t$, find the corresponding function $h(t)$ (again in the notation of Th. 23.3).

23.4   Use the answers to Ex. 23.3 to compute the area of each figure in Ex. 23.1. Note that for a given figure, the choice of the line L can make a significant difference in the ease of computing the area.

23.5   Using the notation of Th. 23.3, let L be the line $x + y = 0$.

  a. What is the equation of the line $L(t)$?
  b. Find the function $h(t)$ for the figure of Ex. 23.1a, and use it to compute the area.
  c. Answer part b for the figure of Ex. 23.1j.
  d. Answer part b for the figure of Ex. 23.1m.

23.6   Let F be the figure of Ex. 23.1a.

  a. Sketch the figure F and the first three steps in the definition of the area of F (in analogy with Fig. 23.2).
  b. Using the notation in the definition of area, compute the quantities N_1, N_2, N_3, A_1, A_2, A_3 for this particular figure.
  c. Find N_k and A_k for arbitrary k for this figure.
  d. Use the answer to part c to obtain the area of F by direct use of the definition.

23.7   Which of the figures in Ex. 23.1 are polynomial figures?

23.8   Sketch the figure bounded by the right-hand loop of the curve $y^2 = x^2(1 - x^2)$ and find its area.

23.9   Sketch the figure bounded by the astroid $x^{2/3} + y^{2/3} = a^{2/3}$, and find its area.

23.10   Note that in the definition of a polynomial figure, it is assumed first that the set of points is a figure, and second, that it is defined by a set of polynomial inequalities. In general, a set of points defined by polynomial inequalities need not be a figure. Show this in the following cases, by sketching the point sets described, and explaining carefully why each one fails to be a figure.

  a. $x^2 + y^2 - 1 \geq 0$
  b. $(x^2 - y^2)^2 \geq 0$
  c. $-(x^2 - y^2)^2 \geq 0$
  d. $-(x^2 + y^2 - 1)^2 \geq 0$
  e. $x^2 - 1 \geq 0,\ 4 - x^2 - y^2 \geq 0$

23.11 *a.* Draw a regular hexagon and sketch a partition of it using seven triangles.
  b. Draw a square and sketch a partition of it using five rectangles.
  c. Find a partition of a square using six squares.
  d. Sketch a partition of a circle using two congruent figures with no straight-line segments in their boundaries.

23.12   Go over the elementary discussion of area which we used to introduce the subject, and note each place that one of the four properties given in Th. 23.1 was used.

*23.13   Carry out the following steps toward proving Th. 23.2.

   a. Using properties 2, 3, and 4 in Th. 23.2, show that if F_{mn} is the figure defined by Eqs. (23.3), then $A(F_{mn}) = (1/2^k)^2$.

   b. Given a figure F, consider for each integer k the partition of F into the squares F_{mn} of part a that lie inside F, and a remaining figure F_k, which is bounded by the boundary of F and parts of the boundaries of the squares F_{mn}. Draw a sketch indicating the figure F_k.

   c. Using property 2 in Th. 23.2, together with part a, show that

$$A(F) = A(F_k) + A_k,$$

where A_k is defined by Eq. (23.4).

   d. Deduce that $A(F) \geq A$, where A is the area of F. Note that in order to complete the proof of Th. 23.2, it is necessary to show that

$$\lim_{k \to \infty} A(F_k) = 0.$$

It is here that some limitation is needed on the regularity of the boundary such as the condition we imposed in the definition of a figure.

23.14   The last step in the proof of Th. 23.3 requires finding the function $h(t)$ and evaluating the integral (23.7) for the square $F: 0 \leq x \leq 1, 0 \leq y \leq 1$, using an arbitrary straight line L. Carry out this computation for the following choices of the line L.

   a. $y = x$
   b. $x + y = 0$
   *c. $y = mx$,   where $0 < m < 1$

*23.15   Let F be the square $0 \leq x \leq 1, 0 \leq y \leq 1$. Let F_1 consist of all those points (x, y) of F for which x is a rational number. Let F_2 consist of all those points of F for which x is irrational. (Thus F_1 and F_2 are both made up out of vertical line segments.)

   a. Show that $F = F_1 \cup F_2$ and that F_1 and F_2 have no points in common. (Thus, if F_1 and F_2 were figures, they would form a partition of F.)

   b. Show that if the definition of area is applied to each of the sets F_1 and F_2, then the value obtained is zero. (In the terminology referred to in footnote 4 of this chapter, the "inner area" of F_1 and of F_2 is zero.) Since the area of F is equal to 1, property 2 of Th. 23.1 fails to hold in this generality.

*23.16   Let F be an arbitrary bounded set of points in the plane. Let M_k be the number of squares of the form defined by Eq. (23.3), which meet the figure F in at least one point. Let $B_k = M_k(1/2^k)^2$.

   a. Show that $B_{k+1} \leq B_k$ for each k.
   b. Show that $M_k \geq N_k$ for each k.
   c. Deduce that the numbers B_k tend to a limit B, and that $B \geq A$, where A is defined by Eq. (23.5).

(*Note:* the number B is called the *outer area* of the set F. If $B = A$, their common value is called the *area* of F. In other words, every bounded set of points in the plane has a well-defined outer area and inner area. When the two coincide, their common value is defined to be the area. It is not hard to show that for a given set F, the outer area equals the inner area if and only if the boundary of F has zero area. It can also be shown that a piecewise smooth curve has zero area. Thus, for the sets we have denoted as "figures," the quantity that we have defined to be area (and which is generally known as the inner area) coincides with the outer area and with the value of area defined more generally by the method indicated here.)

24 The double integral

Having at our disposal a definition and the basic properties of area, we can now define the double integral. Our definition for an arbitrary figure is a natural extension of the one given at the beginning of the previous section for the case of a rectangle.

Let $f(x, y)$ be a bounded function defined at all points of a figure F. Let $F_1, \ldots, F_n$ be an arbitrary partition of F, and form the sum

(24.1) $$\sum_{i=1}^{n} f(\xi_i, \eta_i) A_i,$$

where (ξ_i, η_i) is any point of F_i, and A_i is the area of F_i. We wish to form the limit of this expression as the partition of F becomes finer and finer. To make this notion precise, we introduce the following terminology.

Definition 24.1 The *diameter* of a figure F is the maximum distance between two points of F.

Definition 24.2 The *norm* of a partition $F_1, \ldots, F_n$ of a figure F is the maximum diameter of the figures $F_1, \ldots, F_n$.

Definition 24.3 The expression (24.1) is said to *tend to a limit l as the norm of the partition tends to zero* if for every $\epsilon > 0$ there exists a $\delta > 0$ such that

$$\left| \sum_{i=1}^{n} f(\xi_i, \eta_i) A_i - l \right| < \epsilon$$

for every partition whose norm is less than δ.

Definition 24.4   If the expression (24.1) tends to a limit l as the norm of the partition tends to zero, then $f(x, y)$ is said to be *integrable over F*, and the limit l is called the *double integral* of f over F and is denoted by

$$\iint_F f\,dA.$$

Before investigating properties of the double integral, let us give two interpretations in the case that f is a positive continuous function.

Geometric Interpretation: Volume

By a procedure exactly analogous to that used in the previous section for defining the area of a two-dimensional figure, we can define the notion of *volume* for a three-dimensional body and show that it has properties similar to those described in Ths. 23.1 and 23.3. In particular, the volume of a cylindrical body of constant cross section is equal to its height multiplied by the area of its base. Thus each term in the sum (24.1) represents the volume of the cylindrical body with base F_i and height $f(\xi_i, \eta_i)$. This sum is therefore an approximation to the volume of the body lying over the figure F and under the surface $z = f(x, y)$ (see Fig. 24.1). The precise volume of this body is obtained by taking finer and finer partitions of F and is thus equal to the limiting value of the sum (24.1), which is the double integral $\iint_F f\,dA$.

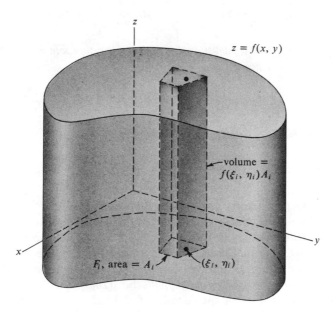

FIGURE 24.1   Volume under a surface

Physical Interpretation: Mass

If the figure F represents a thin plate of constant density c ($=$ mass per unit area), then the total mass of the plate is equal to cA, where A is the area of F. If a thin plate in the shape of the figure F has a variable density given by a function $f(x, y)$, then each term in the sum (24.1) represents an approximation to the mass of the figure F_i; namely, it represents the mass that the figure F_i would have if it had constant density equal to $f(\xi_i, \eta_i)$. The total mass is therefore approximated by the sum (24.1), and it is precisely equal to the limiting value $\iint_F f \, dA$.

Just as in the case of the ordinary integral for functions of a single variable, one rarely applies the definition directly to compute the value of an integral. The definition is used to deduce certain basic properties, and from these properties are derived methods of computation. For the double integral, the basic properties are stated in the following theorem.

Theorem 24.1 *The double integral has the following properties.*

1. *If f is integrable over F, and $f \geq 0$, then $\iint_F f \, dA \geq 0$.*
2. *If f and g are both integrable over F, and if λ, μ are any two constants, then $\lambda f + \mu g$ is also integrable over F, and*

$$(24.2) \qquad \iint_F (\lambda f + \mu g) \, dA = \lambda \iint_F f \, dA + \mu \iint_F g \, dA.$$

3. *If $F_1, \ldots, F_n$ form a partition of F, and if f is integrable over F, then f is integrable over each F_i, and*

$$(24.3) \qquad \iint_F f \, dA = \iint_{F_1} f \, dA + \cdots + \iint_{F_n} f \, dA.$$

4. *The function $f \equiv 1$ on F is integrable over F, and*

$$(24.4) \qquad \iint_F 1 \, dA = \text{Area of } F.$$

PROOF. Properties 1 and 2 are an immediate consequence of the fact that the corresponding properties hold for each of the partial sums of (24.1), and they must therefore continue to hold in the limit. We do not prove property 3 in detail, but we note that if each of the figures $F_1, \ldots, F_n$ is partitioned, then the combined figures of all the partitions form a partition of F. Using partitions of F formed in such a way yields Eq. (24.3). Finally, property 4 follows from the fact that for the function $f \equiv 1$, each of the partial sums (24.1) equals the area of F, by property 2 in Th. 23.1. ◆

We next show that these four properties serve to characterize uniquely the double integral.

Theorem 24.2   *Suppose that to each figure F and to each function f integrable over F there is assigned a number $I_F(f)$, in such a way that the following conditions are verified.*

1.   *If $f \geq 0$ on F, then $I_F(f) \geq 0$.*
2.   *$I_F(\lambda f + \mu g) = \lambda I_F(f) + \mu I_F(g)$, for any constants λ, μ.*
3.   *If $F_1, \ldots, F_n$ form a partition of F, then*

$$I_F(f) = I_{F_1}(f) + \cdots + I_{F_n}(f).$$

4.   *$I_F(1) = $ area of F.*

Then the following condition also holds:

(24.5) $$f \geq g \text{ on } F \Rightarrow I_F(f) \geq I_F(g),$$

and furthermore, for all F and f,

(24.6) $$I_F(f) = \iint_F f \, dA.$$

PROOF.   To derive (24.5), note first that choosing $\lambda = 1$, $\mu = -1$ in condition 2 yields $I_F(f - g) = I_F(f) - I_F(g)$. Now $f \geq g$ means $f - g \geq 0$, and applying condition 1 to the function $f - g$ gives $I_F(f) - I_F(g) = I_F(f - g) \geq 0$, which proves (24.5). To prove (24.6), let $F_1, \ldots, F_n$ be any partition of F. Let A_i be the area of F_i, and let

$$\left. \begin{array}{l} M_i = \text{maximum} \\ m_i = \text{minimum} \end{array} \right\} \text{ of } f \quad \text{on} \quad F_i, \qquad i = 1, \ldots, n.$$

Then

(24.7) $$m_i \leq f \leq M_i \quad \text{on} \quad F_i.$$

By condition 4, we have

$$m_i A_i = m_i I_{F_i}(1)$$

and by condition 2,

$$m_i I_{F_i}(1) = I_{F_i}(m_i).$$

By (24.5) and (24.7), we have

$$I_{F_i}(m_i) \leq I_{F_i}(f) \leq I_{F_i}(M_i)$$

and, again using conditions 2 and 4,

$$I_{F_i}(M_i) = M_i I_{F_i}(1) = M_i A_i.$$

Combining all these yields

$$m_i A_i \le I_{F_i}(f) \le M_i A_i.$$

If we sum over all F_i in the partition and use condition 3, we obtain

$$\sum_{i=1}^{n} m_i A_i \le I_F(f) \le \sum_{i=1}^{n} M_i A_i.$$

However, since both m_i and M_i represent the values of the function f at points of F_i, both the left- and right-hand sides of this inequality are sums of the form (24.1). By the definition of the integral, both sides tend in the limit to $\iint_F f \, dA$. Since the quantity $I_F(f)$ is bounded above and below by quantities that converge to the same limit, it must equal this limit. This proves the theorem. ◆

Remark 1 The significance of this theorem is that it reduces all properties of the double integral to four basic ones. The first property is known as *positivity*, and the second as *linearity*. These are basic properties of integration, and hold equally for ordinary integrals and line integrals. Some analog of property 3 is also common to all forms of integration. Property 4 is, of course, specially related to double integrals. Since many different definitions are possible for the double integral, it is important to know that as long as these four basic properties are satisfied, the value obtained from any alternative definition must coincide with ours.

Remark 2 It is useful to observe that the conclusion of the theorem continues to hold even if the quantity $I_F(f)$ is not defined for *all* integrable f on F, but only for a subset. This subset must be large enough so that the statement of the properties makes sense. For example, the subset must include the function that is identically equal to 1 on any figure, and with each pair of f, g, it must include the linear combination $\lambda f + \mu g$. An example of such a subset is the set of all continuous functions on each figure F. One can in fact prove that every function f continuous on F is integrable over F, but we do not give the proof here. If a number $I_F(f)$ is assigned to each continuous f on F, in such a manner that the four given conditions are satisfied, then this number must coincide with the double integral. We see an example of how this may be done in Th. 24.3. Another useful class of functions are the *piecewise continuous* functions, obtained by "patching together" a number of functions, each of which is defined and continuous on one figure in a partition of F. Such functions are also integrable over F.

Remark 3 The proof of Th. 24.2 as it stands is valid only if the functions f are taken to be continuous, since we assumed that each function f had a maximum and minimum on each figure F_i. This is true of continuous functions, as we have noted in Th. 8.3. To prove Th. 24.2 for arbitrary integrable functions f would require an elaboration of the above argument.

The next theorem provides the method most often used for the actual computation of a double integral. Stating the theorem for an arbitrary figure F necessitates the introduction of some complex notation, which is apt to appear forbidding at first sight. It may be advisable to read through the theorem and its proof without concentrating on details, and then to study the subsequent remarks and examples. Familiarity with the way the theorem is applied in various cases greatly helps to clarify the general statement.

Theorem 24.3   *Let F be an arbitrary figure. Let c and d represent the minimum and maximum values of the y coordinate on F. For each value of t, $c \leq t \leq d$, the part of the line $y = t$ lying in F consists of a number of intervals of the form $a_k(t) \leq x \leq b_k(t)$ (see Fig. 24.2). Let $f(x, y)$ be an integrable function on F, and suppose that the expression*

$$(24.8) \qquad \varphi_{f,F}(t) = \sum_k \int_{a_k(t)}^{b_k(t)} f(x, t)\, dx$$

exists for every t, $c \leq t \leq d$. Suppose further that the integral

$$(24.9) \qquad I_F(f) = \int_c^d \varphi_{f,F}(t)\, dt$$

exists. Then

$$(24.10) \qquad I_F(f) = \iint_F f\, dA.$$

PROOF.   We must show that the quantity $I_F(f)$ defined by (24.8) and (24.9) satisfies the conditions of Th. 24.2. We shall carry out the proof only for the case of polynomial figures. We note first of all that if F is a polynomial figure, then for each t there will be only a finite number of intervals $a_k(t) \leq x \leq b_k(t)$.

Property 1.   If $f \geq 0$, then by (24.8) $\varphi_{f,F}(t) \geq 0$ for all t, and by (24.9), $I_F(f) \geq 0$.

Property 2.   If $h(x, y) = \lambda f(x, y) + \mu g(x, y)$, then by (24.8), for $c \leq t \leq d$ we have $\varphi_{h,F}(t) = \lambda \varphi_{f,F}(t) + \mu \varphi_{g,F}(t)$, and by (24.9), $I_F(h) = \lambda I_F(f) + \mu I_F(g)$.

Property 3.   If $F_1, \ldots, F_n$ form a partition of F into polynomial figures, then for $c \leq t \leq d$ we have $\varphi_{f,F}(t) = \varphi_{f,F_1}(t) + \cdots + \varphi_{f,F_n}(t)$, except for at most a finite number of values of t for which the line $y = t$ contains a line

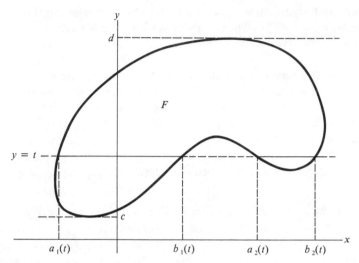

FIGURE 24.2 Notation for Th. 24.3

segment that is a common boundary of two different F_i. It follows from (24.9) that $I_F(f) = I_{F_1}(f) + \cdots + I_{F_n}(f)$.

Property 4. If $f \equiv 1$, then for $c \le t \le d$, we have

$$\varphi_{1,F} = \sum_k (b_k(t) - a_k(t)) = h(t),$$

in the notation of Th. 23.3. It follows from that theorem that

$$I_F(1) = \int_c^d h(t)\, dt = \text{area of } F.$$

Thus the quantity $I_F(f)$ satisfies the hypotheses of Th. 24.2, and Eq. (24.10) follows. ◆

Remark 1 In the case that $f(x, y)$ is continuous on F, the integrals (24.8) and (24.9) always exist, and (24.10) holds.

Remark 2 Substituting the expression (24.8) for $\varphi_{f,F}(t)$ in (24.9), we may write (24.10) in the form

(24.11) $$\iint_F f\, dA = \int \left[\int f(x, y)\, dx \right] dy,$$

where the limits of integration on the right are determined by the figure F in the way indicated in the theorem. The right-hand side of (24.11) is referred to as an *iterated integral* of a function of two variables. We first hold y fixed and integrate with respect to x, obtaining a number that depends on y. This gives us a function of y, which we then integrate with respect to y. The theorem states that this process of repeated ordinary integration yields the same value as the double integral.

Remark 3   Let f be a positive continuous function, and consider the three-dimensional body B lying above the figure F and under the surface $z = f(x, y)$. The quantity $\varphi_{f,F}(t)$ then represents the area of the section cut out of the body B by the vertical plane $y = t$ (see Fig. 24.3). In view of our interpretation of the double integral as the volume of the body B, Th. 24.3 states that the volume may be obtained by integrating the cross-sectional area. This is the precise analog for volume of Th. 23.3 for area, which stated that area is equal to the integral of cross-sectional width.

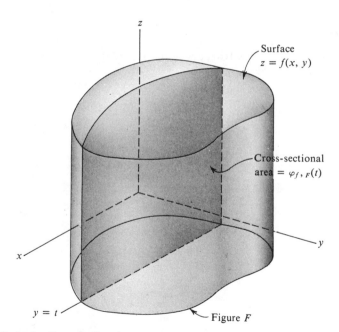

FIGURE 24.3   Cross-sectional area

Remark 4   When computing the integral of a given function $f(x, y)$ over a given figure F we simplify the notation by omitting the subscripts f, F and writing $\varphi(t)$ for $\varphi_{f,F}(t)$.

Example 24.1

Let F be the rectangle $a \leq x \leq b$, $c \leq y \leq d$. Evaluate $\iint_F xy^2 \, dA$.

In this case each horizontal line $y = t$, $c \leq t \leq d$, intersects F in the interval $a \leq x \leq b$. Thus, if $f(x, y) = xy^2$, then

$$\varphi(t) = \int_a^b f(x, t) \, dx = \int_a^b xt^2 \, dx = t^2 \int_a^b x \, dx = t^2 \frac{b^2 - a^2}{2},$$

and

$$\iint_F xy^2 \, dA = \int_c^d \varphi(t) \, dt = \frac{b^2 - a^2}{2} \int_c^d t^2 \, dt = \frac{1}{6} (b^2 - a^2)(d^3 - c^3).$$

Remark   In the notation of (24.11), we would write

$$\iint_F xy^2 \, dA = \int_c^d \left[\int_a^b xy^2 \, dx \right] dy = \int_c^d \tfrac{1}{2}(b^2 - a^2) y^2 \, dy$$
$$= \tfrac{1}{6}(b^2 - a^2)(d^3 - c^3).$$

Example 24.2

Find the volume of the body lying above the square $0 \leq x \leq 1$, $0 \leq y \leq 1$, and under the surface $z = xy^2$ (Fig. 24.4).

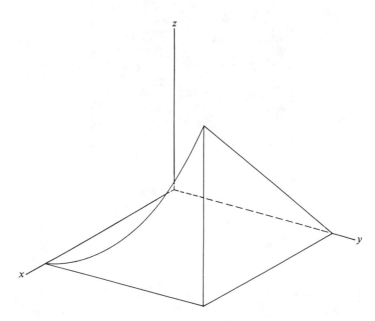

FIGURE 24.4   Volume under the surface $z = xy^2$, over a square

If F denotes the square indicated, then the volume is given by $\iint_F xy^2 \, dA$. We find its value simply by substituting $a = 0, b = 1, c = 0, d = 1$ in Example 24.1; the answer is $\frac{1}{6}$.

Example 24.3

Find the volume of the right isosceles tetrahedron cut out of the first octant by the plane $x + y + z = 1$.

The volume in question is bounded above by the plane $z = 1 - x - y$ and below by the triangle in the x, y plane described by $x \geq 0, y \geq 0, x + y \leq 1$ (Fig. 24.5). If we denote this triangle by F, then the volume is equal to $\iint_F (1 - x - y) \, dA$. For $0 \leq t \leq 1$, the line $y = t$ cuts F in the interval $0 \leq x \leq 1 - t$. Thus

$$\varphi(t) = \int_0^{1-t} (1 - x - t) \, dx = \left[(1 - t)x - \frac{x^2}{2} \right]_0^{1-t} = \frac{1}{2}(1 - t)^2,$$

and

$$\int_0^1 \varphi(t) \, dt = \frac{1}{2} \left[\frac{-(1 - t)^3}{3} \right]_0^1 = \frac{1}{6}.$$

Example 24.4

Evaluate $\iint_F y \, dA$, where F is defined by $y \geq 0, 4 \leq x^2 + y^2 \leq 9$ (Fig. 24.6).

The line $y = t$ intersects the figure F for those values of t satisfying $0 \leq t \leq 3$. When $0 \leq t < 2$, the intersection consists of the two distinct segments

$$-(9 - t^2)^{1/2} \leq x \leq -(4 - t^2)^{1/2} \quad \text{and} \quad (4 - t^2)^{1/2} \leq x \leq (9 - t^2)^{1/2}.$$

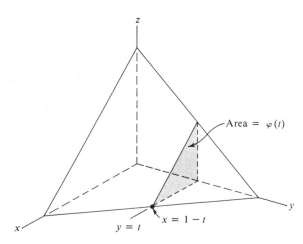

FIGURE 24.5 Volume of a tetrahedron

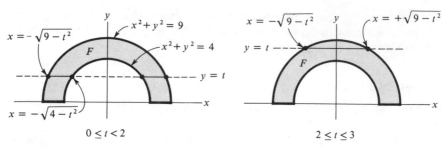

FIGURE 24.6 Integration over upper half-annulus

For $2 \le t \le 3$, there is a single segment $-(9 - t^2)^{1/2} \le x \le (9 - t^2)^{1/2}$. Thus, for $0 \le t < 2$,

$$\varphi(t) = \int_{-(9-t^2)^{1/2}}^{-(4-t^2)^{1/2}} t \, dx + \int_{(4-t^2)^{1/2}}^{(9-t^2)^{1/2}} t \, dx = 2t((9 - t^2)^{1/2} - (4 - t^2)^{1/2}),$$

while for $2 \le t \le 3$,

$$\varphi(t) = \int_{-(9-t^2)^{1/2}}^{(9-t^2)^{1/2}} t \, dx = 2t(9 - t^2)^{1/2}.$$

Thus

$$\iint_F y \, dA = \int_0^3 \varphi(t) \, dt = \int_0^3 2t(9 - t^2)^{1/2} \, dt - \int_0^2 2t(4 - t^2)^{1/2} \, dt$$

$$= -\frac{2}{3}(9 - t^2)^{3/2} \Big|_0^3 - \left[-\frac{2}{3}(4 - t^2)^{3/2} \right] \Big|_0^2 = 18 - \frac{16}{3}$$

$$= \frac{38}{3}.$$

We next state the theorem corresponding to Th. 24.3, with the roles of the x and y axes reversed.

Theorem 24.4 *Let F be an arbitrary figure. Let a and b represent the minimum and maximum values of the x coordinate on F. For each value of t, $a \le t \le b$, the part of the line $x = t$ lying in F consists of a number of intervals of the form $c_k(t) \le y \le d_k(t)$. Let $f(x, y)$ be an integrable function on F, and suppose that the expression*

$$\psi(t) = \sum_k \int_{c_k(t)}^{d_k(t)} f(t, y) \, dy$$

exists for every t, $a \le t \le b$. Then

$$\int_a^b \psi(t) \, dt = \iint_F f \, dA,$$

provided the integral on the left exists.

PROOF. Identical to that of the previous theorem.

Corollary   *In the notation of Ths. 24.3 and 24.4, we have*

$$(24.12) \quad \int_c^d \left[\sum_k \int_{a_k(y)}^{b_k(y)} f(x, y) \, dx \right] dy = \int_a^b \left[\sum_k \int_{c_k(x)}^{d_k(x)} f(x, y) \, dy \right] dx,$$

assuming that all the integrals in question exist; each side of (24.12) *equals the double integral* $\iint_F f \, dA$.

Remark   Equation (24.12) may be written in the abbreviated form

$$\int \left[\int f(x, y) \, dx \right] dy = \int \left[\int f(x, y) \, dy \right] dx,$$

always bearing in mind that the limits of integration are determined by the particular figure. This equation asserts that *the value of an iterated integral does not depend on the order of integration.* It is sometimes referred to as Fubini's theorem.

Example 24.5

Evaluate the integral in Example 24.4, using the method of Th. 24.4.
   In this case, the line $x = t$ intersects F for $-3 \le t \le 3$, and the intersection consists of a single segment (see Fig. 24.7). This segment is

$$0 \le y \le (9 - t^2)^{1/2}, \qquad\qquad \text{for} \quad -3 \le t \le -2$$
$$(4 - t^2)^{1/2} \le y \le (9 - t^2)^{1/2}, \qquad \text{for} \quad -2 \le t \le 2$$
$$0 \le y \le (9 - t^2)^{1/2}, \qquad\qquad \text{for} \quad 2 \le t \le 3.$$

Thus

$$\psi(t) = \begin{cases} \displaystyle\int_0^{(9-t^2)^{1/2}} y \, dy = \frac{1}{2} y^2 \Big|_0^{(9-t^2)^{1/2}} = \frac{1}{2}(9 - t^2), & -3 \le t \le -2 \\[2ex] \displaystyle\int_{(4-t^2)^{1/2}}^{(9-t^2)^{1/2}} y \, dy = \frac{1}{2} y^2 \Big|_{(4-t^2)^{1/2}}^{(9-t^2)^{1/2}} = \frac{1}{2}(9 - t^2) - \frac{1}{2}(4 - t^2), & -2 \le t \le 2 \\[2ex] \displaystyle\int_0^{(9-t^2)^{1/2}} y \, dy = \frac{1}{2} y^2 \Big|_0^{(9-t^2)^{1/2}} = \frac{1}{2}(9 - t^2), & 2 \le t \le 3. \end{cases}$$

Finally,

$$\iint_F y \, dA = \int_{-3}^3 \psi(t) \, dt = \int_{-3}^3 \frac{1}{2}(9 - t^2) \, dt - \int_{-2}^2 \frac{1}{2}(4 - t^2) \, dt$$

$$= \frac{1}{2}\left[9t - \frac{t^3}{3} \right]_{-3}^3 - \frac{1}{2}\left[4t - \frac{t^3}{3} \right]_{-2}^2 = 18 - \frac{16}{3} = \frac{38}{3}.$$

Thus, Eq. (24.12) is verified in this particular case.

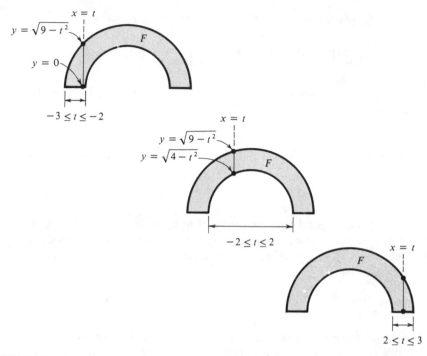

FIGURE 24.7 Integration over half annulus: alternative approach

Remark Once the procedure involved in evaluating an iterated integral is clearly understood, there is no need to introduce the auxiliary variable t. It is sufficient to consider one of the two variables to be fixed, and integrate with respect to the other, thus obtaining the expression inside the brackets on either side of (24.12). A second integration is then carried out with respect to the variable originally held fixed.

We conclude this section with a brief discussion of one of the many applications of double integration.

Definition 24.5 For an arbitrary figure F, the quantities

$$M_x(F) = \iint_F x \, dA, \qquad M_y(F) = \iint_F y \, dA$$

are called the *moments* of F with respect to the y and x axes, respectively.

The physical meaning of these quantities may be described as follows. Suppose that the figures F_1 and F_2 are cut out of cardboard and placed on opposite sides of a seesaw attached along the y axis (Fig. 24.8). Then the side that goes down is the one whose figure has the greater x moment $|M_x(F)|$. The two sides balance if and only if the moments are equal in magnitude $|M_x(F_1)| = |M_x(F_2)|$. Since the x coordinate is positive on one side of the y axis and negative on the other side, the condition for balance may be written as $M_x(F_2) = -M_x(F_1)$, or $M_x(F_1) + M_x(F_2) = 0$.

Suppose that a single figure F is placed so that it crosses the y axis, and we wish to know whether it will balance by itself. We may consider it to be divided by the y axis into two figures, F_1, F_2, and according to the above discussion it balances if and only if $M_x(F) = M_x(F_1) + M_x(F_2) = 0$. A similar discussion applies to the y moment and balancing along the x axis.

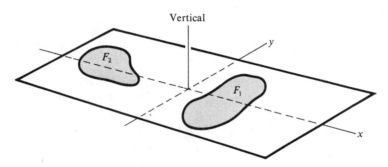

FIGURE 24.8 Moments and "balancing"

Definition 24.6 For any figure F, the point $(\bar{x}, \bar{y})$ defined by

$$\bar{x} = \frac{M_x(F)}{A(F)} = \frac{\iint_F x\, dA}{\iint_F 1\, dA}, \qquad \bar{y} = \frac{M_y(F)}{A(F)} = \frac{\iint_F y\, dA}{\iint_F 1\, dA}$$

is called the *centroid* of F.

Lemma 24.1 *The centroid of a figure is a point that depends only on the geometry of the figure, and not on the choice of coordinates; more precisely, if new coordinates are chosen by a translation or rotation of the axes, and if the centroid is computed in these coordinates, then the point obtained is the same as in the original coordinates.*

PROOF. Suppose first that new coordinates are chosen by a translation of the axes $X = x - h$, $Y = y - k$. Then

$$M_X(F) = \iint_F X\, dA = \iint_F (x - h)\, dA = M_x(F) - hA(F),$$

and

$$\bar{X} = \frac{M_X(F)}{A(F)} = \frac{M_x(F)}{A(F)} - h = \bar{x} - h.$$

Similarly, $\bar{Y} = \bar{y} - k$. Thus $\bar{X}$, $\bar{Y}$ are the coordinates with respect to the new axes of the same point whose coordinates with respect to the original axes were $\bar{x}$, $\bar{y}$.

In particular, by making a preliminary translation of axes if necessary, we may assume that the centroid of F lies at the origin; that is, $(\bar{x}, \bar{y}) = (0, 0)$ or $\iint_F x \, dA = \iint_F y \, dA = 0$ (see Ex. 24.15). Suppose next that new co-ordinates are chosen by a rotation of axes $X = x \cos \alpha - y \sin \alpha$, $Y = x \sin y + y \cos \alpha$. Then

$$M_X(F) = \iint_F X \, dA = \cos \alpha \iint_F x \, dA - \sin \alpha \iint_F y \, dA = 0,$$

and similarly, $M_Y(F) = 0$. Thus $(\bar{X}, \bar{Y}) = (0, 0)$, and the centroid with respect to the rotated axes is again the origin. This proves the lemma. ◆

Concerning the physical interpretation of the centroid $(\bar{x}, \bar{y})$, we note that if we choose a new Y axis to pass through the centroid, by setting $X = x - \bar{x}$, then

$$M_X(F) = M_x(F) - \bar{x}A(F) = 0,$$

so that the figure F balances about the Y axis ($X = 0$), which is to say, about the line $x = \bar{x}$. Similarly, it balances about the line $y = \bar{y}$. Thus, one conse-quence of the above lemma is the following statement (which may not seem obvious by physical intuition): "if an arbitrary figure balances about each of two perpendicular lines, then it balances about every line through their intersection."

Example 24.6

Find the centroid of the figure $F: y \geq 0$, $4 \leq x^2 + y^2 \leq 9$.

It is clear from the symmetry of F about the y axis, and may be verified directly, that $M_x(F) = 0$, and therefore $\bar{x} = 0$. In Example 24.4 we have computed for this figure $M_y(F) = \iint_F y \, dA = \frac{38}{3}$. The area $A(F)$ is half of the area between two concentric circles of radius 2 and 3; thus $A(F) = \frac{1}{2}(9\pi - 4\pi) = \frac{5}{2}\pi$. Hence,

$$\bar{y} = \frac{\frac{38}{3}}{\frac{5}{2}\pi} = \frac{76}{15\pi} \sim 1.6.$$

An interesting feature of this example is that it draws our attention to the fact that the centroid of a figure may well lie outside the figure itself (see Fig. 24.9).

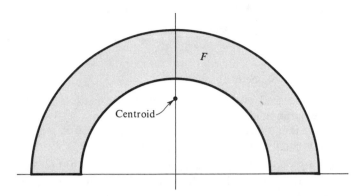

FIGURE 24.9   Centroid of a half annulus

Exercises

24.1   Let F be the rectangle $-1 \leq x \leq 3$, $2 \leq y \leq 5$. Evaluate the following integrals.

a. $\iint_F x^2 y \, dA$ **c.** $\iint_F xy\sqrt{x^2 + y^2} \, dA$
b. $\iint_F (x^2 + y^2) \, dA$ **d.** $\iint_F x e^{xy} \, dA$

24.2   Let F be the triangle $x \geq 0$, $y \geq 0$, $x + 2y \leq 1$. Evaluate the following integrals.

a. $\iint_F xy \, dA$ **b.** $\iint_F (x^2 + y^2) \, dA$

24.3   Let F be the disk $x^2 + y^2 \leq 4$. Evaluate the following integrals.

a. $\iint_F x^2 y \, dA$ **b.** $\iint_F (x^2 + y^2) \, dA$

*24.4   Let F be the trapezoid $x \geq 0$, $y \geq 0$, $1 \leq x + y \leq 2$, and let $f(x, y) = x + y$.

a. Sketch F and find the function $\varphi_{f,F}(t)$ of Eq. (24.8).
b. Find $\iint_F f \, dA$.

24.5   Find the volume of the following three-dimensional bodies.

a. Lying over the square $0 \leq x \leq 1$ and under the paraboloid $z = x^2 + y^2$.
b. Lying under the surface $z = xy$ and over the part of the disk $x^2 + y^2 \leq a^2$ in the first quadrant.
c. The tetrahedron cut out of the first octant by the plane $x/a + y/b + z/c = 1$, $a, b, c > 0$.

In Exs. 24.6–24.8 use the fact that a three-dimensional volume is equal to the integral of its cross-sectional area.

24.6 *a.* Find the area of the cross section of the ellipsoid

$$\frac{x^2}{a^2} + \frac{y^2}{b^2} + \frac{z^2}{c^2} \le 1 \qquad\qquad a, b, c > 0,$$

with the plane $y = t$.

b. Find the volume of this ellipsoid.

24.7 *a.* Find the area of the cross section by the plane $z = t$ of the body bounded by the paraboloid $z = x^2 + y^2$.

b. Show that the cylindrical body $x^2 + y^2 \le 4$, $0 \le z \le 4$ is divided into two parts of equal volume by the paraboloid $z = x^2 + y^2$.

24.8 Let F be a figure lying in the right half-plane $x \ge 0$. Let B be the solid of revolution obtained by revolving the figure F about the y axis.

a. Show that the cross-sectional area of B with the plane $y = t$ is given by the function $\varphi_{f,F}(t)$ of Eq. (24.8), where $f(x, y) = 2\pi x$. (*Hint:* note that the cross section consists of a number of annular domains generated by the intervals $a_k(t) \le x \le b_k(t)$.)

b. Show that the volume of B is equal to

$$\iint_F 2\pi x \, dA.$$

24.9 Use the formula of Ex. 24.8b to find the volume of the following solids of revolution.

a. The ellipsoid $x^2/a^2 + y^2/b^2 + z^2/a^2 \le 1$.

b. The torus obtained by revolving about the y axis the disk $(x - b)^2 + y^2 \le a^2$, where $b > a > 0$.

c. The solid obtained by revolving about the y axis the figure $0 \le x \le 1/y$, $1 \le y \le b$.

d. The limit of the volume obtained in part c as $b \to \infty$.

24.10 *a.* Prove the *Theorem of Pappus*: If a figure F, which lies on one side of a line L, is revolved about the line L, then the volume of the resulting solid of revolution is equal to the area of F multiplied by the distance traveled by the centroid, that is,

$$V = 2\pi R \, A(F),$$

where R is the distance from the centroid of F to the line L. (*Hint:* assume L is the y axis and use Ex. 24.8b.)

b. Use the theorem of Pappus to find the volume of the torus in Ex. 24.9b.

c. Use the theorem of Pappus to find the centroid of the figure $x^2/a^2 + y^2/b^2 \le 1$, $x \ge 0$. (*Hint:* use the result of Ex. 24.9a.)

24.11 Prove that the centroid of an arbitrary triangle lies at the intersection of the medians. (*Hint:* one method would be to place the triangle so that a given median lies on the y axis, with the foot of the median at the origin. The vertices would then be (a, c), $(0, b)$, $(-a, -c)$. Show that the centroid then lies on the y axis.)

24.12   Let a figure F represent a thin plate of variable density $\rho(x, y)$. The following physical quantities are represented in terms of double integrals.

mass:	$m = \iint_F \rho \, dA$
x moment:	$M_x = \iint_F x\rho \, dA$
y moment:	$M_y = \iint_F y\rho \, dA$
center of gravity:	

$$(\bar{x}, \bar{y}), \qquad \text{where} \quad \bar{x} = M_x/m, \quad \bar{y} = M_y/m$$

moment of inertia about a line L:

$$I_L = \iint_F d^2\rho \, dA,$$

where $d(x, y)$ is the distance from the point (x, y) to the line L
radius of gyration about L:

$$R_L, \qquad \text{where} \quad mR_L^2 = I_L.$$

(*Note:* in case L is the y axis, we may use instead of I_L the notation $I_x = \iint_F x^2\rho \, dA$; if L is the x axis, we write $I_y = \iint_F y^2\rho \, dA$. If the density $\rho(x, y)$ is constant, then the center of gravity coincides with the centroid of F, and the quantities I_x, I_y are equal to the product of the density and $\iint_F x^2 \, dA$, $\iint_F y^2 \, dA$. These integrals are called the *second x moment* and *second y moment* of the figure F.)

   a. Find the mass of the figure $x^2 + y^2 \le a^2$, $y \ge 0$ if $\rho(x, y) = y$.
   b. Find M_x for the figure $x \ge 0$, $y \ge 0$, $x + y \le 1$, if $\rho(x, y) = x^2 + y^2$.
   c. Find M_y for the figure $0 \le x \le a$, $0 \le y \le b$, if $\rho(x, y) = x(x^2 + y^2)^{1/2}$.
   d. Find M_x and M_y for the figure $x^2 + y^2 \le a^2$, $x \ge 0$, $y \ge 0$, where the density is proportional to the distance to the origin. (*Hint:* use Th. 24.3 or 24.4, depending on which is more convenient.)
   e. Find the center of gravity of the half-disk of part a, where the density is proportional to the height above the x axis.
   f. Find the center of gravity of the figure $x^2 \le y$, $y^2 \le x$ if $\rho(x, y) = ky$.
   g. Find the radius of gyration about the x axis of the figure in part f, with the given density.
   h. Find the moment of inertia about the y axis of the figure $x \ge 0$, $y \le 2$, $1 + x^3 \le e^y$, if $\rho(x, y) = (1 + y^3)/(1 + x^3)$.

24.13   Let F be a rectangle $a \le x \le b$, $c \le y \le d$. Let $f(x, y) = g(x)h(y)$. Show that

$$\iint_F f \, dA = \left(\int_a^b g(x) \, dx \right) \left(\int_c^d h(y) \, dy \right).$$

24.14 **a.** Let F be a figure that is symmetric about the y axis. Suppose that $f(x, y)$ satisfies the condition $f(-x, y) = -f(x, y)$. Show that $\iint_F f \, dA = 0$.
   b. Show that if a figure F is symmetric about some straight line L, then the centroid of F lies on L.

24.15 Let P and Q be any two points in the plane. Show that an arbitrary rotation about the point P may be expressed as the composition of a translation taking P into Q, a rotation about Q and a translation taking Q back to P.

24.16 Let F be the rectangle $a \le x \le b$, $c \le x \le d$. Let $h(x, y) \in \mathscr{C}^2$ in F, and let $f(x, y) = h_{xy}(x, y)$. Show that

$$\iint_F f \, dA = h(a, c) + h(b, d) - h(b, c) - h(a, d).$$

(Note the resemblance to the fundamental theorem of calculus for functions of a single variable.)

24.17 For each of the integrands $f(x, y)$ in Ex. 24.1, find a function $h(x, y)$ such that $h_{xy} = f$. Use Ex. 24.16 to check the answers to Ex. 24.1.

Remark At the end of Sect. 7, we considered ordinary integrals of a function of two variables with respect to one of the variables. A second integration with respect to the other variable yields an iterated integral, which may also be interpreted as a double integral, using Ths. 24.3 and 24.4. These relationships are the subject of Ex. 24.18–24.25.

24.18 Let $f(x, y)$ be continuous in the triangle $F: x \le b, y \ge a, y \le x$. Define

$$g(x) = \int_a^x f(x, y) \, dy, \qquad h(y) = \int_y^b f(x, y) \, dx.$$

Show that

$$\int_a^b g(x) \, dx = \int_a^b h(y) \, dy.$$

(*Hint:* show that both sides are equal to $\iint_F f \, dA$.)

24.19 Use the result of Ex. 24.18 to evaluate $\int_0^1 h(y) \, dy$, where

$$h(y) = \int_y^1 e^{x^2} \, dx.$$

24.20 Given a continuous function h of one variable, define a function g by

$$g(b) = \int_a^b \left[\int_a^x h(y) \, dy \right] dx.$$

a. Show that $g'(b) = \int_a^b h(y) \, dy$, $g''(b) = h(b)$, and $g(a) = g'(a) = 0$.

b. Show that $g(b) = \int_a^b (b - y)h(y) \, dy$. (*Hint:* apply the method of Ex. 24.18.)

Note: the result of Ex. 24.20 may be stated as follows: *if the indefinite integral $\int_a^x h(t) \, dt$ of a function h is integrated again, the resulting function g can be expressed as a single integral.* Using a somewhat ambiguous notation (whose meaning is made precise in Ex. 24.20) we may write this in the form

$$\int_a^x \left[\int_a^x h(t) \, dt \right] dx = \int_a^x (x - t)h(t) \, dt.$$

This process may be repeated. Denote the function g of Ex. 24.20 by g_1, and perform repeated indefinite integration; that is, define g_n by

$$g_n(x) = \int_a^x g_{n-1}(t) \, dt.$$

(Again, using dubious notation, this could be written as

$$g_n(x) = \int_a^x \left[\int_a^x \cdots \int_a^x \left[\int_a^x h(t) \, dt \right] dx \cdots \right] dx, \qquad (n+1 \text{ integrations})$$

or in, more precise notation,

$$g_n(x) = \int_a^x \left[\int_a^{t_n} \cdots \int_a^{t_2} \left[\int_a^{t_1} h(t) \, dt \right] dt_1 \cdots \right] dt_n.)$$

Then using induction and the reasoning of Ex. 24.20, it can be shown that $g_n(x)$ may be expressed as a single integral,

$$g_n(x) = \frac{1}{n!} \int_a^x (x-t)^n h(t) \, dt.$$

See the following exercise; also compare with Ex. 7.23.

24.21 Let

$$g_n(x) = \frac{1}{n!} \int_a^x (x-y)^n h(y) \, dy, \qquad g_{n+1}(b) = \int_a^b g_n(x) \, dx.$$

Show that

$$g_{n+1}(b) = \frac{1}{(n+1)!} \int_a^b (b-y)^{n+1} h(y) \, dy.$$

24.22 Certain figures may be expressed in both of the forms:

$$a(y) \le x \le b(y), \qquad c \le y \le d$$

or

$$c(x) \le y \le d(x), \qquad a \le x \le b.$$

For example, the unit disk may be described by

$$-\sqrt{1-y^2} \le x \le \sqrt{1-y^2}, \qquad -1 \le y \le 1$$

and also by

$$-\sqrt{1-x^2} \le y \le \sqrt{1-x^2}, \qquad -1 \le x \le 1.$$

Each of the following figures is expressed in one of the above forms. Sketch the figure and express it in the other form.

a. $0 \le x \le 1 - y/2, \qquad 0 \le y \le 2$
b. $0 \le x \le 2\sqrt{1-y^2}, \qquad 0 \le y \le 1$
c. $x^2 \le y \le x, \qquad 0 \le x \le 1$
d. $-\sqrt{4-x^2} \le y \le \sqrt{4-x^2}, \qquad 1 \le x \le 2$
e. $\dfrac{x-1}{e-1} \le y \le \log x, \qquad 1 \le x \le e$

24.23 If F is a figure that can be represented in both of the forms given in Ex. 24.22, then the double integral over F of a function $f(x, y)$ can be expressed in two ways as an iterated integral; that is, by the Corollary to Th. 24.4, we have

$$\int_c^d \left[\int_{a(y)}^{b(y)} f(x, y)\, dx \right] dy = \int_a^b \left[\int_{c(x)}^{d(x)} f(x, y)\, dy \right] dx.$$

This equation summarizes the method for *change of order of integration*, when two integrations are carried out successively. Namely, express the iterated integral as a double integral over some figure, and use that figure to express the double integral as an iterated integral in the other order.

Use this process to evaluate the following iterated integrals. Sketch the figure F in each case, expressing it in the two forms referred to in Ex. 24.22.

a. $\displaystyle \int_0^1 \left[\int_0^{(1-x^2)^{1/2}} (1 - y^2)^{3/2}\, dy \right] dx$

b. $\displaystyle \int_0^1 \left[\int_{y^{1/2}}^{y^{1/5}} \sqrt{1 - x^3}\, dx \right] dy$

c. $\displaystyle \int_\pi^{2\pi} \left[\int_{y-\pi}^{\pi} \frac{\sin x}{x}\, dx \right] dy$

d. $\displaystyle \int_0^1 \left[\int_{x^2}^1 x^3 \cos y^3\, dy \right] dx$

Remark The following section contains a number of important results that are based entirely on the equality of the iterated integrals

$$\int_c^d \left[\int_a^b f(x, y)\, dx \right] dy = \int_a^b \left[\int_c^d f(x, y)\, dy \right] dx.$$

This is a special case of the Corollary to Th. 24.4. However, it is possible to give a direct proof in this case without using the theory of the double integral. Exercises 24.24 and 24.25 give two different methods of proof. The first one is a very simple argument that can be applied immediately in many cases. The second is more difficult, but also more direct, and applies to arbitrary continuous functions.

24.24 Show that if there exists a function $h(x, y) \in \mathscr{C}^2$ such that $f(x, y) = h_{xy}(x, y)$, then (using the equality of the mixed derivatives h_{xy}, h_{yx})

$$\int_c^d \left[\int_a^b f(x, y)\, dx \right] dy = \int_a^b \left[\int_c^d f(x, y)\, dy \right] dx.$$

(*Hint:* compare with Ex. 24.16.)

*24.25 Given a rectangle $R: a \le x \le b,\ c \le x \le d$, define

$$I_R(f) = \int_c^d \left[\int_a^b f(x, y)\, dx \right] dy$$

$$J_R(f) = \int_a^b \left[\int_c^d f(x, y)\, dy \right] dx$$

and $A(R) = (b - a)(d - c) = $ area of R.

a. Show that if $m \leq f(x, y) \leq M$ on R, then

$$mA(R) \leq I_R(f) \leq MA(R), \qquad mA(R) \leq J_R(f) \leq MA(R)$$

and

$$(m - M) \cdot A(R) \leq J_R(f) - I_R(f) \leq (M - m) A(R).$$

b. Show that if a rectangle R is divided into a number of smaller rectangles $R_{jk}: x_{j-1} \leq x \leq x_j, \ y_{k-1} \leq y \leq y_k$ by lines parallel to the x and y axes, then

$$I_R(f) = \sum_{j,k} I_{R_{jk}}(f), \qquad J_R(f) = \sum_{j,k} J_{R_{jk}}(f).$$

c. Let $f(x, y)$ be continuous on a rectangle R. Then $f(x, y)$ is uniformly continuous on R (see the Remark following Lemma 7.2). This means that for any $\epsilon > 0$, the rectangle R can be divided into sufficiently small rectangles R_{jk} such that the maximum M and minimum m of f on each R_{jk} satisfy $M - m < \epsilon$. Using this fact, together with parts a and b above, show that for any $\epsilon > 0$,

$$|J_R(f) - I_R(f)| < \epsilon A(R).$$

Deduce that $J_R(f) = I_R(f)$.

*24.26 Prove the following theorem: if $f(x, y)$ is continuous on a figure F, if $f(x, y) \geq 0$ at each point, and if $\iint_F f \, dA = 0$, then $f(x, y) \equiv 0$. (*Hint:* proceed by contradiction. If $f(x_0, y_0) = \eta > 0$ for some interior point (x_0, y_0) of F, then there exists an $r > 0$ such that $f(x, y) > \eta/2$ for $(x - x_0)^2 + (y - y_0)^2 < r^2$ (see Ex. 5.16a). It follows that $\iint_F f \, dA \geq \pi r^2 \eta/2 > 0$. This shows that $f(x, y) = 0$ at every interior point of F. By continuity, $f(x, y) = 0$ at every boundary point of F.)

24.27 Show that if $|f(x, y)| \leq M$ on a figure F, and if the area of F is denoted by $A(F)$, then

$$\left| \iint_F f \, dA \right| \leq M \, A(F).$$

24.28 **a.** Show that if a figure F lies inside a rectangle $R: a \leq x \leq b, c \leq y \leq d$, then the centroid of F lies in R.

b. More generally, if F lies on one side of a line $Ax + By + C = 0$ then so does the centroid. (*Hint:* the hypothesis means that either $Ax + By + C \geq 0$ or else $Ax + By + C \leq 0$ throughout F.)

c. If F lies in a disk, so does the centroid of F. (*Hint:* if not, use part b to get a contradiction.)

25 Green's theorem for a rectangle

In Sect. 22 we have seen how the fundamental theorem of calculus generalizes to line integrals. Our next objective is to obtain an analog of the fundamental theorem in the case of double integrals.

As our point of departure; we take the second form of the fundamental theorem, Eq. (22.2) which we may write in the form

(25.1)
$$\int_a^b f'(x)\, dx = f(b) - f(a).$$

The left-hand side of this equation consists of the integral over an interval of the derivative of a function of one variable. The corresponding expression in two variables consists of a double integral over a rectangle of a partial derivative of a function of two variables. Specifically, consider the rectangle

(25.2)
$$R: a \le x \le b,\ c \le y \le d,$$

and let $q(x, y) \in \mathscr{C}^1$ in a domain that includes the rectangle R. We wish to evaluate

$$\iint_R \frac{\partial q}{\partial x}\, dA.$$

Applying Th. 24.3, we first integrate along each horizontal line $y = t$, obtaining a function of t

$$\varphi(t) = \int_a^b q_x(x, t)\, dx.$$

For a fixed value of t, the partial derivative $q_x(x, t)$ reduces to the ordinary derivative with respect to x of the function $f(x) = q(x, t)$, and by Eq. (25.1),

$$\varphi(t) = \int_a^b q_x(x, t)\, dx = q(b, t) - q(a, t).$$

Then

(25.3)
$$\iint_R q_x\, dA = \int_c^d \varphi(t)\, dt = \int_c^d q(b, t)\, dt - \int_c^d q(a, t)\, dt.$$

In exactly the same fashion, applying Th. 24.4 to the partial derivative $\partial p/\partial y$, where $p(x, y) \in \mathscr{C}^1$ in some domain including R, we find

(25.4)
$$\iint_R p_y\, dA = \int_a^b p(t, d)\, dt - \int_a^b p(t, c)\, dt.$$

Equations (25.3) and (25.4) constitute, in primitive form, the two-variable generalization of Eq. (25.1). A more useful version is obtained by observing that the right-hand sides of (25.3) and (25.4) are in fact line integrals over parts of the boundary of the rectangle R. In fact, the boundary of R can be described as the piecewise smooth curve C consisting of the four line segments (Fig. 25.1)

(25.5)
$$
\begin{array}{lll}
C_1: x = t,\ y = c, & \qquad & a \le t \le b, \\
C_2: x = b,\ y = t, & \qquad & c \le t \le d, \\
C_3: x = b - t,\ y = d, & \qquad & 0 \le t \le b - a, \\
C_4: x = a,\ y = d - t, & \qquad & 0 \le t \le d - c.
\end{array}
$$

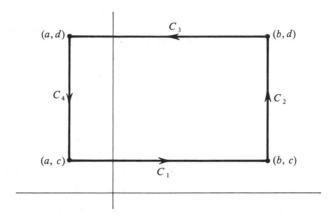

FIGURE 25.1   The curve C defined by Eqs. (25.5)

Note that these segments are oriented so that when traversed in succession they form the closed curve C, beginning and ending at (a, c).

Now the right-hand side of (25.3) consists of two line integrals of the form $\int q\, dy$ taken over the two vertical sides of R, each side being traversed from bottom to top. Since side C_4, as described in (25.5), is traversed from top to bottom, it follows from Lemma 21.2 that the value of the line integral has the opposite sign. Thus, (25.3) can be written as

$$(25.6) \qquad \iint_R q_x\, dA = \int_{C_2} q\, dy + \int_{C_4} q\, dy.$$

Since $dy/dt \equiv 0$ on the horizontal sides C_1 and C_3, we have

$$\int_{C_1} q\, dy = \int_{C_3} q\, dy = 0.$$

It follows that

$$(25.7) \qquad \int_C q\, dy = \int_{C_1} q\, dy + \int_{C_2} q\, dy + \int_{C_3} q\, dy + \int_{C_4} q\, dy = \iint_R q_x\, dA.$$

In the same way, we have

$$\int_{C_2} p\, dx = \int_{C_4} p\, dx = 0,$$

and observing the orientation of the sides C_1 and C_3, we may write (25.4) in the form

$$(25.8) \qquad \iint_R p_y\, dA = -\int_{C_3} p\, dx - \int_{C_4} p\, dx = -\int_C p\, dx.$$

Equations (25.7) and (25.8) together yield the desired two-variable generalization of (25.1). We may state the result as follows.

Theorem 25.1   Green's Theorem for a Rectangle   *Let $p(x, y)$, $q(x, y) \in \mathcal{C}^1$ in a domain that includes the rectangle R defined by (25.2). Let the boundary of R be the closed curve C described by (25.5). Then*

(25.9)
$$\iint_R q_x \, dA = \int_C q \, dy, \qquad \iint_R p_y \, dA = -\int_C p \, dx$$

and hence

(25.10)
$$\int_C p \, dx + q \, dy = \iint_R (q_x - p_y) \, dA.$$

Remark   The two equations in (25.9) are the direct generalizations of Eq. (25.1). In a rough way, these equations state the following: *to evaluate the double integral over a rectangle of a partial derivative of a function, it is sufficient to know the values of the function on the boundary.* Clearly, Eq. (25.1) yields an analogous statement for ordinary derivatives, replacing "rectangle" by "interval" and "boundary" by "endpoints." For most applications, however, it is not the individual equations in (25.9) that occur, but the combined form (25.10).

Example 25.1

Evaluate $\int_C x \, dy$, where C is described by (25.5).

   Method 1.   By direct computation, using the definition of the line integral,

$$\int_C x \, dy = \int_{C_2} x \, dy + \int_{C_4} x \, dy = \int_c^d b \, dt - \int_0^{d-c} a \, dt = b(d - c) - a(d - c).$$

   Method 2.   Applying (25.10), with $p = 0$, $q = x$ we obtain

$$\int_C x \, dy = \iint_R 1 \, dA = \text{Area of } R = (b - a)(d - c).$$

We are now ready to take up again the thread of our discussion at the end of Sect. 22. We prove the following result, which is a partial converse of Lemma 19.1.

Theorem 25.2   *Let $\langle p(x, y), q(x, y) \rangle$ be a $\mathcal{C}^1$ vector field in a disk $D : (x - x_1)^2 + (y - y_1)^2 < r^2$. If*

(25.11)
$$p_y = q_x$$

throughout D, then there exists a function $f(x, y) \in \mathcal{C}^2$ in D such that

(25.12)
$$f_x = p, \qquad f_y = q.$$

   PROOF.   We have seen in Lemmas 22.3 and 22.4 how to construct a pair of functions $g(x, y)$, $h(x, y)$ in D such that $g_x = p$ and $h_y = q$. We shall show

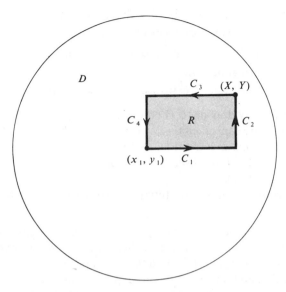

FIGURE 25.2   Notation for the proof of Th. 25.2

that if (25.11) holds, then the functions $g(x, y)$ and $h(x, y)$ are in fact equal, and denoting their common value by $f(x, y)$, we have the desired result (25.12).

We recall first the definition of g and h. Let (X, Y) be any point of D, and consider the line segments (25.5), where $(a, c) = (x_1, y_1)$ and $(b, d) = (X, Y)$ (Fig. 25.2). Then from Lemma 22.4,

$$h(X, Y) = \int_{C_1} p\, dx + q\, dy + \int_{C_2} p\, dx + q\, dy.$$

Similarly from Lemma 22.3,

$$g(X, Y) = -\int_{C_4} p\, dx + q\, dy - \int_{C_3} p\, dx + q\, dy.$$

But the rectangle R defined by (25.2) is included in the disk D, and we may apply Th. 25.1. The integrand on the right-side of (25.10) is identically zero by (25.11), and hence

$$h(X, Y) - g(X, Y) = \int_C p\, dx + q\, dy = 0.$$

Thus, for an arbitrary point (X, Y) in D, $h(X, Y) = g(X, Y)$ and the theorem is proved.   ◆

Let us now collect and summarize our various results for conservative vector fields.

Let $\mathbf{v}(x, y) = \langle p(x, y), q(x, y) \rangle \in \mathscr{C}^1$ in a domain D. Consider the following properties.

1. There exists $f(x, y) \in \mathscr{C}^2$ in D such that $\mathbf{v} = \nabla f$.
2. $\int_C p \, dx + q \, dy$ is independent of path.
3. $\int_C p \, dx + q \, dy = 0$, for all closed curves C in D.
4. $p_y = q_x$ throughout D.

We have the following relations between these properties.

a. $1 \Rightarrow 4$ for any domain D (Lemma 19.1).
b. 1, 2, and 3 are equivalent for any domain D (Th. 22.1 and Lemma 22.5).
c. In the case that D is a disk, $4 \Rightarrow 1$ (Th. 25.2).
d. In the case that D is a disk, all four properties are equivalent (combining relations a, b, c).
e. For certain domains D, the four properties are *not* equivalent (Example 22.3).

These facts are basic in the theory of vector fields. There is an important distinction between property 4 and the other three properties that may further clarify the situation. Property 4 is a *local* property, in the sense that its validity at each point of the domain depends only on the values of the functions p and q in a neighborhood of that point and is not affected by the values in the rest of the domain. The first three properties, on the other hand are *global* properties in the sense that they can only be verified if the functions p and q are known throughout the domain D. We may then add to our list of relations

f. $4 \Rightarrow 1$ locally.
g. All four properties are locally equivalent.

The precise meaning of these statements is the following. Suppose that property 4 holds in D, and let (x_1, y_1) be any point of D. If D' is a disk centered at (x_1, y_1) and included in D (Fig. 25.3), then by Th. 25.2, there exists a function $f(x, y)$ defined in D' satisfying $\mathbf{v} = \nabla f$. Thus, such a function may br defined in some neighborhood of each point of D, but it may be impossibl‹ to define a single function in all of D satisfying property 1 (see Example 25.: below). This is the meaning of relation f. Similarly, relation g means tha properties 2 and 3 also hold, provided that the curves C referred to in thes‹ properties are required to lie in the neighborhood D'.

Example 25.2

Let

$$p = -\frac{y}{x^2 + y^2}, \qquad q = \frac{x}{x^2 + y^2}$$

in the domain D consisting of the whole plane except the origin. We have seen in the discussion of Example 22.3 that in this case property 4 holds, but properties 1, 2, 3 do not. On the other hand, we now know by Th. 25.2 that in

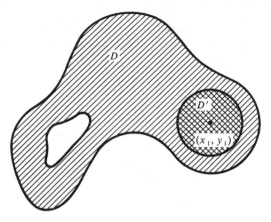

FIGURE 25.3   Localization to a neighborhood of (x_1, y_1)

some neighborhood of each point of D, there must exist a function $f(x, y)$ satisfying

(25.13) $$f_x = -\frac{y}{x^2 + y^2}, \qquad f_y = \frac{x}{x^2 + y^2}.$$

In order to find such a function, let us consider for example the second equation in (25.13) for a fixed value of x as a one-variable problem:

$$\frac{d}{dy} f(x_0, y) = \frac{x_0}{x_0^2 + y^2}.$$

It follows that

$$f(x_0, y) = \tan^{-1} \frac{y}{x_0} + c(x_0),$$

where $c(x_0)$ denotes a constant depending on x_0. This means that the function $f(x, y)$ satisfying (25.13), whose local existence we know, must be of the form

$$f(x, y) = \tan^{-1} \frac{y}{x} + g(x),$$

where $g(x)$ is some function of x. Furthermore, since $f(x, y) \in \mathscr{C}^2$, $g(x)$ is differentiable, and it follows that

$$f_x = \frac{\partial}{\partial x} \left(\tan^{-1} \frac{y}{x} \right) + g'(x) = -\frac{y}{x^2 + y^2} + g'(x).$$

Comparing this with (25.13), we see that $g'(x)$ must be zero, and hence $g(x)$ is constant. We thus have a solution of (25.13) in the form

$$f(x, y) = \tan^{-1} \frac{y}{x} + c.$$

Geometrically, the function $\tan^{-1}(y/x)$ represents the polar angle θ between the positive x axis and the ray from the origin to the point (x, y) (Fig. 25.4). The additive constant c may be interpreted geometrically by saying that the function $f(x, y)$ represents the angle between the ray from the origin to (x, y) and any fixed ray through the origin.

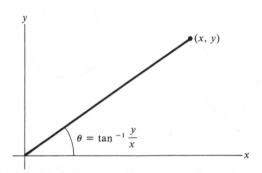

FIGURE 25.4 Geometric interpretation of the function $\tan^{-1} y/x$

Example 25.2 illustrates clearly the distinction between local and global potential functions alluded to in relation f above. Namely, given any point of D, that is to say, any point other than the origin, then in some neighborhood of that point one can choose a single-valued branch of $\tan^{-1}(y/x)$, and that will be a potential function of the vector field. Furthermore, up to an additive constant, it is the unique potential function in the neighborhood. On the other hand, it is impossible to find a function $f(x, y) \in \mathscr{C}^2$ in all of D that satisfies (25.13), since that would imply a continuous single-valued choice of the polar angle θ in the whole plane except the origin. But if we choose a determination of θ (or f) near some point, and then continue to observe its value as we traverse a circle going once counterclockwise around the origin, then the final value is 2π more than the starting value. We are thus faced with the choice of making θ multivalued or discontinuous, neither of which fits the definition of a potential function.

Let us conclude with the observation that this discussion explains the value 2π found by a direct computation in Example 22.3 for the line integral

$$\int_C -\frac{y}{x^2 + y^2} \, dx + \frac{x}{x^2 + y^2} \, dy,$$

where C is the circle $x = \cos t$, $y = \sin t$, $0 \le t \le 2\pi$. Namely, in a neighborhood of any point on C, there is a well-defined potential function $f = \theta$, and if we choose a short arc C_1 of C, which lies in the neighborhood, then the line integral over C_1 is equal to the change in the potential function (Fig. 25.5):

$$\int_{C_1} -\frac{y}{x^2 + y^2} \, dx + \frac{x}{x^2 + y^2} \, dy = \theta_2 - \theta_1,$$

which is the angle subtended by that arc. If we divide the circle into a number of such arcs, the total integral around C is the sum of these angles, which is 2π.

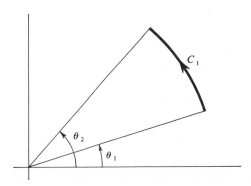

FIGURE 25.5   Geometric interpretation of $\int_{C_1} -y/(x^2 + y^2) \, dx + x/(x^2 + y^2) \, dy$

Exercises

25.1   Let C be the piecewise smooth curve consisting of the four successive line segments C_1, C_2, C_3, C_4 described in Eq. (25.5) (see Fig. 25.1). Compute each of the following line integrals in two ways, first by a direct computation, and then by Green's theorem.

    a. $\displaystyle\int_C y \, dx$ *c.* $\displaystyle\int_C x \, dx + y \, dy$

    b. $\displaystyle\int_C x^2 \, dy$ *d.* $\displaystyle\int_C ye^x \, dx + (x + e^x) \, dy$

25.2   Obtain the answer to Ex. 25.1c by a third method, using a potential function for the vector field $\langle x, y \rangle$.

25.3   Show that the answer to Ex. 25.1b is equal to twice the x moment of the rectangle R bounded by C. Write down the coordinates of the centroid of R, and check the answer to Ex. 25.1b.

25.4   Let C be the curve of Ex. 25.1, where $(a, c) = (-1, -1)$, $(b, d) = (1, 1)$. Use Green's theorem to evaluate the following line integrals.

    a. $\displaystyle\int_C \log \frac{2 + y}{1 + x^2} \, dx + \frac{x(3y + 7)}{y + 2} \, dy$

b. $\int_C (\sinh^{-1} x + y e^{xy}) \, dx + (\tanh y + x e^{xy}) \, dy$

c. $\int_C (xy^2 - x^2 y^4) \, dx + (x^2 y + 3x^3 y^3) \, dy$

(*Hint:* use Ex. 24.13 or 24.14a.)

25.5   Let F be the unit square $0 \le x \le 1, \ 0 \le y \le 1$. For each of the following functions $f(x, y)$, express $\iint_F f \, dA$ as a line integral over the boundary of F, and evaluate that integral.

 a. $f(x, y) = xy$
 b. $f(x, y) = 3x^2 y + y^3$
 c. $f(x, y) = e^x \sin \pi y - e^y \cos \pi x$

25.6   **a.** Evaluate by direct computation

$$\int_C -\frac{y}{x^2 + y^2} \, dx + \frac{x}{x^2 + y^2} \, dy,$$

where C is the curve of Ex. 25.4.
   b. If Green's theorem (Eq. (25.10)) is applied to the integral in part a, then since $q_x = p_y$ in this case, the value of the double integral would appear to be zero. This contradicts the result obtained by direct computation. What is the source of difficulty?
   c. Find the answer to part a by using a local potential function on each of the line segments constituting the curve C.

25.7   Let C be the curve of Ex. 25.1. For any point (X, Y) not on C, define $\varphi(X, Y)$ by

$$\varphi(X, Y) = \int_C -\frac{y - Y}{(x - X)^2 + (y - Y)^2} \, dx + \frac{x - X}{(x - X)^2 + (y - Y)^2} \, dy.$$

   ***a.** Find $\varphi(X, Y)$ for a point (X, Y) inside C. (*Hint:* find a local potential function on each side of C, and use it to evaluate the integral over that side.)
   b. Find $\varphi(X, Y)$ for a point (X, Y) outside C. (*Hint:* use either the method of part a, or Green's theorem.)

***25.8**   Let $u(x, y) \in \mathscr{C}^2$ in a domain that includes a rectangle R. Let the boundary of R be the curve C consisting of the four line segments described in Eq. (25.5). At each point of C other than the corners, let $\mathbf{n}$ denote the *exterior normal* to R; that is, the unit vector perpendicular to C, pointing toward the exterior of R. Denote by $\partial u / \partial n$, the directional derivative of u in the direction of $\mathbf{n}$. Show that

$$\iint_R (u_{xx} + u_{yy}) \, dA = \int_C \frac{\partial u}{\partial n} \, ds.$$

25.9   Find a potential function $f(x, y)$ for each of the following vector fields $\langle p, q \rangle$, using the method given in the text to solve Eq. (25.13).

a. $\langle p, q \rangle = \left\langle \dfrac{x}{x^2 + y^2}, \dfrac{y}{x^2 + y^2} \right\rangle$

b. $\langle p, q \rangle = \langle x^2 - y^2, -2xy \rangle$

c. $\langle p, q \rangle = \left\langle \dfrac{xy + 1}{x^2}, \log x + \dfrac{1}{y} \right\rangle$

d. $\langle p, q \rangle = \langle 2 \sec^2 x \tan x \tan y, \sec^2 x \sec^2 y \rangle$

25.10   Apply the method of Ex. 25.9 to obtain the answers to Ex. 22.8.

25.11   Let $v(x, y)$ be a "horizontal" vector field in a domain D; $v(x, y) = \langle p(x, y), 0 \rangle$, where $p(x, y) \in \mathscr{C}^1$.

   a. Show that if D is a disk, then v is conservative if and only if $p(x, y)$ depends only on x. (*Hint:* see Ex. 7.27.)
   ***b.** Show that it is possible to find a domain D and a function $p(x, y) \in \mathscr{C}^1$ in D such that $p(x, y)$ cannot be expressed as a function of x alone, and nevertheless $\langle p(x, y), 0 \rangle$ is conservative in D. (*Hint:* see Ex. 7.29.)

25.12   Show that harmonic vector fields in a disk coincide with the gradient vector fields of harmonic functions.

25.13   Show that a $\mathscr{C}^1$ vector field $\mathbf{w} = \langle u(x, y), v(x, y) \rangle$ in a disk D whose divergence is identically zero always has a stream function. (*Hint:* apply Th. 25.2 to the vector field $\langle p, q \rangle = \langle -v, u \rangle$, and use Ex. 19.12.)

25.14   Let F be the figure bounded by the lines $x = 0$, $y = 0$, $x/a + y/b = 1$. Let C be the boundary of F, where C consists of the three successive line segments: C_1, from $(0, 0)$ to $(a, 0)$, C_2, from $(a, 0)$ to $(0, b)$, and C_3, from $(0, b)$ to $(0, 0)$. Show that Green's theorem holds in the following form.

   a. Given $q(x, y) \in \mathscr{C}^1$ in a domain which includes the figure F, verify each of the following steps.

$$\int_0^{a - (a/b)t} q_x(x, t) \, dt = q\left(a - \frac{a}{b} t, t\right) - q(0, t)$$

$$\iint_F q_x \, dA = \int_0^b \left[\int_0^{a - (a/b)t} q_x(x, t) \, dx \right] dt = \int_0^b q\left(a - \frac{a}{b} t, t\right) dt - \int_0^b q(0, t) \, dt$$

$$= \int_{C_2} q \, dy + \int_{C_3} q \, dy = \int_C q \, dy.$$

   b. If $p(x, y) \in \mathscr{C}^1$ in a domain that includes F, show by a reasoning analogous to part a that $\iint_F p_y \, dA = -\int_C p \, dx$.
   c. Deduce that

$$\iint_F (q_x - p_y) \, dA = \int_C p \, dx + q \, dy.$$

25.15 Let C be the curve of Ex. 25.14. Use Ex. 25.14c to evaluate

a. $\displaystyle\int_C x \, dy$

b. $\displaystyle\int_C y \, dx$

c. $\displaystyle\int_C \cos e^x \, dx + e^{\cos y} \, dy$

*25.16 Let C and $\varphi(X, Y)$ be defined as in Ex. 25.7.

a. Show that for any point (X, Y) not on C, the partial derivatives φ_X, φ_Y may be obtained by "differentiating under the integral sign," and that the result in this case takes the following form:

if $\qquad p = -\dfrac{y - Y}{(x - X)^2 + (y - Y)^2}, \qquad q = \dfrac{x - X}{(x - X)^2 + (y - Y)^2}$

then

$$\varphi_X = \int_C -p_x \, dx - q_x \, dy, \qquad \varphi_Y = \int_C -p_y \, dx - q_y \, dy.$$

(*Hint:* apply Lemma 7.2 to the integrals over each side of C.)

b. Using the fact that $q_x = p_y$ at each point of the curve C in part a, show that $\nabla\varphi = 0$ at each point (X, Y) not in C.

c. Use part b together with the answer to Ex. 25.6a, to obtain the answer to Ex. 25.7a.

25.17 *a.* Consider the curve $C: x = \cos kt,\ y = \sin kt,\ 0 \le t \le 2\pi$, where k is an integer, positive or negative. Evaluate

$$\int_C -\frac{y}{x^2 + y^2} \, dx + \frac{x}{x^2 + y^2} \, dy.$$

b. Let C be a closed piecewise smooth curve that does not pass through the origin. Explain why you would expect the integral in part a to equal $2\pi n$ for some integer n. (*Hint:* use the reasoning given in the text at the end of this section; divide the curve into short arcs, and describe geometrically the value of the integral over each of these arcs.)

*25.18 Let C be a piecewise smooth closed curve. For any point (X, Y) not on C let

$$n(C; X, Y) = \frac{1}{2\pi} \int_C -\frac{y - Y}{(x - X)^2 + (y - Y)^2} \, dx + \frac{x - X}{(x - X)^2 + (y - Y)^2} \, dy.$$

Show that $n(C; X, Y)$ has the following properties.

a. If D is any domain which does not meet the curve C, then $n(C; X, Y)$ is constant on D. (*Hint:* see Ex. 25.16.)

b. $n(C; X, Y)$ is an integer. (*Hint:* see Exs. 25.7a and 25.17b.)

c. $n(C; X, Y)$ may be described as "the number of times the curve C goes around the point (X, Y) in the counterclockwise direction."

Note: the number n $(C;X, Y)$ is called the *winding number* of the curve C about the point (X, Y). Using the notation described in Ex. 21.25 we could write, for fixed (X, Y),

$$n(C; X, Y) = \frac{1}{2\pi} \int_C d\theta,$$

where (see Fig. 25.6)

$$\theta = \tan^{-1}\frac{y - Y}{x - X}.$$

Although θ is not single-valued, the derivatives θ_x, θ_y are single-valued, and the integral $\int_C d\theta$ is well-defined. Intuitively, it represents the total variation of the angle θ as the point (x, y) traverses the curve C. For an excellent discussion of winding numbers and their applications, see Part II of [7].

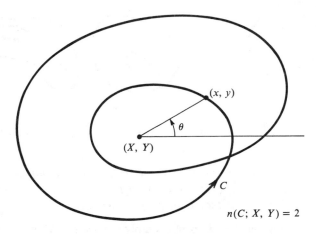

FIGURE 25.6   A curve C which winds twice about a point (X, Y)

Addendum to Section 25   Simple Connectivity

Remark   It is quite clear that the proof of Th. 25.2 can be adapted for domains D that are more general than a disk. The widest class of domains for which the theorem holds is the class of *simply-connected domains*. Simple connectivity is a very important geometric concept, which may be applied not only to plane domains, but to surfaces in space, three-dimensional bodies, and to still more general situations. However, in the case of the plane there are a number of different, but equivalent characterizations of simple connectivity. The following definition is specially adapted to plane domains, making use of the winding number $n(C; X, Y)$ defined in Ex. 25.18.

Definition 25.1   A plane domain D is *simply-connected* if the winding number $n(C; X, Y)$ is equal to zero for every piecewise smooth closed curve

C lying in *D* and for all points (X, Y) not in *D*. A plane domain which is not simply connected is called *multiply-connected*.

Intuitively, a simply-connected domain is one "without holes." If a domain *D* has a hole in it, then choosing a point (X, Y) inside the hole and a curve *C* going around the hole gives $n(C; X, Y) \neq 0$. For example, if *D* is the annulus $1 < x^2 + y^2 < 9$, then choose $(X, Y) = (0, 0)$ and $C: x \cos t$, $y = \sin t$, $0 \leq t \leq 2\pi$, then $n(C; X, Y) = 1$; hence *D* is multiply-connected.

A fact that is useful to bear in mind is that a domain bounded by a single closed curve is simply-connected, whereas a domain bounded by two or more closed curves is multiply-connected (Fig. 25.7).

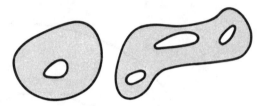

Simply–connected domain

Multiply–connected domains

FIGURE 25.7 Examples of simple connectivity and of multiple connectivity

Theorem 25.3 *Given a domain D, the necessary and sufficient condition that every $\mathscr{C}^1$ vector field $\langle p(x, y), q(x, y) \rangle$ in D satisfying $p_y \equiv q_x$ should have a potential function in D is that D be simply-connected.*

PROOF I. Suppose that *D* is not simply-connected. Then there exists a closed curve *C* in *D* and a point (X, Y) not in *D* such that $n(C; X, Y) \neq 0$. Since the fixed point (X, Y) is not in *D*, it follows that the functions

$$p(x, y) = -\frac{y - Y}{(x - X)^2 + (y - Y)^2}, \qquad q(x, y) = \frac{x - X}{(x - X)^2 + (y - Y)^2}$$

are $\mathscr{C}^1$ functions in *D*. By direct computation $p_y \equiv q_x$. However,

$$2\pi n(C; X, Y) = \int_C p \, dx + q \, dy \neq 0$$

and by Th. 22.2, the vector field $\langle p, q \rangle$ cannot have a potential function in *D*.

II. Suppose that *D* is simply-connected. Let $\langle p, q \rangle$ be an arbitrary $\mathscr{C}^1$ vector field in *D* satisfying $p_y \equiv q_x$. To define a potential function $f(x, y)$ we fix a point (x_1, y_1) in *D*, and then for an arbitrary point (x_2, y_2) we set

$$f(x_2, y_2) = \int_C p \, dx + q \, dy,$$

where C is a curve in D that starts at (x_1, y_1), ends at (x_2, y_2), and consists of a finite number of horizontal and vertical line segments. (The existence of such a curve is easily proved.) If we can show that the value of this integral depends only on the point (x_2, y_2) and not on the particular choice of path C, then by choosing first a path ending in a horizontal segment and then a path ending in a vertical segment, it follows by the reasoning of Lemmas 22.3 and 22.4 that the function $f(x, y)$ satisfies $f_x = p$, $f_y = q$ and is therefore a potential function. Suppose then that C_1, C_2 are two paths having the properties described. Let $\tilde{C}$ be the closed curve in D consisting of C_1 followed by C_2 described in the opposite direction. We wish to show that

$$\int_{C_1} p \, dx + q \, dy - \int_{C_2} p \, dx + q \, dy = \int_{\tilde{C}} p \, dx + q \, dy = 0.$$

The idea of the proof is to show that the curve $\tilde{C}$ can be broken up into a number of line segments, some of which are described twice in opposite directions making the line integrals cancel out, and the rest of which may be grouped together to form the boundaries of rectangles lying in D (Fig. 25.8). Applying Green's theorem to each of these rectangles (using the hypothesis that $q_x - p_y \equiv 0$ in D) and adding over all the rectangles gives the desired result $\int_{\tilde{C}} p \, dx + q \, dy = 0$. A complete proof that $\tilde{C}$ can be decomposed in the manner indicated requires a careful analysis, which is omitted here. We refer to the proof of Theorem 15 in Chapter 4, Section 4.3 of [2].

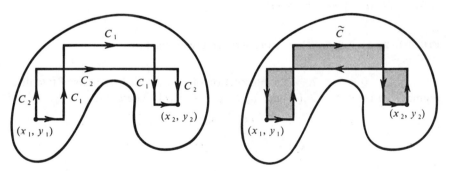

FIGURE 25.8 Decomposition of the curve $\tilde{C}$ into a collection of boundaries of rectangles

We note in conclusion that the above discussion is based on the treatment of simple connectivity given in the book of Ahlfors [2]. The equivalence of the definition given here with other characterizations of simply-connected plane domains is proved in Section 4.2 of Chapter 4 and in Section 1.5 of Chapter 8 of Ahlfors' book.

26 Green's theorem for arbitrary figures; applications

Our proof of Green's theorem, which we gave for the special case of a rectangle, turns out to hold equally well for general figures. More precisely, the idea of the proof is unchanged, but the details must be modified.

Before discussing the general case, let us illustrate the argument for a specific figure (see also Ex. 25.14). Consider the figure

$$F: y \geq 0, \qquad 4 \leq x^2 + y^2 \leq 9.$$

Let $q(x, y) \in \mathscr{C}^1$ in a domain that includes F. We wish to evaluate $\iint_F q_x \, dA$. To do so, we use the method and notation of Th. 24.3. The horizontal line $y = t$ intersects F for $0 \leq t \leq 3$ (Fig. 26.1). There are two cases to consider.

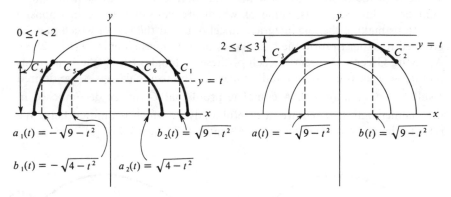

FIGURE 26.1 Proof of Green's theorem for a semi-annulus

Case 1. $0 \leq t < 2$. The intersection of $y = t$ with F consists of two line segments $a_1(t) \leq x \leq b_1(t)$ and $a_2(t) \leq x \leq b_2(t)$, where

$$a_1(t) = -\sqrt{9 - t^2}, \qquad b_1(t) = -\sqrt{4 - t^2};$$
$$a_2(t) = \sqrt{4 - t^2}, \qquad b_2(t) = \sqrt{9 - t^2}.$$

Thus

$$\varphi(t) = \sum_{k=1}^{2} \int_{a_k(t)}^{b_k(t)} q_x(x, t) \, dx = \sum_{k=1}^{2} [q(b_k(t), t) - q(a_k(t), t)]$$

and

$$(26.1) \quad \int_0^2 \varphi(t) \, dt = \int_0^2 q(b_1(t), t) \, dt - \int_0^2 q(a_1(t), t) \, dt$$

$$+ \int_0^2 q(b_2(t), t) \, dt - \int_0^2 q(a_2(t), t) \, dt.$$

We thus have four integrals each of which may be considered as a line integral along a part of the boundary of F. Consider, for example.

$$\int_0^2 q(b_2(t), t) \, dt = \int_0^2 q(\sqrt{9 - t^2}, t) \, dt = \int_{C_1} q \, dy,$$

where the curve $C_1 : x = \sqrt{9 - t^2}$, $y = t$, $0 \le t \le 2$, is an arc of the outer circle. Similarly,

$$-\int_0^2 q(a_2(t), t) \, dt = -\int_0^2 q(\sqrt{4 - t^2}, t) \, dt = -\int_{\tilde{C}_6} q \, dy = \int_{C_6} q \, dy,$$

where the curve $\tilde{C}_6 : x = \sqrt{4 - t^2}$, $y = t$, $0 \le t \le 2$, is an arc of the inside boundary oriented in the direction of increasing y, and C_6 is the same arc with the opposite orientation (Fig. 26.1). Thus, with C_4 and C_5 denoting the (oriented) boundary arcs indicated in Fig. 26.1, Eq. (26.1) takes the form

$$(26.2) \qquad \int_0^2 \varphi(t) \, dt = \int_{C_5} q \, dy + \int_{C_4} q \, dy + \int_{C_1} q \, dy + \int_{C_6} q \, dy.$$

Case 2. $2 \le t \le 3$.   The intersection of $y = t$ with F consists of a single segment. We have $a(t) = -\sqrt{9 - t^2}$, $b(t) = \sqrt{9 - t^2}$,

$$\varphi(t) = \int_{a(t)}^{b(t)} q_x(x, t) \, dx = q(b(t), t) - q(a(t), t)$$

and

$$(26.3) \qquad \int_2^3 \varphi(t) \, dt = \int_2^3 q(b(t), t) \, dt - \int_2^3 q(a(t), t) \, dt$$

$$= \int_{C_2} q \, dy + \int_{C_3} q \, dy,$$

where C_2 and C_3 are the boundary arcs indicated in Fig. 26.1.

Let us denote by $\tilde{C}_1$ the entire outer semicircle of F, oriented in the counterclockwise direction, and let $\tilde{C}_3$ denote the inner semicircle oriented in the clockwise direction (Fig. 26.2). The curve $\tilde{C}_1$ consists of the four arcs C_1, C_2, C_3, C_4, while $\tilde{C}_3$ consists of C_5 and C_6. Thus, adding Eqs. (26.2) and (26.3), and applying Th. 24.3, we have

$$(26.4) \qquad \iint_F q_x \, dA = \int_0^3 \varphi(t) \, dt = \int_0^2 \varphi(t) \, dt + \int_2^3 \varphi(t) \, dt$$

$$= \int_{\tilde{C}_1} q \, dy + \int_{\tilde{C}_3} q \, dy.$$

Finally,   let $\tilde{C}_2 : x = t$, $y = 0$, $-3 \le t \le -2$,   and $\tilde{C}_4 : x = t$, $y = 0$, $2 \le t \le 3$, denote the two horizontal line segments indicated in Fig. 26.2.

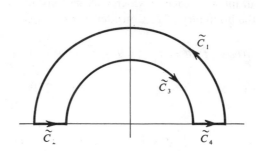

FIGURE 26.2 The oriented boundary of a semi-annulus

Then

$$\int_{\tilde{C}_2} q \, dy = \int_{\tilde{C}_4} q \, dy = 0,$$

and it follows from (26.4) that

(26.5)
$$\iint_F q_x \, dA = \int_C q \, dy,$$

where C is the piecewise smooth boundary curve of F, consisting of the successive arcs $\tilde{C}_1, \tilde{C}_2, \tilde{C}_3, \tilde{C}_4$. An analogous argument using Th 24.4 shows that

(26.6)
$$\iint_F p_y \, dA = -\int_C p \, dx$$

for any function $p(x, y) \in \mathscr{C}^1$ in a domain that includes F. Equations (26.5) and (26.6), or their combined form

(26.7)
$$\iint_F (q_x - p_y) \, dA = \int_C p \, dx + q \, dy,$$

constitute Green's theorem for the figure F. Once the reasoning used in this special case is understood, there should be no difficulty in seeing how it extends to an arbitrary figure. In order to state the general theorem, it is useful to introduce the following notation.

Definition 26.1 A curve $C: x(t), y(t), a \le t \le b$, lying on the boundary of a figure F is said to have *positive orientation* with respect to F if the vector $\langle -y'(t), x'(t) \rangle$ is directed toward the interior of F.

Remark The vector $\langle -y'(t), x'(t) \rangle$ is perpendicular to the tangent vector $\langle x'(t), y'(t) \rangle$, and is obtained from the latter by a rotation through $90°$

in the counterclockwise direction. Thus, positive orientation with respect to F may be described as the direction of motion along C such that the interior of F lies to the left (Fig. 26.3). Note that a figure F is typically bounded by one outer curve and a number of inner curves. Along the outer curve positive orientation means traversing the curve in the counterclockwise direction, whereas along the inner curves, positive orientation implies traversing in the clockwise direction.

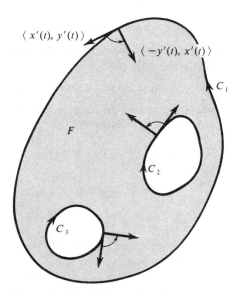

FIGURE 26.3 Positive orientation of boundary curves

Definition 26.2 The *oriented boundary* of a figure F, denoted by ∂F, consists of the totality of boundary curves of F, each of them assigned a positive orientation with respect to F.

Remark The only context in which we use the oriented boundary is in forming a line integral $\int_{\partial F} p\, dx + q\, dy$, where p and q are continuous functions on the boundary of F. Thus

$$(26.8) \qquad \int_{\partial F} p\, dx + q\, dy = \sum_{k=1}^{n} \int_{C_k} p\, dx + q\, dy,$$

where the boundary of F consists of the curves $C_1, \ldots, C_n$, each assigned a positive orientation with respect to F. As we know from Lemma 21.1, the value of this integral does not depend on the parameter chosen to represent the curves C_k, providing the orientation is preserved.

Example 26.1

Evaluate $\int_{\partial F} y\, dx$ for the figure $F: 4 \le x^2 + y^2 \le 9$.

The boundary of F consists of two circles (Fig. 26.4), which may be represented parametrically as

$$C_1: x = 3 \cos t, \, y = \quad 3 \sin t, \qquad\qquad 0 \le t \le 2\pi;$$

$$C_2: x = 2 \cos t, \, y = -2 \sin t, \qquad\qquad 0 \le t \le 2\pi.$$

Note that along the outer curve C_1, the vector

$$\langle -y'(t),\, x'(t) \rangle = \langle -3 \cos t,\, -3 \sin t \rangle = \langle -x,\, -y \rangle$$

is directed toward the origin, hence toward the interior of F, while along the inner curve C_2, the vector

$$\langle -y'(t),\, x'(t) \rangle = \langle 2 \cos t,\, -2 \sin t \rangle = \langle x,\, y \rangle$$

is directed outwards, hence again toward the interior of F. Thus both C_1 and C_2 are positively oriented with respect to F. Now

$$\int_{C_1} y\, dx = \int_0^{2\pi} (3 \sin t)(-3 \sin t)\, dt$$

$$= -9 \int_0^{2\pi} \sin^2 t\, dt = -9\pi,$$

and

$$\int_{C_2} y\, dx = \int_0^{2\pi} (-2 \sin t)(-2 \sin t)\, dt = 4 \int_0^{2\pi} \sin^2 t\, dt = 4\pi.$$

Thus

$$\int_{\partial F} y\, dx = \int_{C_1} y\, dx + \int_{C_2} y\, dx = -5\pi.$$

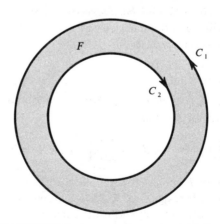

FIGURE 26.4 The positively oriented boundary of an annulus

Theorem 26.1   Green's Theorem   *Let $p(x, y), q(x, y) \in \mathscr{C}^1$ in a domain that includes a figure F. Then*

$$(26.9) \qquad \iint_F (q_x - p_y)\, dA = \int_{\partial F} p\, dx + q\, dy.$$

Remark   Equation (26.9) is equivalent to the two separate equations

$$(26.10) \qquad \iint_F q_x\, dA = \int_{\partial F} q\, dy, \qquad \iint_F p_y\, dA = -\int_{\partial F} p\, dx.$$

PROOF.   We first evaluate $\iint_F q_x\, dA$, using the method and notation of Th. 24.3. Let us indicate the reasoning for the case where F is a polynomial figure.[7] The line $y = t$ intersects F in a finite number of intervals $a_k(t) \le x \le b_k(t)$. We set

$$\varphi(t) = \sum_k \int_{a_k(t)}^{b_k(t)} q_x(x, t)\, dx = \sum_k [q(b_k(t), t) - q(a_k(t), t)].$$

Then

$$(26.11) \qquad \iint_F q_x\, dA = \int_c^d \varphi(t)\, dt$$

$$= \sum_k \int_c^d q(b_k(t), t)\, dt - \sum_k \int_c^d q(a_k(t), t)\, dt.$$

Now for each k, the integral $\int_c^d q(b_k(t), t)\, dt$ can be broken into a finite number of intervals such that in each interval, say $t_j \le t \le t_{j+1}$, the equations $x = b_k(t)$, $y = t$, define a curve $\tilde{C}_j$ lying on the boundary of F. Then

$$\int_{t_j}^{t_{j+1}} q(b_k(t), t)\, dt = \int_{\tilde{C}_j} q\, dy.$$

Since the point $(b_k(t), t)$ is a boundary point of F lying at the right-hand endpoint of a line segment contained in F, the interior of F lies to the left if $\tilde{C}_j$ is traversed in the direction of increasing y. Thus, each of the $\tilde{C}_j$ so defined is positively oriented with respect to F.

Similarly, each of the integrals $\int_c^d q(a_k(t), t)\, dt$ may be broken down into a sum, such that $\bar{C}_i : x = a_k(t)$, $y = t$, $t_i \le t \le t_{i+1}$, is a curve on the boundary of F. Hence

$$\int_{t_i}^{t_{i+1}} q(a_k(t), t)\, dt = \int_{\bar{C}_i} q\, dy = -\int_{\tilde{C}_i} q\, dy,$$

where $\tilde{C}_i$ consists of the curve $\bar{C}_i$ with the opposite orientation. Since F lies to the right of $\bar{C}_i$ as $\bar{C}_i$ is traversed in the direction of increasing y, $\bar{C}_i$ has negative orientation with respect to F, and $\tilde{C}_i$ therefore has positive orienta-

---

[7] For a different method of proof, which applies to an arbitrary figure, see [28], Sections 14.1–14.3.

tion. If we form the sum over all k, the right-hand side of (26.11) represents a sum of line integrals of the form $\int_{\tilde{C}_j} q \, dy$, where each $\tilde{C}_j$ is a curve lying on the boundary of F and having positive orientation with respect to F. Furthermore the totality of these $\tilde{C}_j$ represents the complete oriented boundary of F, except for possible horizontal segments along which $\int q \, dy = 0$. Thus, the right-hand side of (26.11) is equal to $\int_{\partial F} q \, dy$, and (26.11) reduces to the first equation in (26.10). A similar reasoning using Th. 24.4 yields the second equation in (26.10). Combining these two equations, we obtain (26.9), and the theorem is proved. ◆

Example 26.2

Evaluate $\int_{\partial F} y \, dx$ for the figure $F: 4 \leq x^2 + y^2 \leq 9$.

We have already found the value -5π for this integral by a direct calculation. Using Green's theorem, with $p = y, q = 1$, we find $q_x \equiv 0, p_y \equiv 1$, and

$$\int_{\partial F} y \, dx = \iint_F (-1) \, dA = -\text{Area of } F.$$

But the area of F is equal to the area inside a circle of radius 3 and outside a circle of radius 2; that is, $9\pi - 4\pi$ or 5π.

Example 26.3

Evaluate

(26.12) $$\int_{\partial F} -\frac{y}{x^2 + y^2} \, dx + \frac{x}{x^2 + y^2} \, dy,$$

where F is the square $-1 \leq x \leq 1, -1 \leq y \leq 1$.

In this case ∂F is a single closed curve C consisting of four line segments (Fig. 26.5). We can find the value of (26.12) by computing the integral over each line segment and adding (see Ex. 25.6a). Green's theorem cannot be applied directly, because the functions p, q in this case have a singularity at the origin. Thus the figure F, which contains the origin, does not lie in a domain where $p(x, y), q(x, y) \in \mathscr{C}^1$. We can, however, simplify the computation greatly by using the figure

$$\tilde{F}: -1 \leq x \leq 1, -1 \leq y \leq 1, x^2 + y^2 \geq r^2,$$

where r is any fixed number between 0 and 1 (Fig. 26.5). The boundary of $\tilde{F}$ consists of two parts. The outer curve is the same as the boundary of F, and the inner curve is a circle of radius r. The functions $p = -y/(x^2 + y^2)$, $q = x/(x^2 + y^2) \in \mathscr{C}^1$ in the domain D consisting of the whole plane except for the origin. Since the figure $\tilde{F}$ lies in this domain we may apply Green's theorem. As we have observed before, $q_x - p_y \equiv 0$ in this case, and therefore

$$\int_{\partial \tilde{F}} p \, dx + q \, dy = \iint_{\tilde{F}} (q_x - p_y) \, dA = 0.$$

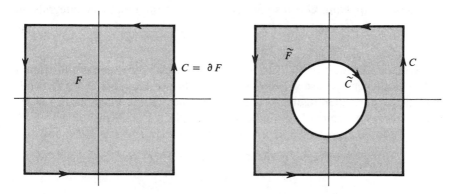

FIGURE 26.5   Use of Green's theorem to reduce integral over a square to an integral over a circle

Denoting by $\tilde{C}$ the inner circle, $x = r \cos t$, $y = -r \sin t$, $0 \le t \le 2\pi$, positively-oriented with respect to $\tilde{F}$, and once again $C = \partial F$, we obtain

$$\int_C p\,dx + q\,dy + \int_{\tilde{C}} p\,dx + q\,dy = \int_{\partial \tilde{F}} p\,dx + q\,dy = 0$$

or

$$\int_{\partial F} -\frac{y}{x^2 + y^2}\,dx + \frac{x}{x^2 + y^2}\,dy$$

$$= -\int_{\tilde{C}} \frac{-y}{x^2 + y^2}\,dx + \frac{x}{x^2 + y^2}\,dy$$

$$= -\int_0^{2\pi} \left[\frac{r \sin t}{r^2}(-r \sin t) + \frac{r \cos t}{r^2}(-r \cos t)\right] dt$$

$$= 2\pi.$$

Thus we have found the integral in question, not by a direct integration over the original curve, but by reducing it to an integral over a simpler curve by virtue of Green's theorem.

The above examples illustrate two of the many applications of Green's theorem. We next consider the way in which Example 26.2 generalizes to arbitrary figures.

Lemma 26.1   *Let F be an arbitrary figure, and A(F) its area; then*

(26.13) $$A(F) = \int_{\partial F} -y\,dx = \int_{\partial F} x\,dy = \frac{1}{2}\int_{\partial F} x\,dy - y\,dx.$$

PROOF. Applying Green's theorem to each of the boundary integrals on the right yields $\iint_F 1 \, dA = A(F)$. ◆

Remark In addition to Example 26.2, we have already seen an illustration of this result for the case where F is a rectangle (see Example 25.1). Depending on the particular figure F, one or the other of the integrals in (26.13) may be more convenient to evaluate. For example, if F is the disk $x^2 + y^2 \leq r^2$, then ∂F can be represented as $x = r \cos t$, $y = r \sin t$, $0 \leq t \leq 2\pi$, and

$$-yx'(t) = r^2 \sin^2 t, \qquad xy'(t) = r^2 \cos^2 t,$$
$$\tfrac{1}{2}[xy'(t) - yx'(t)] = \tfrac{1}{2}r^2.$$

Each of these expressions integrated from 0 to 2π yields the value πr^2, but the last one is clearly the simplest.

We are now able to prove a theorem that provides a basic link between the double integral and the theory of differentiable transformations. The Jacobian of a transformation, which first appeared as the determinant of an associated linear transformation, turns up here in a totally different manner.

Theorem 26.2 *Let $u(x, y)$, $v(x, y)$ define a diffeomorphism G of a domain D onto a domain $\tilde{D}$. Let F be a figure lying in D, and let its image be a figure $\tilde{F}$; then*

$$(26.14) \qquad \text{Area of } \tilde{F} = \iint_F \left| \frac{\partial(u, v)}{\partial(x, y)} \right| dA.$$

PROOF. We apply Lemma 26.1 to the figure $\tilde{F}$. For this purpose, let us denote by $C_1, \ldots, C_n$ the oriented curves which constitute ∂F. Let

$$(26.15) \qquad\qquad C_k : x(t), y(t), \qquad\qquad a \leq t \leq b$$

be any one of these curves, and let

$$(26.16) \qquad\qquad \tilde{C}_k : u(x(t), y(t)), v(x(t), y(t)), \qquad\qquad a \leq t \leq b$$

be its image under the transformation G. Using the chain rule we may transform any line integral over $\tilde{C}_k$ into a line integral over C_k. In particular,

$$(26.17) \quad \int_{\tilde{C}_k} u \, dv = \int_a^b uv'(t) \, dt$$

$$= \int_a^b u(v_x x'(t) + v_y y'(t)) \, dt = \int_{C_k} uv_x \, dx + uv_y \, dy.$$

Now setting $p = uv_x$, $q = uv_y$, we find

$$q_x = uv_{xy} + u_x v_y, \qquad p_y = uv_{xy} + u_y v_x.$$

and

$$q_x - p_y = u_x v_y - u_y v_x = \frac{\partial(u, v)}{\partial(x, y)}.$$

Applying Green's formula (26.9) to the figure F, we find

$$(26.18) \qquad \sum_{k=1}^{n} \int_{C_k} uv_x \, dx + uv_y \, dy = \int_{\partial F} uv_x \, dx + uv_y \, dy$$

$$= \iint_{F} \frac{\partial(u, v)}{\partial(x, y)} \, dA.$$

As we have remarked in the proof of Th. 18.2, for a diffeomorphism, the Jacobian is either everywhere positive or everywhere negative. If the Jacobian is positive, then the transformation G preserves orientation. This implies that since the figure F lies to the left when C_k is traversed in the direction of increasing t, the figure $\tilde{F}$ also lies to the left when $\tilde{C}_k$ is traversed in the direction of increasing t. In other words, for positive Jacobian, the fact that the curve C_k defined by (26.15) is positively oriented with respect to F implies that the curve $\tilde{C}_k$ defined by (26.16) is positively oriented with respect to $\tilde{F}$. Thus the curves $\tilde{C}_1, \ldots, \tilde{C}_n$ constitute $\partial \tilde{F}$. Applying Lemma 26.1 to $\tilde{F}$ and using Eqs. (26.17) and (26.18) yields

$$\text{Area of } \tilde{F} = \int_{\partial \tilde{F}} u \, dv = \sum_{k=1}^{n} \int_{\tilde{C}_k} u \, dv = \iint_{F} \frac{\partial(u, v)}{\partial(x, y)} \, dA.$$

This proves (26.14) for the case that $\partial(u, v)/\partial(x, y) > 0$. In the opposite case, where $\partial(u, v)/\partial(x, y) < 0$, orientation is reversed, so that the curves $\tilde{C}_k$ defined by (26.16) are negatively oriented with respect to $\tilde{F}$. In this case

$$\text{Area of } \tilde{F} = \int_{\partial \tilde{F}} u \, dv = -\sum_{k=1}^{n} \int_{\tilde{C}_k} u \, dv = \iint_{F} -\frac{\partial(u, v)}{\partial(x, y)} \, dA.$$

Since

$$\frac{\partial(u, v)}{\partial(x, y)} < 0, \qquad \left| \frac{\partial(u, v)}{\partial(x, y)} \right| = -\frac{\partial(u, v)}{\partial(x, y)},$$

and (26.14) holds again. ◆

Corollary *Let*

$$G: \begin{cases} u = ax + by \\ v = cx + dy \end{cases}$$

be a nonsingular linear transformation, and let $\Delta = \det G = ad - bc$. Then, if F is any figure in the x, y plane and $\tilde{F}$ is its image in the u, v plane,

$$(26.19) \qquad |\Delta| = \frac{\text{area of } \tilde{F}}{\text{area of } F}.$$

PROOF. Since a nonsingular linear transformation is a diffeomorphism of the whole x, y plane onto the u, v plane, we may apply the theorem. But $\partial(u, v)/\partial(x, y) = \Delta$, a constant, and (26.14) reduces to

$$\text{Area of } \tilde{F} = \iint_F |\Delta| \, dA = |\Delta| \iint_F 1 \, dA = |\Delta| \text{ (area of } F\text{).} \quad \blacklozenge$$

Remark We have already indicated the geometric interpretation on (26.19) of the determinant in Sect. 14, for the case where F was a disk, and at the end of Sect. 15, for the case where F was a triangle. One way to make Eq. (26.14) plausible is to note that the integral on the right-hand side is the limit of sums of the form

$$(26.20) \qquad \qquad \sum \left| \frac{\partial(u, v)}{\partial(x, y)} \right|_{(\xi_i, \eta_i)} A_i$$

taken over partitions of F. If a figure F_i in the partition is taken to be a small triangle, then the differential of G at the point (ξ_i, η_i) is a linear transformation that approximates G on F_i and whose determinant is equal to

$$\frac{\partial(u, v)}{\partial(x, y)} \bigg|_{(\xi_i, \eta_i)}.$$

Thus the corresponding term in the sum (26.20) equals the area of the image under the differential of G at (ξ_i, η_i) and is approximately equal to the area of the image under G. Summing over i, we obtain a quantity that approximates the area of the image of all of F, and Eq. (26.14) states that the sums (26.20) tend precisely to this area in the limit.

Example 26.4

Let G be the transformation

$$G : \begin{cases} u = \frac{1}{2}(x^2 - y^2) \\ v = -xy. \end{cases}$$

Let F be the figure $x \geq 0$, $y \geq 0$, $x^2 + y^2 \leq 1$.
   To compute the integral in (26.14), we note

$$\begin{aligned} u_x &= x & u_y &= -y \\ v_x &= -y & v_y &= -x, \end{aligned}$$

and

$$\frac{\partial(u, v)}{\partial(x, y)} = -x^2 - y^2.$$

Thus

$$\iint_F \left| \frac{\partial(u, v)}{\partial(x, y)} \right| dA = \iint_F (x^2 + y^2) \, dA = \int_0^1 \left[\int_0^{(1 - y^2)^{1/2}} (x^2 + y^2) \, dx \right] dy.$$

But

$$\int_0^{(1-y^2)^{1/2}} (x^2 + y^2)\, dx = \left[\frac{x^3}{3} + y^2 x\right]_0^{(1-y^2)^{1/2}} = \frac{(1 - y^2)^{1/2}}{3}(1 + 2y^2),$$

and we find (for example, by the trignometric substitution $y = \sin t$):

$$\int_0^1 (1 - y^2)^{1/2}\, dy = \frac{\pi}{4}, \qquad \int_0^1 y^2 (1 - y^2)^{1/2}\, dy = \frac{\pi}{16}.$$

Thus

(26.21)
$$\iint_F (x^2 + y^2)\, dA = \frac{1}{3}\left(\frac{\pi}{4} + \frac{\pi}{8}\right) = \frac{\pi}{8}.$$

Let us now examine the figure F and its image $\tilde{F}$ under the transformation G (Fig. 26.6). Introducing polar coordinates

$$x = r \cos \theta \qquad\qquad y = r \sin \theta$$
$$u = R \cos \varphi \qquad\qquad v = R \sin \varphi,$$

we find that G takes the form

$$u = \tfrac{1}{2}(x^2 - y^2) = \tfrac{1}{2} r^2 (\cos^2 \theta - \sin^2 \theta) = \tfrac{1}{2} r^2 \cos 2\theta,$$
$$v = -xy = -r^2 \cos \theta \sin \theta = -\tfrac{1}{2} r^2 \sin 2\theta,$$

or

$$R = \tfrac{1}{2} r^2, \qquad\qquad \varphi = -2\theta.$$

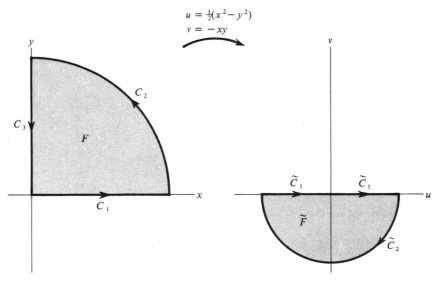

FIGURE 26.6 The image of a quarter disk under the transformation $G: u = \tfrac{1}{2}(x^2 - y^2)$, $v = -xy$

Thus each radius $\theta = \theta_0$, $0 \le r \le 1$, of the quarter circle F maps one-to-one onto the ray $\varphi = -2\theta_0$, $0 \le R \le \frac{1}{2}$, and the figure F maps one-to-one onto the semicircle $\tilde{F}$: $-\pi \le \varphi \le 0$, $0 \le R \le \frac{1}{2}$. The area of $\tilde{F}$ is $\frac{1}{2}\pi(\frac{1}{2})^2 = \frac{1}{8}\pi$, which, in view of (26.21), confirms Eq. (26.14) for this example.

In conclusion, let us examine ∂F and $\partial \tilde{F}$. First of all, ∂F consists of a single closed piecewise smooth curve, obtained by describing in succession the following parts:

$$C_1 : x = t, \, y = 0, \qquad\qquad 0 \le t \le 1,$$
$$C_2 : x = \cos t, \, y = \sin t, \qquad 0 \le t \le \tfrac{1}{2}\pi,$$
$$C_3 : x = 0, \, y = 1 - t, \qquad\qquad 0 \le t \le 1.$$

Their images under G are, respectively,

$$\tilde{C}_1 : u = \tfrac{1}{2}t^2, \, v = 0, \qquad\qquad 0 \le t \le 1,$$
$$\tilde{C}_2 : u = \tfrac{1}{2}\cos 2t, \, v = -\tfrac{1}{2}\sin 2t, \qquad 0 \le t \le \tfrac{1}{2}\pi,$$
$$\tilde{C}_3 : u = -\tfrac{1}{2}(1 - t)^2, \, v = 0, \qquad\qquad 0 \le t \le 1.$$

Thus $\tilde{C}_1$ and $\tilde{C}_3$ are each segments of the x axis, traversed from left to right, and $\tilde{C}_2$ is a semicircle traversed in the clockwise direction (see Fig. 26.6). We observe that C_1, C_2, C_3 are positively oriented with respect to F, but $\tilde{C}_1$, $\tilde{C}_2$, $\tilde{C}_3$ are negatively oriented with respect to F. This reflects the fact that the transformation G has negative Jacobian, $-(x^2 + y^2)$. Thus $\partial \tilde{F}$ is the closed curve consisting of the three parts $\tilde{C}_1$, $\tilde{C}_2$, $\tilde{C}_3$ traversed in the opposite direction.

Exercises

26.1 Let C be the curve $x = a \cos t$, $y = b \sin t$, $0 \le t \le 2\pi$. Use Green's theorem to evaluate each of the following integrals.

a. $\displaystyle \int_C y^2 e^x \, dx$

b. $\displaystyle \int_C y \cos x \, dx + (\cos y + \sin x) \, dy$

c. $\displaystyle \int_C \log \frac{1 + y^2}{1 + x^2} \, dx + \frac{x(1 + y)^2}{1 + y^2} \, dy$

d. $\displaystyle \int_C y \, dx - x \, dy$

26.2 Sketch each of the following figures F, and express in parametric form the curve or curves that constitute the oriented boundary ∂F.

a. $F: \dfrac{x^2}{4} + \dfrac{y^2}{9} \le 1$ **d.** $F: x^2 \le y \le x^{1/2}$

b. $F: 1 \le \dfrac{x^2}{4} + \dfrac{y^2}{9} \le 4$ **e.** $F: y^2 \le x \le 1$

c. $F: 2x \le x^2 + y^2 \le 9$ **f.** $F: x + y \le 1, \, x - y \ge -1, \, y \ge 0$

26.3   Use Green's theorem to evaluate

$$\int_{\partial F} (3x^5 y^3 - x^2 y^3 - 3xy + 5x^2)\, dx + (x^6 y - x^3 y^2 + x^2 - 3y)\, dy$$

for each of the figures F in Ex. 26.2.

26.4   Use Lemma 26.1 to compute the area of the figures bounded by the following curves.

  a. The ellipse $x = a \cos t,$ $y = b \sin t,$ $0 \le t \le 2\pi.$
  b. The astroid $x = a \cos^3 t,$ $y = a \sin^3 t,$ $0 \le t \le 2\pi.$
  c. The loop of the strophoid

$$x = a\,\frac{1 - t^2}{1 + t^2}, \qquad y = at\,\frac{1 - t^2}{1 + t^2}, \qquad -1 \le t \le 1.$$

(*Hint:* note that $y = tx$, and use the last expression in (26.13).)
  d. The loop of the folium of Descartes

$$x = \frac{3at}{1 + t^3}, \qquad y = \frac{3at^a}{1 + t^3}, \qquad 0 \le t < \infty.$$

(*Hint:* use the same method as in part c. The resulting integral is an "improper integral" and is evaluated as a limit. See the discussion immediately following Eq. (28.12).)
  e. The cardioid $r = a(1 - \cos \theta),$ $0 \le \theta \le 2\pi,$ where r and θ are polar coordinates. (*Hint:* see Ex. 21.25a.)
  f. One "leaf" of the "rose" $r = a \sin 2\theta,$ $0 \le \theta \le \tfrac{1}{2}\pi.$ (*Hint:* see part e.)

26.5   Let $f(x)$ be a positive continuously differentiable function defined for $a \le x \le b$. Let F be the figure defined by

$$a \le x \le b, \qquad 0 \le y \le f(x).$$

  a. Show that if $p(x, y)$ is continuous on the boundary of F, then

$$\int_{\partial F} p\, dx = - \int_a^b p(x, f(x))\, dx.$$

  b. Using part a and Lemma 26.1, show that the area of F is equal to

$$\int_a^b f(x)\, dx.$$

26.6   Show that the centroid $(\bar{x}, \bar{y})$ of a figure F is given by

$$\bar{x} = \frac{\tfrac{1}{2} \int_{\partial F} x^2\, dy}{\int_{\partial F} x\, dy}, \qquad \bar{y} = \frac{\tfrac{1}{2} \int_{\partial F} y^2\, dx}{\int_{\partial F} y\, dx}.$$

26.7 For each of the following figures F and transformations G, sketch F and its image $\tilde{F}$ under G, and verify Eq. (26.14).

  a. $F : x \ge 0,\ x^2 + y^2 \le 1;$ $G : u = x^2,\ v = y$
  b. $F : -3 \le x \le -2,\ 1 \le y \le 2;$ $G : u = x^2,\ v = y^3$
  c. $F : 0 < a \le x \le b,\ 0 \le y \le \pi;$ $G : u = x \cos y,\ v = x \sin y$
  d. $F : a \le x \le b,\ 0 \le y \le \pi;$ $G : u = e^x \cos y,\ v = e^x \sin y$

26.8 Let $C: x(t), y(t), a \le t \le b$, be a regular curve lying on the boundary of a figure F, and suppose that C has positive orientation with respect to F. Show that at each point of C,

$$\langle y'(t), -x'(t) \rangle = \frac{ds}{dt} \mathbf{N},$$

where s is the parameter of arc length along C, and $\mathbf{N}$ is the unit vector perpendicular to C directed toward the exterior of F.

26.9 Let $u(x, y) \in \mathscr{C}^2$ and $v(x, y) \in \mathscr{C}^2$ in a domain D. Let F be a figure lying in D. At each boundary point of F where the boundary curve is regular, define the *normal derivative* $\partial v / \partial n$ to be the directional derivative of $v(x, y)$ in the direction of the unit normal $\mathbf{N}$ directed toward the exterior of F. Prove the following identities.

*a. $\iint_F u \Delta v \, dA + \iint_F (\nabla u \cdot \nabla v) \, dA = \int_{\partial F} u \frac{\partial v}{\partial n} \, ds$

Hint: apply Green's theorem with $p = -uv_y$, $q = uv_x$, and use Ex. 26.8.)

b. $\iint_F (u \Delta v - v \Delta u) \, dA = \int_{\partial F} \left(u \frac{\partial v}{\partial n} - v \frac{\partial u}{\partial n} \right) ds$

c. $\iint_F u \Delta v \, dA = \int_{\partial F} u \frac{\partial v}{\partial n} \, ds$

(*Note:* parts a and b are known as *Green's identities*.)

26.10 Using the notation of Ex. 26.9, show that the following statements hold.

a. If $v(x, y)$ is harmonic in D, and if $u(x, y)$ is equal to zero at each point of the boundary of F, then

$$\iint_F (\nabla u \cdot \nabla v) \, dA = 0.$$

b. If $u(x, y)$ is harmonic in D, then

$$\int_{\partial F} \frac{\partial u}{\partial n} \, ds = 0.$$

(*Note:* if $u(x, y)$ represents temperature, then $\partial u / \partial n$ describes the rate of flow of heat across the boundary at each point. The integral $\int_{\partial F} (\partial u / \partial n) \, ds$ represents the total flow across the boundary. The vanishing of this integral means that the total flow across the boundary is zero, or that the amount of heat entering the figure F is equal to the amount leaving F.)

26.11 Suppose $u(x, y)$ is harmonic in a domain D. Setting $v(x, y) = u(x, y)$ in Green's identity, Ex. 26.9a, yields

$$\iint_F |\nabla u|^2 \, dA = \int_{\partial F} u \frac{\partial u}{\partial n} \, ds.$$

By Ex. 24.26, if the integral on the left is equal to zero, then $\nabla u \equiv 0$ on F. Use this fact to prove the following statements.

a. If $u(x, y) = 0$ at each point of the boundary of F, then $u(x, y) \equiv 0$ in F. (See Corollary 3 to Th. 12.3 for a different proof.)

b. If $\partial u/\partial n = 0$ at each point of the boundary of F, then $u(x, y)$ is constant on F.

c. If $u(x, y)$ and $v(x, y)$ are both harmonic in D, and if $\partial u/\partial n = \partial v/\partial n$ at each point of the boundary of F, then $u(x, y)$ and $v(x, y)$ differ by a constant.

d. Suppose $u(x, y)$ and $v(x, y)$ both satisfy *Poisson's equation* in D:

$$\Delta u = f, \qquad \Delta v = f,$$

where $f(x, y)$ is a given function. If $u(x, y) = v(x, y)$ at each point of the boundary of F, then $u(x, y) \equiv v(x, y)$ in F.

*26.12 Let $h(x, y) \in \mathscr{C}^2$ in a domain D. Suppose that the disk

$$F: (x - x_0)^2 + (y - y_0)^2 \leq r_0^2$$

lies in D. The quantity

$$m(r) = \frac{1}{2\pi} \int_0^{2\pi} h(x_0 + r \cos \theta, y_0 + r \sin \theta)\, d\theta$$

is called the *mean value* of $h(x, y)$ on the circle $(x - x_0)^2 + (y - y_0)^2 = r^2$. (See the general discussion of mean values at the beginning of Sect. 27.) Using Ex. 26.9c and Lemma 7.2, one can prove the following relation

$$\frac{d}{dr} m(r)\Big|_{r=r_0} = \frac{1}{2\pi r_0} \iint_F \Delta h\, dA.$$

Carry out the details by verifying each step in the following string of equalities. Set

$$g(r, \theta) = h(x_0 + r \cos \theta, y_0 + r \sin \theta),$$

so that

$$m(r) = \frac{1}{2\pi} \int_0^{2\pi} g(r, \theta)\, d\theta;$$

then

$$2\pi m'(r_0) = \int_0^{2\pi} g_r(r_0, \theta)\, d\theta = \frac{1}{r_0} \int_0^{2\pi} g_r(r_0, \theta) r_0\, d\theta$$

$$= \frac{1}{r_0} \int_{\partial F} \frac{\partial h}{\partial n}\, ds = \frac{1}{r_0} \iint_F \Delta h\, dA.$$

26.13 Using the result of Ex. 26.12, prove the following theorem, known as the *mean-value property* of harmonic functions: *if $h(x, y)$ is harmonic in a domain D, then for every point (x_0, y_0) in D,*

$$h(x_0, y_0) = \frac{1}{2\pi} \int_0^{2\pi} h(x_0 + r \cos \theta, y_0 + r \sin \theta)\, d\theta$$

for every value of r such that the disk $(x - x_0)^2 + (y - y_0)^2 \leq r^2$ lies in D.

(*Hint:* in the notation of Ex. 26.12, if $h(x, y)$ is harmonic, then $d/dr\ m(r) = 0$, and $m(r)$ is constant. But as a consequence of Lemma 7.2, $m(r)$ is a continuous function of r, and $\lim_{r \to 0} m(r) = m(0) = h(x_0, y_0)$.)

*26.14 The mean-value property described in Ex. 26.13 is a property unique to harmonic functions. This may be stated in the following manner: *if $h(x, y) \in \mathscr{C}^2$ in a domain D, and if for every point (x_0, y_0) in D the value of h at (x_0, y_0) equals its mean value over all small circles about (x_0, y_0), then $h(x, y)$ is harmonic in D.*

Prove this statement. (*Hint:* proceed by contradiction. Suppose that at some point (x_0, y_0) in D, $\Delta h \neq 0$; say $\Delta h(x_0, y_0) = \eta > 0$. Then for some $r > 0$, $\Delta h \geq \eta/2 > 0$ for $(x - x_0)^2 + (y - y_0)^2 \leq r^2$. By Ex. 26.12, $m'(r) > 0$. This contradicts the assumption $m(r) \equiv h(x_0, y_0)$.)

*26.15 Let $h(x, y)$ be continuous on the circle $C: (x - x_0)^2 + (y - y_0)^2 = r^2$, and suppose $h(x, y) \leq M$ on C. Show that if the mean value of $h(x, y)$ on C is equal to M, then $h(x, y) \equiv M$ on C. (*Hint:* let

$$g(r, \theta) = h(x_0 + r \cos \theta, y_0 + r \sin \theta),$$

and let $f(r, \theta) = M - g(r, \theta)$. Then $f(r, \theta)$ is continuous for $0 \leq \theta \leq 2\pi$, and $f(r, \theta) \geq 0$. To show that $\int_0^{2\pi} f(r, \theta)\, d\theta = 0$ implies $f(r, \theta) \equiv 0$, proceed by contradiction. If $f(r, \theta_0) = \eta > 0$, then $f(r, \theta) \geq \eta/2$ for $|\theta - \theta_0| < \delta$, and

$$\int_0^{2\pi} f(r, \theta)\, d\theta \geq \int_{\theta_0 - \delta}^{\theta_0 + \delta} f(r, \theta)\, d\theta \geq 2\delta \frac{\eta}{2} > 0.)$$

26.16 Prove the *strong maximum principle* for harmonic functions: *if $h(x, y)$ is harmonic in a domain D, if $h(x, y) \leq M$ throughout D, and if $h(x_0, y_0) = M$ for some point (x_0, y_0) in D, then $h(x, y) \equiv M$ in D.* Proceed in two steps.

    a. Show that if $h(x, y)$ has a local maximum M at (x_0, y_0) then $h(x, y) \equiv M$ on every small circle about (x_0, y_0). (*Hint:* use Ex. 26.15 together with the mean value property of harmonic functions.)

    *b. Show that if there were some point (x_1, y_1) in D such that $h(x_1, y_1) < M$, this would lead to a contradiction as follows. Join (x_0, y_0) to (x_1, y_1) by a curve $C: x(t), y(t), 0 \leq t \leq 1$. Then the function $h_C(t)$ satisfies $h_C(0) = M$, $h_C(1) < M$. Let b be the largest value of t such that $h_C(b) = M$. Then $0 \leq b < 1$, and applying part a to the point $(x_2, y_2) = (x(b), y(b))$ leads to a contradiction.

26.17 Let $f(x, y) \in \mathscr{C}^2$ in a domain D, and let $\langle p, q \rangle = \nabla f$. Give two different proofs that for every figure F lying in D, $\int_{\partial F} p\, dx + q\, dy = 0$.

26.18 Let $p(x, y), q(x, y) \in \mathscr{C}^1$ in a domain D, except at a point (x_0, y_0). Let C_1, C_2 be circles in D defined by

$$C_k: x = x_0 + r_k \cos t,\ y = y_0 + r_k \sin t, \qquad 0 \leq t \leq 2\pi;\ k = 1, 2.$$

Suppose that all points between these two circles also lie in D. Show that if $p_y = q_x$ at every point of D except (x_0, y_0), then

$$\int_{C_1} p\, dx + q\, dy = \int_{C_2} p\, dx + q\, dy.$$

(*Hint:* apply Green's theorem to the figure F bounded by C_1 and C_2. Pay careful attention to the orientation of ∂F and of C_1, C_2.)

26.19 Under the same hypotheses as Ex. 26.18, let F be a figure lying in D. Show that the following statements hold.

   a. If (x_0, y_0) does not lie in F, then $\int_{\partial F} p \, dx + q \, dy = 0$.
   b. If (x_0, y_0) is an interior point of F, then

$$\int_{\partial F} p \, dx + q \, dy = \int_{C_1} p \, dx + q \, dy,$$

where r_1 is chosen sufficiently small so that the set of points (x, y) satisfying $(x - x_0)^2 + (y - y_0)^2 \le r^2$ lies in the interior of F. (*Hint:* apply Green's theorem to the figure $\tilde{F}$ bounded by ∂F and C_1.)

26.20 Let C be a piecewise smooth curve that forms the oriented boundary of a figure F. Show that the winding number $n(C; x_0, y_0)$ of C about the point (x_0, y_0) satisfies

$$n(C; x_0, y_0) = \begin{cases} 0 & \text{if } (x_0, y_0) \text{ is not in } F \\ 1 & \text{if } (x_0, y_0) \text{ is an interior point of } F. \end{cases}$$

(*Hint:* see Ex. 25.18 for the definition of winding number, and use Ex. 26.19.)

26.21 Let G be a differentiable transformation of a domain D in the x, y plane into a domain E of the u, v plane. Suppose $P(u, v)$, $Q(u, v) \in \mathscr{C}^1$ in E. Set

$$\tilde{P}(x, y) = P(u(x, y), v(x, y)), \qquad \tilde{Q}(x, y) = Q(u(x, y), v(x, y))$$

and

$$p(x, y) = \tilde{P}u_x + \tilde{Q}v_x, \qquad q(x, y) = \tilde{P}u_y + \tilde{Q}v_y.$$

Let C be a curve in D, and let $\tilde{C}$ be its image under G. Show that

$$\int_{\tilde{C}} P \, du + Q \, dv = \int_C p \, dx + q \, dy.$$

(Note that Eq. (26.17) is a special case of this formula. Use the same reasoning here as was used to derive (26.17).)

26.22 Using the notation of Ex. 26.21, show that if $C = \partial F$, where F is a figure lying in D, then

$$\int_{\tilde{C}} P \, du + Q \, dv = \iint_F (Q_u - P_v) \frac{\partial(u, v)}{\partial(x, y)} \, dA.$$

(*Hint:* see Ex. 17.15a.)

26.23 Using the notation of Ex. 26.21, let (x_0, y_0) be a point of D, and let F be the figure $(x - x_0)^2 + (y - y_0)^2 \le r^2$, where r is chosen sufficiently small so that F lies in D. Let $C = \partial F$, and let $\tilde{C}$ be the image of C under the transformation G. Let (U, V) be any point in the u, v plane such that $G(x, y) \ne (U, V)$ for all points (x, y) in F (that is, (U, V) is a point that is not in the image of the figure F). Show that the winding number $n(\tilde{C}; U, V) = 0$. (*Hint:* let

$$P(u, v) = -\frac{v - V}{(u - U)^2 + (v - V)^2},$$

$$Q(u, v) = \frac{u - U}{(u - U)^2 + (v - V)^2}.$$

Then $P(u, v) \in \mathscr{C}^1$, $Q(u, v) \in \mathscr{C}^1$ in the domain E consisting of the whole plane minus the point (U, V), and $P_v = Q_u$ in E. Since

$$n(\tilde{C}; U, V) = \frac{1}{2\pi} \int_{\tilde{C}} P \, du + Q \, dv,$$

the result follows from Ex. 26.22.)

Remark Ex. 26.23 gives a useful criterion for determining when a pair of equations

$$f(x, y) = U$$

$$g(x, y) = V$$

can be solved for x and y, when U and V are given. Namely, suppose C is a circle in the x, y plane, and $\tilde{C}$ is the image of C under the transformation

$$G: u = f(x, y), v = g(x, y).$$

If $\tilde{C}$ does not pass through the point (U, V) and if the winding number of $\tilde{C}$ about the point (U, V) is different from zero, then the above equations must be satisfied by some point (x, y) lying inside C (since otherwise we could conclude from Ex. 26.23 that the winding number of $\tilde{C}$ about (U, V) must equal zero). Exercise 26.24 shows how this reasoning may be used to guarantee the existence of a solution to a pair of simultaneous quadratic equations. It may be difficult to solve such a pair of equations explicitly, or even to determine whether a solution exists, since the equation may represent two ellipses, or an ellipse and a hyperbola, which may or may not intersect. In Ex. 26.28 the same reasoning is used to prove a part of the inverse mapping theorem, which may be thought of as guaranteeing the existence of a solution of a pair of simultaneous equations. Exercise 26.29 completes the proof of the inverse mapping theorem. Exercises 26.25–26.27 lead up to this result, but they are also of interest in their own right for an understanding of some of the more subtle properties of differentiable mappings.

26.24 Let G be the mapping

$$u = 3x^2 - 2xy + y^2$$

$$v = 4x^2 + 2xy + 2y^2.$$

Let $\tilde{C}$ be the image of the unit circle $C: x = \cos t, y = \sin t, 0 \leq t \leq 2\pi$.

a. Show that $\tilde{C}$ may be written in the form

$$u = 2 + \sqrt{2} \cos (2t + \tfrac{1}{4}\pi), \qquad v = 3 + \sqrt{2} \sin (2t + \tfrac{1}{4}\pi), \qquad 0 \leq t \leq 2\pi.$$

b. Show that $n(\tilde{C}; 2, 3) = 2$.

c. Deduce that the equations

$$3x^2 - 2xy + y^2 = 2$$

$$4x^2 + 2xy + 2y^2 = 3$$

have a simultaneous solution (x, y) satisfying $x^2 + y^2 < 1$. (*Hint:* use the reasoning indicated in the Remark just above.)

26.25 Let G be a linear transformation with determinant $\Delta \neq 0$. Let C be a circle $x = r \cos t, y = r \sin t, 0 \leq t \leq 2\pi$, and let $\tilde{C}$ be the image of C under G. Show that

$$n(\tilde{C}; 0, 0) = \begin{cases} 1 & \text{if } \Delta > 0 \\ -1 & \text{if } \Delta < 0. \end{cases}$$

(*Hint:* introducing polar coordinates R, φ on the curve $\tilde{C}$ (which is an ellipse centered at the origin), we have by Eq. (14.11) that $d\varphi/dt > 0$ if $\Delta > 0$ and $d\varphi/dt < 0$ if $\Delta < 0$. But

$$\int_0^{2\pi} \frac{d\varphi}{dt} \, dt = 2n\pi$$

for some integer n, and $n \neq 0$, since the integrand is either positive for all t or negative for all t. If $|n| \geq 2$, then there would be some $t_0, 0 < t_0 < 2\pi$, such that

$$\left| \int_0^{t_0} \frac{d\varphi}{dt} \, dt \right| = 2\pi,$$

which would mean that the (distinct) points $(\cos t_0, \sin t_0)$ and $(\cos 0, \sin 0)$ of C would map onto the same point of $\tilde{C}$, contradicting the fact that a nonsingular linear transformation is injective. Thus $n = \pm 1$.)

*26.26 Prove *Rouché's theorem:* Let (u_0, v_0) be a given point, and let

$$C: u(t), v(t), \qquad a \leq t \leq b$$

$$\hat{C}: \hat{u}(t), \hat{v}(t), \qquad a \leq t \leq b$$

be two closed curves that do not pass through (u_0, v_0). Suppose that for each value of t, the corresponding points on C and $\hat{C}$ are closer together than the distance from the point on C to (u_0, v_0); that is,

$$[u(t) - \hat{u}(t)]^2 + [v(t) - \hat{v}(t)]^2 < [u(t) - u_0]^2 + [v(t) - v_0]^2$$

(see Fig. 26.7). Then $n(\hat{C}: u_0, v_0) = n(C; u_0, v_0)$.
(*Hint:* choose $\varphi_0, \hat{\varphi}_0$ such that

$$u(a) = u_0 + R \cos \varphi_0, \qquad v(a) = v_0 + R \sin \varphi_0,$$

$$\hat{u}(a) = u_0 + R \cos \hat{\varphi}_0, \qquad \hat{v}(a) = v_0 + R \sin \hat{\varphi}_0.$$

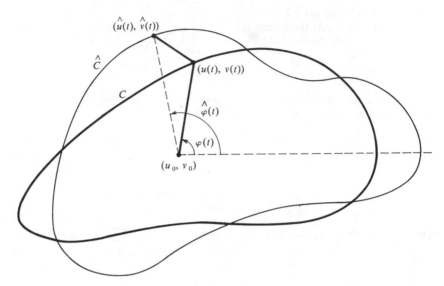

FIGURE 26.7 Illustration of Rouché's theorem

For any τ, $a \le \tau \le \theta$, define

$$\varphi(\tau) = \varphi_0 + \int_a^\tau \left[-\frac{v(t) - v_0}{(u(t) - u_0)^2 + (v(t) - v_0)^2} \, u'(t) \right.$$

$$\left. + \frac{u(t) - u_0}{(u(t) - u_0)^2 + (v(t) - v_0)^2} \, v'(t) \right] dt$$

and $\hat{\varphi}(\tau)$ similarly. The integrand is $d/dt \tan^{-1}[(v(t) - v_0)/(u(t) - u_0)]$, and these integrals simply define in a continuous way the angles $\varphi(t)$ and $\hat{\varphi}(t)$ along the curve (see Fig. 26.7). Furthermore,

$$\varphi(b) - \varphi(a) = 2\pi n(C; u_0, v_0) \qquad \text{and} \qquad \hat{\varphi}(b) - \hat{\varphi}(a) = 2\pi n(\hat{C}; u_0, v_0).$$

Thus

$$\int_a^b \frac{d}{dt} [\varphi(t) - \hat{\varphi}(t)] \, dt = 2\pi k,$$

where $k = n(C; u_0, v_0) - n(\hat{C}; u_0, v_0)$. If $k \ne 0$, then the function $\varphi(t) - \hat{\varphi}(t)$ increases or decreases by at least 2π as t goes from a to b, and since it is a continuous function, it must be equal to an odd multiple of π for some t_0 between a and b. This means that the points $(u(t_0), v(t_0))$ and $(\hat{u}(t_0), \hat{v}(t_0))$ lie on a line through (u_0, v_0) and on opposite sides of (u_0, v_0). But then the distance between these points is greater than the distance from $(u(t_0), v(t_0))$ to (u_0, v_0), contrary to assumption. Thus k must equal zero, which is the desired result.)

Remark One way to visualize Rouché's theorem is the following. Let the point (u_0, v_0) be the position of the sun, and let C and $\hat{C}$ be the paths of the earth and the moon, respectively. If after a certain number of orbits the earth and the moon both return to their original position, then C and $\hat{C}$ will be closed curves. The fact that the distance from the moon to the earth is always less than the distance from the earth to the sun implies that the number of times the moon has gone around the sun must equal the number of times the earth has gone around the sun.

*26.27 Let G be a differentiable transformation, let $G(x_0, y_0) = (u_0, v_0)$, and suppose that the Jacobian of G at (x_0, y_0) is different from zero. Let C be the circle $x = x_0 + r \cos t$, $y = y_0 + r \sin t$, $0 \leq t \leq 2\pi$ and let $\hat{C}$ be its image under G. Show that for all sufficiently small values of r,

$$n(\hat{C}; u_0, v_0) = \begin{cases} 1 & \text{if } \left.\dfrac{\partial(u, v)}{\partial(x, y)}\right|_{(x_0, y_0)} > 0 \\[2em] -1 & \text{if } \left.\dfrac{\partial(u, v)}{\partial(x, y)}\right|_{(x_0, y_0)} < 0. \end{cases}$$

(*Hint:* if we set $(u, v) = G(x, y)$, then by Th. 16.2,

$$\langle u - u_0, v - v_0 \rangle = dG_{(x_0, y_0)}(\langle x - x_0, y - y_0 \rangle) + \langle h(x, y) \rangle,$$

where $\langle h(x, y) \rangle$ is a remainder term satisfying

$$\lim_{(x, y) \to (x_0, y_0)} \frac{|\langle h(x, y) \rangle|}{d(x, y)} = 0; \qquad d(x, y) = \sqrt{(x - x_0)^2 + (y - y_0)^2}.$$

Let $\tilde{C}$ be the ellipse that is the image of C under the transformation

$$u = u_0 + u_x(x_0, y_0)(x - x_0) + u_y(x_0, y_0)(y - y_0),$$

$$v = v_0 + v_x(x_0, y_0)(x - x_0) + v_y(x_0, y_0)(y - y_0).$$

Since the determinant Δ of this transformation is

$$\left.\frac{\partial(u, v)}{\partial(x, y)}\right|_{(x_0, y_0)},$$

which is assumed nonzero, we may apply Ex. 26.25 and deduce that $n(\tilde{C}; u_0, v_0) = \pm 1$ according as $\Delta > 0$ or $\Delta < 0$. If λ_2 is the minimum of

$$|dG_{(x_0, y_0)}(\langle x - x_0, y - y_0 \rangle)|^2 \qquad \text{for } (x - x_0)^2 + (y - y_0)^2 = 1,$$

then the fact that $dG_{(x_0, y_0)}$ is nonsingular implies that $\lambda_2 > 0$. Then

$$|dG_{(x_0, y_0)}(\langle x - x_0, y - y_0 \rangle)| \geq (\lambda_2)^{1/2} d(x, y)$$

for all (x, y). But for $d(x, y)$ sufficiently small we have $|\langle h(x, y) \rangle| < (\lambda_2)^{1/2} d(x, y)$. This means that the inequality

$$|\langle u - u_0, v - v_0 \rangle - dG_{(x_0, y_0)}(\langle x - x_0, y - y_0 \rangle)|$$

$$< |dG_{(x_0, y_0)}(\langle x - x_0, y - y_0 \rangle)|$$

must hold at every point of the circle C for r sufficiently small. But this inequality is precisely the condition on the curves $\tilde{C}$ and $\hat{C}$ which by Ex. 26.26 guarantees that $n(\hat{C}; u_0, v_0) = n(\tilde{C}; u_0, v_0)$. (Note that the ellipse $\tilde{C}$ of this exercise plays the role of the curve C in Ex. 26.26.)

26.28 Let G be a differentiable transformation, let $(u_0, v_0) = G(x_0, y_0)$, and suppose that the Jacobian of G at (x_0, y_0) is different from zero. Then the image under G of any disk $(x - x_0)^2 + (y - y_0)^2 < r^2$ includes all points in some disk $(u - u_0)^2 + (v - v_0)^2 < R^2$. (*Hint:* in the notation of Ex. 26.27, if r is sufficiently small, then the image $\hat{C}$ of the circle C of radius r about (x_0, y_0) satisfies $n(\hat{C}; u_0, v_0) \neq 0$. Choose R so that the disk $(u - u_0)^2 + (v - v_0)^2 < R^2$ does not contain any points of $\hat{C}$. Let (U, V) be any point of this disk. By Ex. 25.18a, $n(\hat{C}; U, V) = n(\hat{C}; u_0, v_0)$. By Ex. 26.23 above, it follows that (U, V) is the image under G of some point (x, y) inside C.)

*26.29 Prove the inverse mapping theorem (Th. 17.3). Specifically, let $G(x, y)$ be a differentiable mapping defined in a domain D by

$$G:\begin{cases} u = f(x, y) \\ v = g(x, y) \end{cases}$$

where $f(x, y), g(x, y) \in \mathscr{C}^1$ in D. Let $(u_0, v_0) = G(x_0, y_0)$ and suppose

$$\frac{\partial(u, v)}{\partial(x, y)}\bigg|_{(x_0, y_0)} \neq 0.$$

Show that there exists a disk $D': (x - x_0)^2 + (y - y_0)^2 < r^2$, such that G maps D' one-to-one onto a domain E', and such that G^{-1} is a differentiable map of E' onto D'. Proceed in the following steps.

a. Choose r sufficiently small so that

$$f_x(x_3, y_3)g_y(x_4, y_4) - f_y(x_3, y_3)g_x(x_4, y_4) \neq 0,$$

where (x_3, y_3) and (x_4, y_4) are any two points of the disk

$$D': (x - x_0)^2 + (y - y_0)^2 < r^2.$$

Show that G maps D' one-to-one onto a domain E'. (*Hint:* by Ex. 17.17, it is always possible to choose r so that the given condition is satisfied, and for any such r, the map G is one-to-one in D'. Note that this condition implies in particular that $\partial(u, v)/\partial(x, y) \neq 0$ throughout D'. It follows from Ex. 26.28 that the image E' of D' includes a disk about each of its points, hence is an open set. Finally the fact that D' is connected and G is continuous implies that E' is connected. Hence E' is a domain.)

b. Show that the inverse map $H: E' \to D'$ which exists by part a is continuous in E'. (*Hint:* given any point (u_1, v_1) in E', let $(x_1, y_1) = H(u_1, v_1)$. H continuous at (u_1, v_1) means that for any $\epsilon > 0$ there exists $\delta > 0$ such that $(u - u_1)^2 + (v - v_1)^2 < \delta \Rightarrow (x - x_1)^2 + (y - y_1)^2 < \epsilon$, where $(x, y) = H(u, v)$. But this is a consequence of Ex. 26.28 applied at the point (x_1, y_1) with $\epsilon = r$ and $\delta = R$, using the fact that $(x, y) = H(u, v) \Leftrightarrow (u, v) = G(x, y)$.)

c. Show that if the inverse map H is given by

$$H:\begin{cases} x = \varphi(u, v) \\ y = \psi(u, v) \end{cases}$$

then $\varphi(u, v) \in \mathscr{C}^1$ and $\psi(u, v) \in \mathscr{C}^1$ in E'. (*Hint:* use an analogous reasoning to the proof of the implicit function theorem (Theorem 8.2). Given any two points (u_1, v_1), (u_2, v_2) in E', let (x_1, y_1), (x_2, y_2) be the corresponding points in the disk D'. By the mean value theorem applied to the functions $f(x, y)$ and $g(x, y)$, there exist points (x_3, y_3) and (x_4, y_4) on the line segment joining (x_1, y_1) to (x_2, y_2) such that

$$u_2 - u_1 = f_x(x_3, y_3)(x_2 - x_1) + f_y(x_3, y_3)(y_2 - y_1)$$

$$v_2 - v_1 = g_x(x_4, y_4)(x_2 - x_1) + g_y(x_4, y_4)(y_2 - y_1).$$

By the condition of part a, these equations can be solved in the form

$$x_2 - x_1 = \frac{g_y(x_4, y_4)(u_2 - u_1) - f_y(x_3, y_3)(v_2 - v_1)}{f_x(x_3, y_3)g_y(x_4, y_4) - f_y(x_3, y_3)g_x(x_4, y_4)}$$

$$y_2 - y_1 = \frac{-g_x(x_4, y_4)(u_2 - u_1) + f_x(x_3, y_3)(v_2 - v_1)}{f_x(x_3, y_3)g_y(x_4, y_4) - f_y(x_3, y_3)g_x(x_4, y_4)}.$$

In order to show the existence of $\varphi_u(u_1, v_1)$, we choose $v_2 = v_1$, and we obtain from the first of the above equations that

$$\frac{\varphi(u_2, v_1) - \varphi(u_1, v_1)}{u_2 - u_1} = \frac{x_2 - x_1}{u_2 - u_1}$$

$$= \frac{g_y(x_4, y_4)}{f_x(x_3, y_3)g_y(x_4, y_4) - f_y(x_3, y_3)g_x(x_4, y_4)}.$$

Using the continuity of H (see part b) and the fact that (x_3, y_3) and (x_4, y_4) lie on the line between (x_1, y_1) and (x_2, y_2) ,we have

$$u_2 \to u_1 \Rightarrow (u_2, v_1) \to (u_1, v_1)$$

$$\Rightarrow (x_2, y_2) = H(u_2, v_1) \to (x_1, y_1) = H(u_1, v_1)$$

$$\Rightarrow (x_3, y_3) \to (x_1, y_1) \quad \text{and} \quad (x_4, y_4) \to (x_1, y_1).$$

By the continuity of the partial derivatives f_x, f_y, g_x, g_y, it follows that

$$\lim_{u_2 \to u_1} \frac{\varphi(u_2, v_1) - \varphi(u_1, v_1)}{u_2 - u_1} = g_y(x_1, y_1) \Big/ \frac{\partial(f, g)}{\partial(x, y)}\Big|_{(x_1, y_1)}.$$

Thus the partial derivative $\varphi_u(u_1, v_1)$ exists, and in a similar manner, so do φ_v, ψ_u, and ψ_v. Furthermore, the expressions for these derivatives show that they depend continuously on x and y, and since x and y are continuous functions of u and v, the partial derivatives $\varphi_u, \varphi_v, \psi_u, \psi_v$ are continuous functions of u and v.)

***26.30** Let G be a differentiable transformation in a domain D. Let the circle $C: x = x_0 + r \cos t, y = y_0 + r \sin t$, together with its interior, be included in D. Let $\tilde{C}$ be the image of C under G, and let (U, V) be a point not on $\tilde{C}$. Suppose that there are a finite number of points inside C which map onto the point (U, V), and suppose that the Jacobian of G is not zero at any of these points.

Specifically, suppose the Jacobian is positive at m points and negative at n points. Then

$$n(\tilde{C}; U, V) = m - n.$$

(*Hint:* for each point (x_k, y_k) inside C such that $G(x_k, y_k) = (U, V)$, let

$$C_k: x = x_k + r \cos t, \quad y = y_k + r \sin t, \qquad 0 \le t \le 2\pi,$$

be a small circle, where r is chosen sufficiently small so that all these circles lie inside of C and no two of them intersect. Let $\tilde{C}_k$ be the image of C_k under G, and let F be the figure consisting of those points which lie on or inside C, and on or outside each C_k. If we define $P(u, v)$, $Q(u, v)$ as in Ex. 26.23 and $p(x, y)$, $q(x, y)$ as in Ex. 26.21, then

$$
\begin{aligned}
n(\tilde{C}; U, V) - \sum_k n(\tilde{C}_k; U, V) &= \frac{1}{2\pi} \int_{\tilde{C}} P\, du + Q\, dv - \sum_k \frac{1}{2\pi} \int_{\tilde{C}_k} P\, du + Q\, dv \\
&= \frac{1}{2\pi} \int_C p\, dx + q\, dy - \sum_k \frac{1}{2\pi} \int_{C_k} p\, dx + q\, dy \\
&= \frac{1}{2\pi} \int_{\partial F} p\, dx + q\, dy = \frac{1}{2\pi} \iint_F (q_x - p_y)\, dA \\
&= \frac{1}{2\pi} \iint_F (Q_u - P_v) \frac{\partial(u, v)}{\partial(x, y)}\, dA \\
&= 0.
\end{aligned}
$$

But by Ex. 26.27, for r sufficiently small $n(\tilde{C}_k; U, V) = 1$ if $\partial(u, v)/\partial(x, y) > 0$ at (x_k, y_k) and $n(\tilde{C}_k; U, V) = -1$ if $\partial(u, v)/\partial(x, y) < 0$ at (x_k, y_k). Thus

$$n(\tilde{C}; U, V) = \sum_k n(\tilde{C}_k; U, V) = \begin{bmatrix} \text{number of points } (x_k, y_k) \text{ where the Jacobian} \\ \text{is positive minus number of points } (x_k, y_k) \\ \text{where the Jacobian is negative.)} \end{bmatrix}$$

Remark There are many important special cases of the result in Ex. 26.30. For example, if the mapping G has positive Jacobian everywhere, then the result may be stated as follows: *the winding number about a point (U, V) of the image of a circle C is equal to the number of points inside C that map onto (U, V).* This is closely related to the so-called *argument principle* in the theory of functions of a complex variable.

It is most instructive to try to understand the underlying reason for the relation between winding numbers and the number of points that map onto a given point. One way is to picture a disk in the x, y plane made out of a rubber sheet, and to visualize the mapping as a process of distorting this disk by various means, such as stretching, twisting, and folding. When this is done, the resulting form of the rubber sheet may lie in several layers over parts of the u, v plane, and the number of points in the sheet that lie over a given point corresponds to the number of points inside the original disk that map onto the given point. The study of general properties of mappings such as those considered here belongs to the branch of mathematics known as *topology*. The specific quantity denoted in Ex. 26.30 by $m - n$ is called the *degree* of a mapping. For further discussion along these lines we refer to [7], Sections 13–18, and [21], Section 5.

27   Mean value theorem for the double integral; vector forms of Green's theorem

The notion of the mean value (or average value) of a variable quantity is useful in many connections. As an example involving functions of one variable, the average speed of a car is defined to be the distance covered divided by the time elapsed. Thus, if $s(t)$ represents distance covered at time t, then the instantaneous velocity at time t is $f(t) = s'(t)$, and the average velocity between time $t = a$ and time $t = b$ is given by:

(27.1)
$$\frac{s(b) - s(a)}{b - a} = \frac{\int_a^b f(t)\, dt}{\int_a^b 1\, dt}.$$

Thus if the function $f(t)$ represents instantaneous velocity, then the average velocity over any interval is equal to the integral of f divided by the integral of the function 1 over the interval. Whether the function $f(t)$ represents velocity or not, the right-hand side of (27.1) is taken to define the *average value of $f(t)$ over the interval $a \le t \le b$*.

For an example involving functions of two variables, we may take a plate cut out in the shape of a figure F, having variable density $f(x, y)$. The mass of the plate is then equal to $\iint_F f\, dA$. We may define the *average density* of the plate to be the value μ such that if the figure F had constant density μ, the total mass would be the same:

$$\iint_F f\, dA = \iint_F \mu\, dA = \mu \iint_F 1\, dA$$

or

(27.2)
$$\mu = \frac{\iint_F f\, dA}{\iint_F 1\, dA}.$$

Thus the average density equals the total mass divided by the area of F. In a similar fashion, the *average height* of a surface $z = f(x, y)$ would be defined to be that constant height, which would yield the same volume. We would then obtain the same value μ, defined by (27.2), interpreted as the total volume divided by the area of the base.

Regardless of the interpretation of the function $f(x, y)$, its average value is defined in the same way.

Definition 27.1   The *mean value* or *average value* of a function $f(x, y)$ over a figure F is the number μ defined by (27.2).

Thus, the mean value of f is that constant whose integral over F yields the same value as the integral of f.

Example 27.1

The mean value of the function $f(x, y) = x$ over a figure F is precisely the x coordinate of the centroid of F. Similarly for $f(x, y) = y$. Thus the centroid of F may be described as the point whose coordinates are the average values over F of the x and y coordinates, respectively.

In general, the mean value of a function need not be a value that the function actually assumes. Thus, a function that is equal to 2 on one half of a figure, and 4 on the other half, has a mean value of 3, although it is not actually equal to 3 at any point. For a continuous function, however, this cannot happen. We have the following basic result.

Theorem 27.1 Mean-Value Theorem *If $f(x, y)$ is continuous on a figure F, then there is a point (x_0, y_0) in F such that*

$$f(x_0, y_0) = \frac{\iint_F f \, dA}{\iint_F 1 \, dA}.$$

PROOF. Let M be the maximum of $f(x, y)$ on F, and let (x_1, y_1) be a point such that $f(x_1, y_1) = M$. Similarly, let m be the minimum of $f(x, y)$ on F, and suppose $f(x_2, y_2) = m$. Then

$$m \iint_F 1 \, dA = \iint_F m \, dA \leq \iint_F f \, dA \leq \iint_F M \, dA = M \iint_F 1 \, dA.$$

Dividing through by $\iint_F 1 \, dA$, and defining μ by (27.2), yields

$$m \leq \mu \leq M.$$

Let C be a curve in F joining (x_1, y_1) to (x_2, y_2). Then $f_C(t)$ is a continuous function on an interval and takes on the values m and M, hence every value between m and M, and in particular the value μ. At the corresponding point (x_0, y_0) on the curve C, $f(x_0, y_0) = \mu$, which proves the theorem. ◆

Remark It is crucial to the above proof, and to the correctness of the theorem, that the definition of a figure was in terms of a domain, which is a *connected* set.

We next give an important application of the mean-value theorem.

Theorem 27.2 *Let $f(x, y)$ be continuous in a domain D, and let (x_0, y_0) be any point of D. Let $F_1, F_2, \ldots$ be a sequence of figures lying in D, which all* *contain (x_0, y_0) and whose diameters tend to zero. Let $A(F_k)$ be the area of F_k; then*

$$(27.3) \qquad f(x_0, y_0) = \lim_{k \to \infty} \frac{\iint_{F_k} f\, dA}{A(F_k)}.$$

Remark   In many applications, each F_k is a circular disk centered at (x_0, y_0).

PROOF.   Applying the mean value theorem to the figure F_k, we find that there is a point (x_k, y_k) in F_k such that $f(x_k, y_k) = \iint_{F_k} f\, dA / A(F_k)$. Since the diameters of F_k tend to zero as k tends to infinity, it follows that $(x_k, y_k) \to (x_0, y_0)$. By the continuity of f at (x_0, y_0), $f(x_0, y_0) = \lim_{k \to \infty} f(x_k, y_k)$. This proves the theorem.   ◆

Corollary   *Given a differentiable transformation*

$$G : \begin{cases} u(x, y) \\ v(x, y) \end{cases}$$

in a domain D, let (x_0, y_0) be a point at which the Jacobian is different from zero. Let $F_1, F_2, \ldots$ be a sequence of figures lying in D, containing the point (x_0, y_0), whose diameters tend to zero. Let $\tilde{F}_k$ be the image of F_k under G, and let $A(F_k), A(\tilde{F}_k)$ denote the areas of F_k and $\tilde{F}_k$; then

$$(27.4) \qquad \left| \frac{\partial(u, v)}{\partial(x, y)} \right|_{(x_0, y_0)} = \lim_{k \to \infty} \frac{A(\tilde{F}_k)}{A(F_k)}.$$

PROOF.   Since the Jacobian of G is different from zero at (x_0, y_0), it follows from the inverse mapping theorem (Theorem 17.3) that there is a neighborhood D' of (x_0, y_0) in which the mapping G has a differentiable inverse. Thus G restricted to D' is a diffeomorphism, and since the diameters of F_k tend to zero, from some point on they must all lie in D'. We may then apply Th. 26.2 to conclude that

$$A(\tilde{F}_k) = \iint_{F_k} \left| \frac{\partial(u, v)}{\partial(x, y)} \right| dA.$$

Consequently, setting $f(x, y) = |\partial(u, v)/\partial(x, y)|$, Eq. (27.3) reduces to (27.4), and the result is proved.   ◆

Remark   The property of the Jacobian expressed in Eq. (27.4) is the fundamental geometric interpretation of the Jacobian, which we have already mentioned at the end of Sect. 16.

We next give applications of Th. 27.2 to the theory of vector fields.

Let $v(x, y) = \langle p(x, y), q(x, y) \rangle \in \mathscr{C}^1$ in a domain D. We noted in Eqs. (21.12) and (21.13a) that a line integral may be written in vector form as

$$\int_C p \, dx + q \, dy = \int_C (v \cdot T) \, ds = \int_C v_T \, ds$$

where v_T is the tangential component of v along the curve C. Now if F is a figure lying in D such that ∂F consists of the single closed curve C, then Green's theorem takes the form

$$(27.5) \qquad \iint_F (q_x - p_y) \, dA = \int_C v_T \, ds.$$

For $v(x, y)$ the velocity field of a fluid flow, the right-side of (27.5) was interpreted as the total rate of flow around the curve C.

Choose any point (x_0, y_0) in D, and choose a sequence of curves C_n such that $C_n = \partial F_n$, where F_n is a figure lying in D and containing (x_0, y_0) and where the diameters of F_n tend to zero. For example, the C_n could be circles centered at (x_0, y_0). Then applying (27.3), with $f(x, y) = q_x - p_y$, we find

$$(27.6) \quad (q_x - p_y)|_{(x_0, y_0)} = \lim_{n \to \infty} \frac{\iint_{F_n} (q_x - p_y) \, dA}{\iint_{F_n} 1 \, dA} = \lim_{n \to \infty} \frac{\iint_{F_n} p \, dx + q \, dy}{\text{area of } F_n}$$

$$= \lim_{n \to \infty} \frac{\text{flow around } C_n}{\text{area of } F_n}.$$

Thus, the quantity $q_x - p_y$ may be described as a measure of the "vorticity" of the flow at each point. As we mentioned at the end of Sect. 22, a vector field is called "irrotational" if $q_x - p_y \equiv 0$.

Regardless of the physical interpretation, Eq. (27.6) has an important consequence. Namely, in view of (27.5), the numerators on the right-hand side of (27.6) may be described in terms of the vector field v, in a fashion that is completely independent of coordinates. Suppose, in fact, that D is a plane domain, and that without choosing any coordinate system in the plane, we assign to each point of D a vector v in the form of a directed line segment. If C_n is a circle of radius r_n, then the tangential component v_T of v is defined at each point of C_n and provides a function whose integral

$$\int_{C_n} v_T \, ds,$$

is on the one hand defined without any reference to a coordinate system, and on the other hand, once a coordinate system is chosen, is equal by (27.5) to $\iint_{F_n} (q_x - p_y) \, dA$. Thus, the right-hand side of (27.6) consists of a sequence of numbers, which are independent of the coordinate system, and their limit, the value of $q_x - p_y$ at (x_0, y_0) must also be independent of the coordinates.

One can show directly that under a rotation of coordinates the quantity $q_x - p_y$ remains invariant (see Ex. 20.19c). However, such a computation provides no insight as to *why* this particular quantity is invariant. It is always more illuminating to prove the invariance of a quantity by obtaining an expression for it that is intrinsic and independent of coordinates. Equation (27.6), together with (27.5), does this for the quantity $q_x - p_y$.

Example 27.2

Let $p = x$, $q = y$ (Fig. 27.1).

If C is a circle with center at the origin, then the vector $\mathbf{v} = \langle x, y \rangle$ is perpendicular to the circle at the point (x, y), and hence $v_T = 0$ at each point. Thus

$$\int_C x\, dx + y\, dy = \int_C v_T\, ds = 0.$$

If C is a circle not centered at the origin, then v_T is not zero at each point, but it is clear from the symmetry of the flow that the line through the origin and the center of the circle divides the circle into two halves such that at each pair of points on the circle which are symmetric about this line, the quantity v_T has the same magnitude, but opposite sign (Fig. 27.1). Thus the total clockwise flow equals the total counterclockwise flow, and again

$$\int_C x\, dx + y\, dy = \int_C v_T\, ds = 0.$$

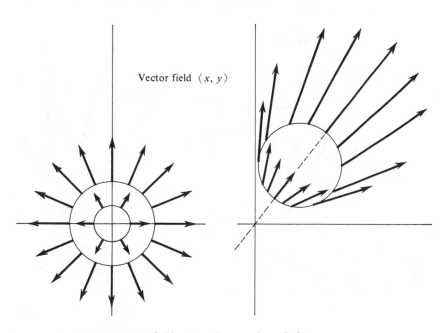

FIGURE 27.1 The vector field $\langle x, y \rangle$ along various circles

If we now choose a sequence of circular disks F_n centered at a point (x_0, y_0), then the flow around ∂F_n is zero for each n, and therefore the right-hand side of (27.6) is equal to zero. That the left-hand side is also zero at every point is easily verified.

We may note that this vector field has the potential function $f(x, y) = \frac{1}{2}(x^2 + y^2)$, and consequently for every closed curve C, $\int_C x \, dx + y \, dy = 0$, even though this may not be obvious geometrically for arbitrary C.

Example 27.3

Let $p = -y$, $q = x$ (Fig. 27.2).

Let C be the circle $x = r \cos t$, $y = r \sin t$, $0 \leq t \leq 2\pi$. Then $\langle x', y' \rangle = \langle -r \sin t, r \cos t \rangle$, and the unit tangent is $\mathbf{T} = \langle -\sin t, \cos t \rangle$. Thus, at each point of C the vector $\mathbf{v} = \langle -y, x \rangle = \langle -r \sin t, r \cos t \rangle$ has the same direction as $\mathbf{T}$, and the tangential component is

$$v_T = \mathbf{v} \cdot \mathbf{T} = r(\sin^2 t + \cos^2 t) = r = |\mathbf{v}|,$$

whence

$$\int_C -y \, dx + x \, dy = \int_C v_T \, ds = r \cdot 2\pi r = 2\pi r^2.$$

Since the area inside C is πr^2, if we choose for the figures F_n a sequence of disks centered at the origin, then each term on the right-hand side of (27.6) is equal to $2\pi r_n^2 / \pi r_n^2 = 2$, and the limit is also equal to 2. As for the left-hand side, we have $q_x = 1$, $p_y = -1$, $q_x - p_y = 2$. Thus Eq. (27.6) is verified for $(x_0, y_0) = (0, 0)$.

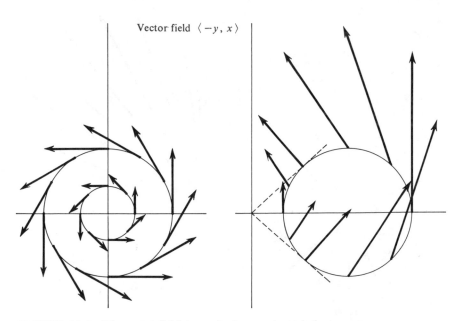

Vector field $\langle -y, x \rangle$

FIGURE 27.2 The vector field $\langle -y, x \rangle$ along various circles

For an arbitrary point (x_0, y_0) the left-hand side of (27.6) is still equal to 2. It is more difficult to compute directly the flow around a circle not centered at the origin, but it is clear geometrically that the value obtained must be a positive number. In fact, the circle is divided into two arcs by the pair of points whose tangent lines pass through the origin (Fig. 27.2). On the arc further from the origin the flow is in the counterclockwise direction, while on the arc nearer the origin it is clockwise. But the further arc is the longer one, and the vector field has greater magnitude. It is therefore clear that the amount of counterclockwise flow is greater than the amount of flow in the clockwise direction; that is, $\int_{v_T} ds > 0$. Using Green's theorem, we find that if C bounds a disk F of radius r, then

$$\int_C v_T \, ds = \int_C -y \, dx + x \, dy = \iint_F 2 \, dA = 2\pi r^2,$$

so that the flow around an arbitrary circle depends only on its radius, and not on the position of its center.

Example 27.4

Let $p = y, q = 0$ (Fig. 27.3).

If we consider the vector field $\mathbf{v} = \langle y, 0 \rangle$ just in the domain $D: 0 < y < 1$, it may be pictured as representing a horizontal flow along a canal, where the velocity is lower near the bottom because of friction. Although this flow does not have the obvious vorticity of the previous example, we nevertheless find that

$$q_x - p_y = -1$$

at every point. This is explained by the fact that on the upper half of every circle, where the velocity is greater, the flow is clockwise, whereas on the lower half it is counterclockwise (Fig. 27.3). Thus the total flow is clockwise, or $\int_C v_T \, ds < 0$.

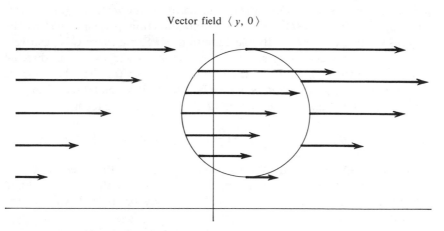

Vector field $\langle y, 0 \rangle$

FIGURE 27.3 A "horizontal" vector field

We may again find the precise value by Green's theorem

$$\int_C v_T \, ds = \iint_F -1 \, dA = -\pi r^2,$$

where F is a disk of radius r, and $C = \partial F$.

We turn next to the quantity $p_x + q_y$, which we introduced in Sect. 19 as the divergence of the vector field $\mathbf{v}$. We note that by Green's theorem:

$$\iint_F p_x \, dA = \int_{\partial F} p \, dy$$

and

$$\iint_F q_y \, dA = -\int_{\partial F} q \, dx,$$

so that

$$(27.7) \qquad \iint_F (p_x + q_y) \, dA = \int_{\partial F} -q \, dx + p \, dy.$$

Suppose now that ∂F consists of a single smooth curve

$$C: x(t), \, y(t), \qquad\qquad a \le t \le b.$$

Then

$$(27.8) \qquad \int_C -q \, dx + p \, dy = \int_a^b (-qx'(t) + py'(t)) \, dt.$$

We observe that the integrand on the right-hand side is equal to the dot product of $\langle p, q \rangle$ and $\langle y'(t), -x'(t) \rangle$. Furthermore, the vector $\langle y'(t), -x'(t) \rangle$ has length $(y'(t)^2 + x'(t)^2)^{1/2} = ds/dt$, and direction perpendicular to the tangent vector $\langle x'(t), y'(t) \rangle$. By the definition of ∂F, the curve C is positively oriented with respect to F, so that the vector $\langle -y'(t), x'(t) \rangle$ is directed toward the interior of F. Thus, the vector $\langle y'(t), -x'(t) \rangle$ has the opposite direction, and is equal to $ds/dt \, \mathbf{N}$, where $\mathbf{N}$ is the unit normal directed toward the *exterior* of F (Fig. 27.4). Equation (27.8) then takes the form

$$(27.9) \quad \int_C -q \, dx + p \, dy = \int_a^b (\mathbf{v} \cdot \mathbf{N}) \frac{ds}{dt} \, dt = \int_C (\mathbf{v} \cdot \mathbf{N}) \, ds = \int_C v_N \, ds,$$

where $v_N = \mathbf{v} \cdot \mathbf{N}$ denotes the *normal component* of $\mathbf{v}$ in the direction of the exterior normal to F. We note again the important fact that the right-hand side of (27.9) is an intrinsic quantity, which may be defined without any reference to coordinates.

Combining (27.7) and (27.9), and introducing the notation

$$\text{div } \mathbf{v} = p_x + q_y$$

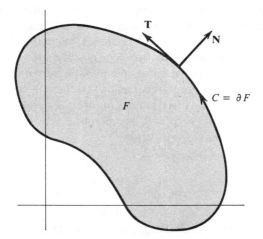

FIGURE 27.4   The exterior normal **N**

for the divergence of the vector field **v**, we obtain

$$(27.10) \qquad \iint_F (\text{div } \mathbf{v}) \, dA = \int_{\partial F} v_N \, ds.$$

This equation is just one more form of Green's theorem, emphasizing the interpretation of the pair of functions p, q as the components of a vector field **v**.

At the end of Sect. 20, we showed that the divergence of a vector field is a quantity that does not depend on the choice of coordinates. We are now in a position to give a direct proof of this fact by obtaining an intrinsic expression for the divergence. We again choose any point (x_0, y_0) and a sequence of figures $F_1, F_2, \ldots$, which contain (x_0, y_0) and whose diameters tend to zero. We apply Eq. (27.3) with $f(x, y) = \text{div } \mathbf{v}$, to obtain

$$(27.11) \qquad \text{div } \mathbf{v}\big|_{(x_0, y_0)} = \lim_{k \to \infty} \frac{\iint_{F_k} (\text{div } \mathbf{v}) \, dA}{\iint_{F_k} 1 \, dA} = \lim_{k \to \infty} \frac{\int_{\partial F_k} v_N \, ds}{\text{area of } F_k}.$$

Again, the terms on the right-hand side are independent of coordinates, hence so is their limit.

Returning to the interpretation of $\mathbf{v}(x, y)$ as the velocity field of a fluid flow, we note that the quantity v_N represents the component of the flow velocity perpendicular to C, so that the integral

$$\int_C v_N \, ds$$

may be regarded as the "total rate of flow *across* C." Equation (27.10) states that the total flow across ∂F is equal to the integral over F of the divergence of **v**. If we take a sequence of circles around a point, form the ratio of flow across these circles to area inside, and take the limit, then by (27.11) we obtain the values of the divergence at that point. Note in particular, that since **N** is the exterior normal to F, a positive value for (27.10) means flow out of F, and a negative value means flow into F. By (27.11), if the divergence is positive at a point, this means that there is flow away from that point, and if it is negative, there is flow *toward* the point. A *divergence-free* vector field, in which $u_x + v_y \equiv 0$, represents a flow that goes past each point, without any "sources" or "sinks."

We now examine Examples 27.2–27.4 in terms of divergence.

Example 27.2a

$$\mathbf{v} = \langle x, y \rangle \quad \text{(Fig. 27.1)}.$$

If C is a circle of radius r centered at the origin, then at each point (x, y) of C the vector $\mathbf{v} = \langle x, y \rangle$ has length $(x^2 + y^2)^{1/2} = r$ and is normal to C, directed toward the exterior. Thus $\mathbf{v} = r\mathbf{N}$, and $v_N = \mathbf{v} \cdot \mathbf{N} = r$. It follows that

$$\int_C v_N \, ds = r \cdot 2\pi r = 2\pi r^2.$$

The area inside this circle is πr^2 and the ratio

$$\int_C v_N \, ds / \text{Area} = 2.$$

Thus, if the figures F_k are chosen to be disks centered at the origin, then each term on the right-hand side of (27.11) is equal to 2. The limit is also 2, and we verify directly for the left-hand side of (27.11) that $p_x + q_y = 2$.

The fact that the divergence is everywhere equal to 2 for this vector field means that the rate of flow out of any given figure F is equal to twice its area.

$$\int_{\partial F} v_N \, ds = \int_{\partial F} -y \, dx + x \, dy = \iint_F 2 \, dA = 2 \times \text{Area of } F.$$

It is clear geometrically from Fig. 27.1 that the amount of flow out of any disk is greater than the amount entering, so that one would expect the divergence to be positive.

Example 27.3a

$$\mathbf{v} = \langle -y, x \rangle \quad \text{(Fig. 27.2)}.$$

In this case the vector field is tangential to any circle C around the origin, so that $v_N = 0$. This implies that the divergence at the origin must be zero, which may also be verified directly. In fact div $\mathbf{v} \equiv 0$ in this case, in accordance with the fact that for an arbitrary closed curve C, by (27.9),

$$\int_C v_N \, ds = \int_C -x \, dx - y \, dy = -\int_C x \, dx + y \, dy = 0,$$

as we have observed at the end of our discussion of Example 27.2.

Example 27.4a

$$\mathbf{v} = \langle y, 0 \rangle \qquad \text{(Fig. 27.3).}$$

In this case it is clear that for an arbitrary circle, the flow into it equals the flow out of it, and it follows from (27.11) that the divergence must be everywhere zero. We may confirm this directly, since both p_x and q_y are everywhere zero.

Exercises

27.1   Find the average height of the surface $z = xy$ over each of the following figures F.

  a. $F: 0 \le x \le 1, 0 \le y \le 1$
  b. $F: x \ge 0, y \ge 0, x^2 + y^2 \le 1$
  c. $F: x \ge 0, y \ge 0, x^2 + y^2 \le 4$
  d. $F: x \ge 0, y \ge 0, 1 \le x^2 + y^2 \le 4$
  e. $F: x \ge 0, y \ge 0, x + y \le a, \quad (a > 0)$
  f. $F: x \ge 0, y \ge 0, x^2/a^2 + y^2/b^2 \le 1, \quad (a > 0, b > 0)$
  g. $F: 0 \le x \le a, 0 \le y \le b, \quad (a > 0, b > 0)$

27.2   *a.* What is the average height of the hemisphere $z = \sqrt{1 - x^2 - y^2}$? (*Hint:* recall the geometric interpretation of average height given in the text.)
      b. What is the average density of the half disk $x \ge 0, x^2 + y^2 \le a^2$, if the density at each point is equal to x^3?

27.3   Find the average value of the function ye^x over the following figures F.

  a. $F: 0 \le x \le 1, 0 \le y \le 1$
  b. $F: x^2 + y^2 \le 1$

27.4   Suppose that $f(x, y)$ is of the form $g(x)h(y)$. If F is a rectangle $a \le x \le b, c \le y \le d$, show that the average value of $f(x, y)$ over F is equal to the product of the average value of $g(x)$ over $a \le x \le b$ and the average value of $h(y)$ over $c \le y \le d$.

27.5   Let F be a figure that is symmetric about the x axis and suppose $f(x, y)$ satisfies $f(x, -y) = -f(x, y)$. Show that the average value of $f(x, y)$ over F is equal to zero.

27.6   For each part of Ex. 27.1, verify the mean-value theorem (Th. 27.1) by finding a specific point (x_0, y_0) in the given figure F such that $x_0 y_0$ equals the average value of xy over F.

27.7   Show that if $f(x, y)$ is a linear function, $ax + by + c$, then the mean value of $f(x, y)$ over any figure F is equal to the value of $f(x, y)$ at the centroid of F.

27.8   In the statement of Th. 27.2, let $f(x, y) = (y + 2)^2$, $(x_0, y_0) = (0, 0)$, and let F_k be the disk $x^2 + y^2 \le 1/k^2$. Evaluate explicitly the expression $\iint_{F_k} f \, dA / A(F_k)$ for each positive integer k, and verify Eq. (27.3).

*27.9 Give a proof of Th. 27.2 without using the mean value theorem, as follows.

    a. Show that the theorem holds if $f(x_0, y_0) = 0$. (*Hint:* given any $\epsilon > 0$, by continuity of $f(x, y)$ there exists a $\delta > 0$ such that $|f(x, y)| < \epsilon$ for (x, y) in the disk D of radius δ about (x_0, y_0). For k sufficiently large, the figure F_k lies in D. By Ex. 24.27, $|\iint_{F_k} f \, dA / A(F_k)| < \epsilon$, and hence the limit on the right-hand side of (27.3) is $\leq \epsilon$. Since this is true for arbitrary small ϵ, the limit must be zero.)

    b. If $f(x_0, y_0) \neq 0$, set $g(x, y) = f(x, y) - f(x_0, y_0)$ and apply part a to $g(x, y)$.

27.10 Suppose, in the Corollary to Th. 27.2, that G is the transformation $u = x^2$, $v = y^3$.

    a. If F_k is the square $x_0 - 1/k \leq x \leq x_0 + 1/k$, $y_0 - 1/k \leq y \leq y_0 + 1/k$, find the area $A(\tilde{F}_k)$ of its image. (Assume that $x_0 > 0$, $y_0 > 0$ and that k is sufficiently large so that F_k lies in the first quadrant.)

    b. Compute explicitly the ratio $A(\tilde{F}_k)/A(F_k)$ for each k, and verify Eq. (27.4).

27.11 Let G be the transformation $u = x \cos y$, $v = x \sin y$. Let F be a rectangle $x_1 \leq x \leq x_2$, $y_1 \leq y \leq y_2$, where $x_1 > 0$.

    a. Sketch the image $\tilde{F}$ of F under the transformation G, and find its area $A(\tilde{F})$.

    b. Verify Eq. (27.4) for this transformation, where each F_k is a rectangle containing (x_0, y_0).

27.12 Let $G:(x, y) \to (u, v)$ and $H:(u, v) \to (z, w)$ be differentiable transformations. Suppose that G is regular at (x_0, y_0) and H is regular at $(u_0, v_0) = G(x_0, y_0)$. In Ex. 17.9, it was shown by a geometrical argument how the sign of the Jacobian of the composed mapping $H \circ G$ at (x_0, y_0) was determined by the signs of the Jacobian of G at (x_0, y_0) and the Jacobian of H at (u_0, v_0). Show how the Corollary to Th. 27.2 may be used to give a geometric derivation of the relation between the absolute values of these Jacobians; namely,

$$\left|\frac{\partial(z, w)}{\partial(x, y)}\right|_{(x_0, y_0)} = \left|\frac{\partial(z, w)}{\partial(u, v)}\right|_{(u_0, v_0)} \left|\frac{\partial(u, v)}{\partial(x, y)}\right|_{(x_0, y_0)}.$$

27.13 The *mean-value theorem for integrals* in one variable states that if $f(t)$ is continuous for $a \leq t \leq b$, then there is a value c, $a < c < b$, such that

$$\int_a^b f(t) \, dt = (b - a)f(c).$$

    a. Show that this is equivalent to the ordinary mean value theorem for derivatives. (*Hint:* set $F(x) = \int_a^x f(t) \, dt$.)

    b. State this result in terms of the average value of $f(t)$.

27.14 Show that the equality of the mixed derivatives (Th. 11.1) is a consequence of the mean value theorem (Th. 27.1) and the fact that the value of an iterated integral does not depend on the order of integration (Corollary to Th. 24.4). (*Hint:* to prove $f_{xy}(x_1, y_1) = f_{yx}(x_1, y_1)$, let F be a rectangle $x_1 \leq x \leq x_2$, $y_1 \leq y \leq y_2$, apply the mean value theorem to $\iint_F f_{xy} \, dA$ and $\iint_F f_{yx} \, dA$, evaluate each of these integrals by a suitable iteration, and take the limit as $x_2 \rightarrow x_1$, $y_2 \rightarrow y_1$.)

27.15 Show that Eq. (27.10) holds for an arbitrary figure F, even though the integrand v_N may have discontinuities at corners of ∂F. (At a corner of the boundary the normal N is not well-defined, but by the definition of a piecewise smooth curve, it tends to a limit from each side of the corner. Thus, the value of v_N at a corner may be defined as either of these limits, since the value of an integral is independent of changes in the value of the integrand at a finite number of points.)

27.16 Show that the identity

$$\iint_F \Delta u \, dA = \int_{\partial F} \frac{\partial u}{\partial n} \, ds$$

is a consequence of Eq. (27.10) (see also Ex. 27.15).

27.17 Show that the expression for Δu must be invariant under rotation of coordinates by obtaining an invariant description of this quantity at any point. (*Hint:* use Ex. 27.16 and find an expression analogous to Eq. (27.11).)

*27.18 Using the methods developed in this section, it is possible to deduce the expression for the Laplacian in polar coordinates (Eq. (20.10)) without actually computing the way the second derivatives transform under change of coordinates. This may be carried out in the following steps. Let $h(x, y) \in \mathscr{C}^2$, and let $g(r, \theta) = h(r \cos \theta, r \sin \theta)$ be the same function expressed in polar coordinates. Let F be the figure defined in polar coordinates by $\theta_1 \leq \theta \leq \theta_2$, $r_1 \leq r \leq r_2$.

   a. Show that

$$\int_{\partial F} \frac{\partial h}{\partial n} \, ds = \int_{C_1} \frac{\partial h}{\partial n} \, ds + \int_{C_2} \frac{\partial h}{\partial n} \, ds + \int_{C_3} \frac{\partial h}{\partial n} \, ds + \int_{C_4} \frac{\partial h}{\partial n} \, ds,$$

where

$$
\begin{aligned}
C_1 &: \theta = \theta_1, r = t, & r_1 \leq t \leq r_2, \\
C_2 &: \theta = t, r = r_2, & \theta_1 \leq t \leq \theta_2, \\
C_3 &: \theta = \theta_2, r = t, & r_1 \leq t \leq r_2, \\
C_4 &: \theta = t, r = r_1, & \theta_1 \leq t \leq \theta_2.
\end{aligned}
$$

(*Hint:* see Ex. 21.14.)

   b. Show that

$$\frac{\partial h}{\partial n} = \begin{cases} -g_\theta(r, \theta_1)/r & \text{on} \quad C_1 \\ g_r(r_2, \theta) & \text{on} \quad C_2 \\ g_\theta(r, \theta_2)/r & \text{on} \quad C_3 \\ -g_r(r_1, \theta) & \text{on} \quad C_4. \end{cases}$$

c. Show that

$$\iint_F \Delta h \, dA = \int_{\theta_1}^{\theta_2} [r_2 g_r(r_2, \theta) - r_1 g_r(r_1, \theta)] \, d\theta + \int_{r_1}^{r_2} \frac{1}{r} [g_\theta(r, \theta_2) - g_\theta(r, \theta_1)] \, dr.$$

(*Hint:* use Ex. 27.16.)

d. Using the mean-value theorem for integrals in one variable (Ex. 27.13), and then the mean-value theorem for derivatives, show that

$$\int_{\theta_1}^{\theta_2} [r_2 g_r(r_2, .\theta) - r_1 g_r(r_1, \theta)] \, d\theta$$

$$= (\theta_2 - \theta_1)[r_2 g_r(r_2, \theta_3) - r_1 g_r(r_1, \theta_3)]$$
$$= (\theta_2 - \theta_1)[r_1(g_r(r_2, \theta_3) - g_r(r_1, \theta_3)) + (r_2 - r_1)g_r(r_2, \theta_3)]$$
$$= (\theta_2 - \theta_1)(r_2 - r_1)[r_1 g_{rr}(r_3, \theta_3) + g_r(r_2, \theta_3)]$$

and

$$\int_{r_1}^{r_2} \frac{1}{r} [g_\theta(r, \theta_2) - g_\theta(r, \theta_1)] \, dr = (r_2 - r_1)\frac{1}{r_4} [g_\theta(r_4, \theta_2) - g_\theta(r_4, \theta_1)]$$

$$= (r_2 - r_1)(\theta_2 - \theta_1)\frac{1}{r_4} g_{\theta\theta}(r_4, \theta_4)$$

where r_3, r_4 are between r_1 and r_2, and θ_3, θ_4 are between θ_1 and θ_2.

e. Show that there is a point (x_0, y_0) in F such that

$$\Delta h(x_0, y_0) = \frac{2}{r_1 + r_2} [r_1 g_{rr}(r_3, \theta_3) + g_r(r_2, \theta_3) + \frac{1}{r_4} g_{\theta\theta}(r_4, \theta_4)].$$

(*Hint:* apply the mean value theorem to the left-hand side of part c, use the answer to Ex. 27.11a for $A(F)$ in polar coordinates, and use part d above.)

f. Show that

$$\Delta h(x_1, y_1) = g_{rr}(r_1, \theta_1) + \frac{1}{r_1} g_r(r_1, \theta_1) + \frac{1}{r_1^2} g_{\theta\theta}(r_1, \theta_1),$$

where $x_1 = r_1 \cos \theta_1$, $x_2 = r_2 \cos \theta_2$. (*Hint:* take the limit in part e as $r_2 \to r_1$, $\theta_2 \to \theta_1$.)

27.19 Show that Green's identity, Ex. 26.9a, is a consequence of Eq. (27.10). (*Hint:* apply Eq. (27.10) to the vector field $\mathbf{w} = u \, \nabla v$.)

27.20 Sketch each of the following vector fields $\mathbf{v}(x, y)$, and compute the integrals $\int_C v_T \, ds$ and $\int_C v_N \, ds$, where C is a circle of radius r centered at a point (x_0, y_0), $\mathbf{T}$ is the unit tangent to C in the counterclockwise direction, and $\mathbf{N}$ is the unit normal directed toward the exterior of C. Choosing a sequence of such circles C_n whose radii tend to zero, verify Eqs. (27.6) and (27.11).

a. $\mathbf{v}(x, y) \equiv \langle 1, 0 \rangle$ (uniform flow)
b. $\mathbf{v}(x, y) = \langle -x, -y \rangle$
c. $\mathbf{v}(x, y) = \langle y, -x \rangle$
d. $\mathbf{v}(x, y) = \langle x, 0 \rangle$
e. $\mathbf{v}(x, y) = \langle x^2, 0 \rangle$
f. $\mathbf{v}(x, y) = \langle -y/(x^2 + y^2), x/(x^2 + y^2) \rangle$.

27.21   Show that if $\mathbf{v}(x, y)$ is a harmonic vector field in a domain D, then $\int_{\partial F} v_T \, ds = 0$, $\int_{\partial F} v_N \, ds = 0$ for every figure F lying in D.

*27.22   The *generalized mean value theorem* for double integrals may be stated as follows: *if $f(x, y)$ and $g(x, y)$ are continuous on a figure F, and if $g(x, y) > 0$ everywhere on F, then there is a point (x_0, y_0) in F such that*

$$\iint_F fg \, dA = f(x_0, y_0) \iint_F g \, dA.$$

  a. Prove this theorem. (*Hint:* use the proof of Th. 27.1 as a model.)
  b. Show by an example that the conclusion may be false if the hypothesis $g(x, y) > 0$ is omitted.

*27.23   Let $f(x, y) \in \mathscr{C}^3$ in a domain D, and suppose that the figure

$$F: (x - x_0)^2 + (y - y_0)^2 \leq R^2$$

lies in D.

  a. Show that

$$\iint_F f \, dA = \pi R^2 f(x_0, y_0) + \tfrac{1}{8}\pi R^4 \, \Delta f(x_0, y_0) + \iint_F R_2 \, dA$$

where $R_2(x, y)$ is the remainder term in the Taylor expansion of $f(x, y)$ about (x_0, y_0) through terms of second order (see Eqs. (12.8), (12.9), (12.10), and (12.11)).
  b. Deduce that

$$\lim_{R \to 0} \frac{\iint_F f \, dA}{\pi R^2} = f(x_0, y_0).$$

  c. Show that

$$\lim_{R \to 0} \frac{\iint_F [f - f(x_0, y_0)] \, dA}{\pi R^4} = \tfrac{1}{8} \Delta f(x_0, y_0).$$

27.24   A mapping G defined in a domain D is said to be *area-preserving* if for every figure F in D, the image $\tilde{F}$ of F has the same area as F. Show that a diffeomorphism $G: (x, y) \to (u, v)$ is area-preserving if and only if $|\partial(u, v)/\partial(x, y)| \equiv 1$.

28   Transformation rule for double integrals

One of the basic methods in the theory of the ordinary integral is that of integration by substitution:

(28.1) $$\int_c^d f(u) \, du = \int_a^b f(u(x)) \frac{du}{dx} \, dx,$$

where u is a differentiable function of x for $a \leq x \leq b$, and $u(a) = c$, $u(b) = d$. Equation (28.1) describes the behavior of a definite integral when the integrating variable is expressed as a differentiable function of another variable. Our next object is to derive a formula analogous to (28.1) for the double integral.

In view of the fact that in the ensuing discussion a single function is expressed in terms of different variables, it is convenient to adopt a modified notation for the double integral. If $f(x, y)$ is integrable over a figure F, we set

$$(28.2) \qquad \iint_F f \, dA = \iint_F f(x, y) \, dx \, dy.$$

The notation on the right is obviously adapted from the iterated integral.

We now state the principal result.

Theorem 28.1 *Let F be a figure lying in a domain D of the x, y plane, and let $\tilde{F}$ be the image of F under a diffeomorphism*

$$G : \begin{cases} u(x, y) \\ v(x, y) \end{cases}$$

of D. Let $f(u, v)$ be integrable over $\tilde{F}$. Then

$$(28.3) \qquad \iint_{\tilde{F}} f(u, v) \, du \, dv = \iint_F f(u(x, y), v(x, y)) \left| \frac{\partial(u, v)}{\partial(x, y)} \right| dx \, dy.$$

Remark Note the analogy between formulas (28.3) and (28.1). When passing from a differentiable function $u(x)$ to a differentiable transformation $u(x, y)$, $v(x, y)$, it is once again the Jacobian which takes over the role of the derivative. The proofs of (28.1) and (28.3), however, have nothing in common. The proof of (28.1) is a simple application of the chain rule, whereas there is no elementary proof of (28.3). We sketch a proof in the following.[8]

PROOF. Let $F_1, \ldots, F_n$ be an arbitrary partition of F, and let $\tilde{F}_k$ be the image of F_k under the transformation G, for $k = 1, \ldots, n$. Then applying Th. 26.2 and the mean value theorem (Th. 27.1),

$$(28.4) \quad \text{Area of } \tilde{F}_k = \iint_{F_k} \left| \frac{\partial(u, v)}{\partial(x, y)} \right| dx \, dy = (\text{Area of } F_k) \times \left| \frac{\partial(u, v)}{\partial(x, y)} \right|_{(x_k, y_k)},$$

where (x_k, y_k) is some point in F_k. Let (u_k, v_k) be the image of (x_k, y_k) under the transformation G, and denote by $A(\tilde{F}_k)$, $A(F_k)$ the areas of $\tilde{F}_k$ and F_k,

---

[8] For a complete proof, as well as an interesting discussion of the advantages and disadvantages of each of the standard proofs, see [32]. See also Ex. 28.14, at the end of this section.

respectively. Then multiplying each side of (28.4) by $f(u_k, v_k)$ and adding over $k = 1, \ldots, n$ yields

$$(28.5) \quad \sum_{k=1}^{n} f(u_k, v_k) A(\tilde{F}_k) = \sum_{k=1}^{n} f(u(x_k, y_k), v(x_k, y_k)) \left| \frac{\partial(u, v)}{\partial(x, y)} \right|_{(x_k, y_k)} A(F_k).$$

If we now choose a sequence of partitions whose norm tends to zero, then by the definition of the double integral the right-hand side of (28.5) tends to the integral on the right of (28.3), whereas the left-hand side of (28.5) tends to the integral on the left of (28.3). This proves the theorem. ◆

There are several remarks to be made concerning the above reasoning. First of all, we assumed implicitly that the $\tilde{F}_k$ form a partition of $\tilde{F}$, and that the norm of this partition tends to zero with the norm of the corresponding partition of F. Only then can we conclude that the left-side of (28.5) tends to the double integral in (28.3). Furthermore, we have not shown that the integrand on the right of (28.3) is an integrable function. However, in the important case that $f(u, v)$ is continuous on $\tilde{F}$, the function $f(u(x, y), v(x, y)) \cdot |\partial(u, v)/\partial(x, y)|$ is continuous on F, and the integral on the right of (28.3) exists.

Example 28.1

If $f(u, v) \equiv 1$, then Th. 28.1 reduces to Th. 26.2.

Example 28.2

Let $\tilde{F}$ be the figure between the lines $u + v = 1$ and $u + v = 2$ in the first quadrant. Evaluate $\iint_{\tilde{F}} (u + v) \, du \, dv$.

This integral may be evaluated directly, but it is a rather long computation (see Ex. 24.4). Examining the figure $\tilde{F}$, we see that it would be advantageous to introduce the quantity $u + v$ as a new independent variable. We set

$$(28.6) \quad \begin{cases} u = x + y \\ v = x - y. \end{cases}$$

Then $u + v = 2x$, $u - v = 2y$. Since the figure $\tilde{F}$ is given explicitly by

$$\tilde{F}: u \geq 0, v \geq 0, \quad 1 \leq u + v \leq 2,$$

it corresponds under the transformation (28.6) to the figure

$$F: x + y \geq 0, \quad x - y \geq 0, \quad 1 \leq 2x \leq 2,$$

or equivalently (Fig. 28.1),

$$F: \quad -x \leq y \leq x, \quad \tfrac{1}{2} \leq x \leq 1.$$

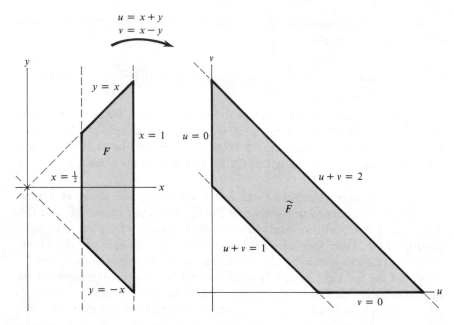

$$u = x + y$$
$$v = x - y$$

FIGURE 28.1 An application of the transformation rule for double integrals

For the transformation (28.6) we have

$$\begin{pmatrix} u_x & u_y \\ v_x & v_y \end{pmatrix} = \begin{pmatrix} 1 & 1 \\ 1 & -1 \end{pmatrix}$$

and

$$\frac{\partial(u, v)}{\partial(x, y)} = -2.$$

Thus Eq. (28.3) takes the form

$$\iint_{\tilde{F}} (u + v) \, du \, dv = \iint_{F} 2x|-2| \, dx \, dy$$

$$= 4 \iint_{F} x \, dx \, dy.$$

But

$$\iint_{F} x \, dx \, dy = \int_{1/2}^{1} \left[\int_{-x}^{x} x \, dy \right] dx,$$

and

$$\int_{-x}^{x} x \, dy = xy \Big|_{-x}^{x} = 2x^2,$$

so that

$$\iint_F x \, dx \, dy = \int_{1/2}^1 2x^2 \, dx = \frac{2x^3}{3}\Big|_{1/2}^1 = \tfrac{2}{3}(1 - \tfrac{1}{8}) = \tfrac{7}{12}$$

and

$$\iint_{\tilde{F}} (u + v) \, du \, dv = 4 \iint_F x \, dx \, dy = \tfrac{7}{3}.$$

There is an important interpretation of Eq. (28.3) that is somewhat different from the version we have presented. Namely, as we have pointed out in the previous chapter, the pair of functions $u(x, y)$, $v(x, y)$ may be considered equally well to define a transformation or a change of coordinates. For example, Eqs. (28.6) amount essentially to a rotation of coordinates, which takes the parallel lines bounding $\tilde{F}$ onto a pair of vertical lines bounding F. Another familiar example is

(28.7) $u = x \cos y, \qquad v = x \sin y.$

We then have

$$\begin{pmatrix} u_x & u_y \\ v_x & v_y \end{pmatrix} = \begin{pmatrix} \cos y & -x \sin y \\ \sin y & x \cos y \end{pmatrix}$$

and

$$\frac{\partial(u, v)}{\partial(x, y)} = x.$$

Equation (28.3) takes the form

(28.8) $$\iint_{\tilde{F}} f(u, v) \, du \, dv = \iint_F f(x \cos y, x \sin y)|x| \, dx \, dy.$$

Equations (28.7) may be interpreted as defining a transformation G, or else as introducing new coordinates. Namely, x and y represent polar coordinates in the u, v plane. In the usual notation, setting $x = r$, $y = \theta$, where it is assumed that $r \geq 0$, (28.8) takes the form

(28.9) $$\iint_{\tilde{F}} f(u, v) \, du \, dv = \iint_F f(r \cos \theta, r \sin \theta) r \, dr \, d\theta.$$

Thus, the function f is expressed in polar coordinates, multiplied by r, and then integrated over the corresponding values of r and θ.

Example 28.3

Calculate the y moment of the figure

$$F: y \geq 0, \, a^2 \leq x^2 + y^2 \leq b^2.$$

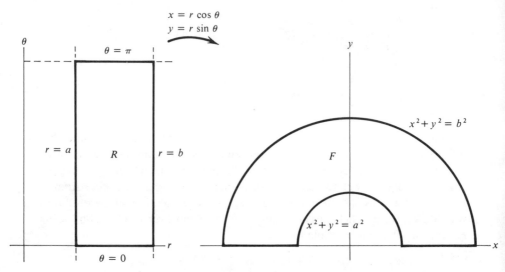

FIGURE 28.2 The rectangle in the plane of polar coordinates corresponding to a semi-annulus in rectangular coordinates

If we introduce polar coordinates, the figure F corresponds to the rectangle $R: a \leq r \leq b$, $0 \leq \theta \leq \pi$ (Fig. 28.2). Thus

$$\iint_F y \, dx \, dy = \iint_R (r \sin \theta) r \, dr \, d\theta = \int_0^\pi \left[\int_a^b r^2 \sin \theta \, dr \right] d\theta$$

$$= \int_0^\pi \left[\frac{b^3 - a^3}{3} \sin \theta \right] d\theta = \tfrac{2}{3}(b^3 - a^3).$$

Polar coordinates are certainly, after rectangular coordinates, the most useful and most frequently encountered. In view of this fact, it may be illuminating to give a direct proof of Eq. (28.9), which does not depend on all the theory that was needed to prove the general formula (28.3). We do so for the figure F of Example 28.3 (see Fig. 28.2).

Let $f(x, y)$ be defined on F, and choose a partition of F by figures

$$F_{ij}: r_{i-1} \leq r \leq r_i, \ \theta_{j-1} \leq \theta \leq \theta_j,$$

where

$$a = r_0 < r_1 < \cdots < r_m = b; \qquad 0 = \theta_0 < \theta_1 < \cdots < \theta_n = \pi$$

(see Fig. 28.3). Since the area of a circular sector of angle α is equal to $\tfrac{1}{2}\alpha r^2$, where r is the radius of the circle, the area of F_{ij} is equal to

$$A_{ij} = \tfrac{1}{2}(\theta_j - \theta_{j-1})r_i^2 - \tfrac{1}{2}(\theta_j - \theta_{j-1})r_{i-1}^2 = \tfrac{1}{2}(\theta_j - \theta_{j-1})(r_i^2 - r_{i-1}^2)$$
$$= \tfrac{1}{2}(r_i + r_{i-1})(r_i - r_{i-1})(\theta_j - \theta_{j-1}).$$

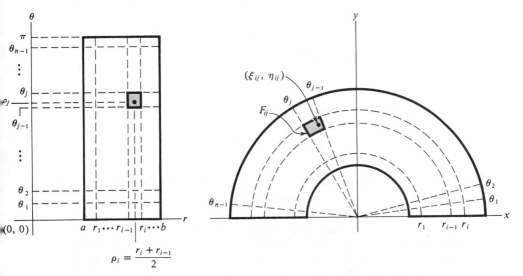

FIGURE 28.3 Direct proof of the transformation rule for polar coordinates

Set $\rho_i = \frac{1}{2}(r_i + r_{i-1})$, and choose φ_j so that $\theta_{j-1} \le \varphi_j \le \theta_j$. Then $r_{i-1} < \rho_i < r_i$, and the point whose polar coordinates are (ρ_i, φ_j) lies in F_{ij}. Finally, let (ξ_{ij}, η_{ij}) be the rectangular coordinates of this point. Then

$$\sum_{\substack{i=1,\ldots,m \\ j=1,\ldots,n}} f(\xi_{ij}, \eta_{ij}) A_{ij}$$

(28.10)

$$= \sum_{\substack{i=1,\ldots,m \\ j=1,\ldots,n}} f(\rho_i \cos \varphi_j, \rho_i \sin \varphi_j) \rho_i (r_i - r_{i-1})(\theta_j - \theta_{j-1}).$$

If we take successively finer partitions of F, then the sum on the left-hand side of (28.10) tends to $\iint_F f \, dA$, by the definition of this integral. On the other hand, if we set

$$g(r, \theta) = f(r \cos \theta, r \sin \theta)r,$$

and if we denote by R the rectangle in the r, θ plane defined by $a \le r \le b$, $0 \le \theta \le \pi$, then the sums on the right of (28.10) are exactly those used in defining the double integral of g over R. Thus, Eq. (28.10) yields in the limit

(28.11) $\displaystyle \iint_F f \, dA = \iint_F f(x, y) \, dx \, dy = \iint_R f(r \cos \theta, r \sin \theta)r \, dr \, d\theta.$

Remark The essential content of Eq. (28.11) is often written in abbreviated form as

$$dA = dx\, dy$$

if x and y are rectangular coordinates, and

$$dA = r\, dr\, d\theta$$

if r, θ are polar coordinates. Similarly, Eq. (28.3) is abbreviated to

$$du\, dv = \left|\frac{\partial(u,\, v)}{\partial(x,\, y)}\right| dx\, dy.$$

The meaning of these "equations" is that in order to compute the double integral of a function with respect to different sets of coordinates, one need only express the function in terms of the given coordinates, multiply by the corresponding factor, and then evaluate the iterated integral. The term "area element" is often used to denote the expression dA and its representation in various coordinates. It should be understood that these expressions are purely a formalism, just as the corresponding abbreviation $du = (du/dx)\, dx$ for Eq. (28.1), unless the quantities dx, dy, and their "product" $dx\, dy$ are suitably defined.

We conclude with an illustration of how some of the ideas introduced in this chapter may be applied to evaluate an important integral. We wish to find the value of

(28.12)
$$\int_0^\infty e^{-x^2}\, dx.$$

This integral arises in many connections, and plays a special role in the theory of probability. It is a so-called *improper* integral, since the domain of integration is an infinite, rather than a finite, interval. Its value is defined to be the number

$$L = \lim_{b \to \infty} L(b),$$

where

(28.13)
$$L(b) = \int_0^b e^{-x^2}\, dx.$$

Elementary methods of integration do not work, because there is no elementary function whose derivative equals e^{-x^2}. We observe, however, that we can evaluate the double integral

$$\iint_F e^{-x^2} e^{-y^2}\, dx\, dy$$

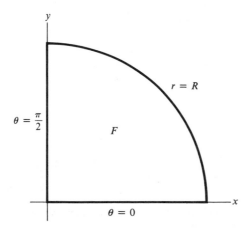

FIGURE 28.4 A quarter disk in polar coordinates

for the figure (Fig. 28.4)

$$F: x \geq 0, \ y \geq 0, \ x^2 + y^2 \leq R^2$$

by introducing polar coordinates.[9] We have

$$\iint_F e^{-x^2} e^{-y^2} dx \, dy = \iint_F e^{-x^2 - y^2} \, dx \, dy = \iint_{\tilde{R}} e^{-r^2} r \, dr \, d\theta,$$

where $\tilde{R}$ is the rectangle $0 \leq r \leq R, \ 0 \leq \theta \leq \tfrac{1}{2}\pi$. But

$$\int_0^R e^{-r^2} r \, dr = -\tfrac{1}{2} e^{-r^2} \Big|_0^R = \tfrac{1}{2}(1 - e^{-R^2}),$$

and

(28.14)
$$\iint_F e^{-x^2} e^{-y^2} \, dx \, dy = \int_0^{\pi/2} \left[\int_0^R e^{-r^2} r \, dr \right] d\theta$$
$$= \tfrac{1}{4}\pi(1 - e^{-R^2}).$$

On the other hand, if we integrate over the square

$$\tilde{F}: 0 \leq x \leq b, \ 0 \leq y \leq b,$$

---

[9] Strictly speaking, Th. 28.1 does not apply in this case since the figure F contains the origin, and the transformation $x = r \cos \theta, \ y = r \sin \theta$ is not a diffeomorphism in any domain containing a point where $r = 0$. However, it is easy to show that the conclusion of Th. 28.1 still holds in this case, either by excluding a small circle of radius ϵ about the origin and taking the limit as ϵ tends to zero, or else by using the special reasoning for polar coordinates, which led to Eq. (28.11).

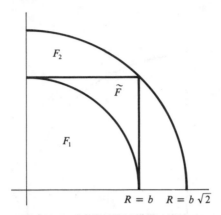

FIGURE 28.5 The use of inscribed and circumscribed circles to estimate an integral over a square

we find

$$(28.15) \quad \iint_{\tilde{F}} e^{-x^2}e^{-y^2}\,dx\,dy = \int_0^b \left[\int_0^b e^{-x^2}e^{-y^2}\,dx\right]dy$$

$$= \int_0^b e^{-y^2}\left[\int_0^b e^{-x^2}\,dx\right]dy$$

$$= \left[\int_0^b e^{-x^2}\,dx\right]\left[\int_0^b e^{-y^2}\,dy\right] = [L(b)]^2$$

using the notation of (28.13). Suppose finally that we consider figures F_1, F_2 of the type F, such that $F_1 \subset \tilde{F} \subset F_2$ (see Fig. 28.5). For example, for F_1 we set $R = b$, and for F_2 we set $R = b\sqrt{2}$. Since the integrand

$$e^{-x^2}e^{-y^2}$$

is positive everywhere, it follows from basic properties of the double integral that

$$\iint_{F_1} e^{-x^2}e^{-y^2}\,dx\,dy \le \iint_{\tilde{F}} e^{-x^2}e^{-y^2}\,dx\,dy \le \iint_{F_2} e^{-x^2}e^{-y^2}\,dx\,dy.$$

By (28.14) and (28.15), this amounts to

$$(28.16) \qquad \tfrac{1}{4}\pi(1 - e^{-b^2}) \le [L(b)]^2 \le \tfrac{1}{4}\pi(1 - e^{-2b^2}) < \tfrac{1}{4}\pi.$$

Thus, using the double integral, we obtain specific upper and lower bounds for the value of the ordinary integral (28.13). In fact, the inequalities in (28.16) are remarkably precise because of the rapidity with which the function e^{-x^2} tends to zero as x tends to infinity. For example, for $b = 10$, using rough estimates such as

$$e > 2, \qquad e^{10} > 2^{10} > 10^3, \qquad \tfrac{1}{4}\pi < 1,$$

we deduce from (28.16) that

$$\tfrac{1}{4}\pi - 10^{-30} < \left[\int_0^{10} e^{-x^2} \, dx\right]^2 < \tfrac{1}{4}\pi.$$

In other words, the square of the integral coincides to the first thirty decimal places with the value $\tfrac{1}{4}\pi$.

To evaluate the original integral (28.12), we let b tend to infinity, and we note from (28.16) that the quantity $[L(b)]^2$ is bounded above by $\tfrac{1}{4}\pi$ for all b, and is bounded below by a quantity that tends to $\tfrac{1}{4}\pi$ as b tends to infinity. Thus

$$L^2 = \lim_{b \to \infty} [L(b)]^2 = \tfrac{1}{4}\pi$$

and

$$\int_0^\infty e^{-x^2} \, dx = L = \tfrac{1}{2}\sqrt{\pi}.$$

Exercises

28.1 Sketch each of the following figures and describe them in terms of polar coordinates.

    a. $x^2 + y^2 \le 1$, $x \le 0$
    b. $y \ge 0$, $x \le 1$, $y \le x$
    c. $x \ge 0$, $y \ge 0$, $x + y \le 1$
    d. $-x \le y \le x$, $2ax \le x^2 + y^2 \le 2bx$ $(b > a > 0)$
    e. $a^2 \le x^2 + y^2 \le 2ay$ $(a > 0)$
    f. $x \ge 0$, $(x^2 + y^2)^2 \le a^2(x^2 - y^2)$ $(a > 0)$

(*Note*: the figure in part f is bounded by the right-hand loop of the "lemniscate of Bernoulli," whose equation is $(x^2 + y^2)^2 = a^2(x^2 - y^2)$. Observe that $x^2 - y^2 \ge 0$ on this figure.)

28.2 Compute the area of each figure F in Ex. 28.1 by expressing the integral $\iint_F 1 \, dA$ in polar coordinates and evaluating this integral.

28.3 Use polar coordinates to find the centroid of the figures in Ex. 28.1a, b.

28.4 Let F be the disk $x^2 + y^2 \le a^2$, $a > 0$. Express each of the following double integrals as an iterated integral in polar coordinates, and evaluate.

    a. $\displaystyle\iint_F x^2 \, dx \, dy$ **b.** $\displaystyle\iint_F (x^2 + y^2)^n \, dx \, dy,$ $n \ge 0$

28.5 Use polar coordinates to compute the volume of each of the following solids.

    a. Under the plane $z = c$ and over the paraboloid $z = x^2 + y^2$, where $c > 0$.

b. Inside the sphere $x^2 + y^2 + z^2 = a^2$, $a > 0$.

c. Under the surface $z = \cos(x^2 + y^2)$, over the disk $x^2 + y^2 \le \frac{1}{2}\pi$ in the x, y plane.

d. Under the surface $z = \cos(x^2 + y^2)$, and over the annulus, $\frac{3}{2}\pi \le x^2 + y^2 \le 2\pi$.

28.6 Use polar coordinates to compute $\iint_F Q(x, y)\, dx\, dy$, where $Q(x, y) = ax^2 + 2bxy + cy^2$ is an arbitrary quadratic form, and F is the disk $x^2 + y^2 \le R^2$.

28.7 Let F be the figure $x^2/a^2 + y^2/b^2 \le 1$, where $a > 0$, $b > 0$. Let $\tilde{F}$ be the image of F under the transformation $u = x/a$, $v = y/b$.

a. Express $\iint_F f(x, y)\, dx\, dy$ as an integral over $\tilde{F}$. (Note that the roles of the x, y and u, v planes are reversed here from the way they appear in Eq. (28.3).)

b. Use the formula in part a to evaluate $\iint_F (x^2 + y^2)\, dx\, dy$.

c. Verify that the expression obtained in part a yields the area of F when $f(x, y) \equiv 1$.

28.8 In each of the following cases there is given a figure F, a transformation G, and a function $f(u, v)$ defined on the image $\tilde{F}$ of F under G. Find $\tilde{F}$ explicitly and verify Th. 28.1 in each case by evaluating both sides of Eq. (28.3). (Use polar coordinates wherever convenient.)

a. $F: |x| \le 1, |y| \le 1$; $G: u = x - y, v = x + y$;
$$f(u, v) = u^2 + v^2$$

b. $F: x^2 + y^2 \le 1$; $G: u = x - y, v = x + y$; $f(u, v) = u^2 + v^2$

c. $F: 0 \le x \le 1, 0 \le y \le 1$; $G: u = x + y^2, v = 3y$;
$$f(u, v) = (9u - v^2)^{1/2}$$

d. $F: 1 \le x^2 + y^2 \le 4, x \ge 0, y \ge 0$; $G: u = x^2 - y^2, v = 2xy$;
$$f(u, v) = 1/(u^2 + v^2)$$

28.9 Let $\tilde{F}$ be the figure $0 \le u + v \le 2$, $0 \le v - u \le 2$. Evaluate

$$\iint_{\tilde{F}} (v^2 - u^2) \exp\left(\tfrac{1}{2}(u^2 + v^2)\right) du\, dv$$

by applying Th. 28.1 to the transformation G of Ex. 28.8a, and transforming to an integral over the corresponding figure F in the x, y plane.

***28.10** Let F be the triangle $x \ge 0$, $y \ge 0$, $x + y \le 2$. Evaluate

$$\iint_F \exp\left(\frac{y - x}{y + x}\right) dx\, dy$$

by making a suitable transformation. (Note that the integrand is not defined at the origin, and in fact it cannot be extended in a continuous manner to the origin. However, if we assign an arbitrary value to the function at the origin, the integral exists and is independent of the particular value chosen.)

28.11   Let $\tilde{F}$ be the parallelogram with vertices $(0, 0)$, $(2, 1)$, $(1, 3)$, $(3, 4)$. Let G be a linear transformation that maps the square $F: 0 \le x \le 1, 0 \le y \le 1$ onto $\tilde{F}$.

   a. Find G explicitly. (*Hint:* write down a general linear transformation, and find the coefficients by setting $G(1, 0) = (2, 1)$, $G(0, 1) = (1, 3)$. Then automatically $G(1, 1) = (3, 4)$.)

   b. Evaluate $\iint_{\tilde{F}} (u + v)\, du\, dv$ by expressing this integral as an integral over F.

28.12   Given $R > 0$, let F be the figure $4x^2 + 9y^2 \le R^2$. Use a suitable transformation to evaluate

$$\iint_F e^{-4x^2 - 9y^2}\, dx\, dy.$$

28.13   Let $Q(x, y) = Ax^2 + 2Bxy + Cy^2$ be an arbitrary positive definite quadratic form.

   a. Show that if $u = ax + by$, $v = cx + dy$ is a linear transformation such that $u^2 + v^2 = Q(x, y)$, then the determinant $\Delta = ad - bc$ satisfies $\Delta^2 = AC - B^2$. (*Hint:* express A, B, C explicitly in terms of a, b, c, d and compute $AC - B^2$.)

   b. Show that, given the quadratic form $Q(x, y)$, it is always possible to find explicitly a nonsingular linear transformation of the type specified in part a. (*Hint:* show that the equations expressing A, B, and C in terms of a, b, c, d, can be solved for a, b, c, d in terms of A, B, C. Note that there is quite a bit of freedom in solving; for example, one may choose $c = 0$, and solve for a, b, d.)

   c. Let F be the figure $Q(x, y) \le R^2$. Evaluate $\iint_F Q(x, y)\, dx\, dy$.

   d. Let F be as in part c. Evaluate

$$\iint_F e^{-Q(x,y)}\, dx\, dy.$$

28.14   Suppose, under the hypotheses of Th. 28.1, that the function $f(u, v)$ can be written in the form $f = h_u$, for some function $h(u, v) \in \mathscr{C}^1$. Show that Th. 28.1 then follows by the same reasoning that was used to prove Th. 26.2; namely, express $\iint_{\tilde{F}} h_u(u, v)\, du\, dv$ as a line integral over $\partial \tilde{F}$, transform this into a line integral over ∂F, and apply Green's theorem to obtain an integral over F. (See Ex. 26.22. Note that this reasoning can only work for $f(u, v)$ continuous, and even then the existence of a function $h(u, v)$ satisfying $h_u = f$ is not clear, as we have seen in Ex. 22.18.)

Remark   It is possible to define *improper double integrals* in a manner analogous to improper integrals for functions of a single variable. Suppose, for example, that the domain D is the entire x, y plane. A sequence of figures F_k is said to *exhaust* D if each figure lies in the succeeding one, and if an arbitrarily large disk $x^2 + y^2 \le R^2$ lies in some F_k for sufficiently large k. (An example would be if each

F_k were chosen to be the disk $x^2 + y^2 \leq k^2$.) The integral of $f(x, y)$ over D is said to *converge* if

1. $\iint_F f \, dA$ exists for every figure F in D.
2. $\lim_{k \to \infty} \iint_{F_k} f \, dA$ exists whenever F_k exhausts D.
3. The limit in condition 2 is independent of the sequence F_k.

If these three conditions hold, then the (improper) integral of f over D is defined by

$$\iint_D f \, dA = \lim_{k \to \infty} \iint_{F_k} f \, dA.$$

Exercises 28.15–28.18 deal with improper double integrals.

*28.15   Suppose that D is the whole plane, $f(x, y) \geq 0$ in D, and $\iint_F f \, dA$ exists for every figure F in D. Show that if for some sequence of figures F_k that exhaust D, $\lim_{k \to \infty} \iint_{F_k} f \, dA$ exists, then for any other sequence of figures $\tilde{F}_k$ that exhaust D, $\lim_{k \to \infty} \iint_{\tilde{F}_k} f \, dA$ exists also, and the two limits are equal. (*Hint:* since $F_k \subset F_{k+1}$ for each k, and $f \geq 0$, it follows that $a_{k+1} \geq a_k$, where $a_k = \iint_{F_k} f \, dA$. Similarly, if $b_k = \iint_{\tilde{F}_k} f \, dA$, then $b_{k+1} \geq b_k$. Furthermore, each F_k is bounded, hence lies in some disk $x^2 + y^2 \leq R^2$, and since the sequence $\tilde{F}_k$ exhausts D, this disk lies in $\tilde{F}_m$ for some m. Hence for each k there exists an m such that $a_k \leq b_m$. Similarly, for each m there exists an n such that $b_m \leq a_n$. Deduce that if the (increasing) sequence a_k tends to a limit l, then the (increasing) sequence b_k tends to the same limit.)

Note: if $f(x, y)$ is continuous in D, then condition 1 in the definition of the improper integral certainly holds. If, in addition, $f(x, y) \geq 0$, then by Ex. 28.15, condition 3 holds, and it is sufficient to verify that the limit in 2 holds for some sequence F_k. This limit is then the value of the improper integral. Use this fact in Exs. 28.16 and 28.17.

28.16   Let D be the entire x, y plane. Evaluate the following expressions.

a. $\displaystyle\iint_D \exp\left(-(x^2 + y^2)\right) dx \, dy$

(*Hint:* use circular disks for the F_k.)

b. $\displaystyle\iint_D \frac{1}{(1 + x^2 + y^2)^2} \, dx \, dy$

(*Hint:* same method as part a.)

c. $\displaystyle\iint_D \exp\left(-4x^2 - 9y^2\right) dx \, dy$

(*Hint:* see Ex. 28.12, and use a suitable sequence of ellipses for the F_k.)

d. $\displaystyle\iint_D \exp\left(-(Ax^2 + 2Bxy + Cy^2)\right) dx \, dy$

where $A > 0$, $AC - B^2 > 0$. (*Hint:* see Ex. 28.13, and use a suitable sequence of ellipses for the F_k.)

e. $\displaystyle\iint_D \frac{1}{(1 + Ax^2 + 2Bxy + Cy^2)^2}\, dx\, dy$

where $A > 0$, $AC - B^2 > 0$. (*Hint:* use the same method as part d.)

f. $\displaystyle\iint_D \frac{1}{1 + x^2 + y^2 + x^2 y^2}\, dx\, dy$

(*Hint:* factor the denominator and use a sequence of squares for the F_k.)

28.17 Show that if $f(x, y)$ is continuous in the whole plane D, and if $0 \le f(r \cos\theta, r \sin\theta) \le 1/r^3$, then $\iint_D f\, dA$ converges.

28.18 Show that if D is the whole plane, then

$$\iint_D \frac{1}{1 + x^2 + y^2}\, dx\, dy$$

does not converge.

28.19 The *gamma function* is defined as an improper integral:

$$\Gamma(x) = \int_0^\infty e^{-t} t^{x-1}\, dt.$$

Specifically, if for any positive numbers a, b, we set $L(a, b) = \int_a^b e^{-t} t^{x-1}\, dt$, then $\Gamma(x) = \lim_{a \to 0,\, b \to \infty} L(a, b)$. This limit exists for all $x > 0$. The function of x so defined is Euler's answer to the problem of finding a continuous function of x with the property that $\Gamma(n + 1) = n!$ for each positive integer n.

    a. Show that $\Gamma(x + 1) = x\Gamma(x)$ for each $x > 0$. (*Hint:* integrate $\int_a^b e^{-t} t^{x-1}\, dt$ by parts, differentiating e^{-t} and integrating t^{x-1} (x being fixed); take the limit as $a \to 0$ and $b \to \infty$.)

    b. Show that $\Gamma(1) = 1$.

    c. Deduce from parts a and b that $\Gamma(n + 1) = n!$ for positive integers n.

    d. Show that $\Gamma(\tfrac{1}{2}) = \sqrt{\pi}$. (*Hint:* make the substitution $t = u^2$ in the definition of $\Gamma(\tfrac{1}{2})$, and use the value found for the integral (28.12).)

References

[1] Agnew, R. P. *Differential Equations*, 2nd ed. (New York: McGraw-Hill Book Company, Inc., 1960).

[2] Ahlfors, Lars V. *Complex Analysis*, 2nd ed. (New York: McGraw-Hill Book Company, Inc., 1966).

[3] Apostol, Tom M. *Calculus*, 2 vols. (New York: Blaisdell Publishing Co., 1961).

[4] Bark, L. S., Ganson, P. P., and Meister, N. A. *Tables of the Velocity of Sound in Sea Water* (New York: MacMillan Company, 1964).

[5] Beaumont, Ross A. *Linear Algebra* (New York: Harcourt, Brace and World, 1965).

[6] Buck, R. C. *Advanced Calculus*, 2nd ed. (New York: McGraw-Hill Book Company, Inc., 1965).

[7] Chinn, W. G., and Steenrod, N. E. *First Concepts of Topology* [No. 18 in the New Mathematical Library] (New York: Random House and the L. W. Singer Company, 1966).

[8] Chrestenson, H. E. *Mappings of the Plane* (San Francisco: W. H. Freeman and Company, 1966).

[9] Christie, D. E. *Vector Mechanics*, 2nd ed. (New York: McGraw-Hill Book Company, Inc., 1964).

[10] Copeland, Arthur H., Sr. *Geometry, Algebra, and Trigonometry by Vector Methods* (New York: MacMillan Company, 1962).

[11] Courant, R. *Differential and Integral Calculus*, Vol. II (New York: Nordeman Publishing Company, 1936).

[12] Czech, J. *The Cathode Ray Oscilloscope* (New York: Interscience Publishers, Inc., 1957).

[13] Davis, Harry F. *Introduction to Vector Analysis*, 2nd ed. (Boston: Allyn and Bacon, 1967).

[14] de Leeuw, Karel. *Calculus* (New York: Harcourt, Brace and World, 1966).

431

[15] Einstein, Albert, *et al. The Principle of Relativity* [A collection of original memoirs on the theory of relativity, with notes by A. Sommerfeld] (New York: Dover Publications, Inc., 1923).

[16] Goffman, Casper. *Calculus of Several Variables* (New York: Harper and Row, 1965).

[17] Guggenheimer, H. W. *Differential Geometry* (New York: McGraw-Hill Book Company, Inc., 1963).

[18] *Handbook of Geophysics*, rev. ed. (New York: MacMillan Company, 1961).

[19] James, Robert C. *Advanced Calculus* (Belmont, Calif.: Wadsworth Publishing Co., 1966).

[20] Kacser, Claude. *Introduction to the Special Theory of Relativity* (Englewood Cliffs, N.J.: Prentice-Hall, 1967).

[21] Krasnoselskij, M. A., *et al. Plane Vector Fields* (New York: Academic Press, 1966).

[22] Lagemann, Robert T. "The Scientist as Artist," *American Scientist* **36**, 415 (1948).

[23] Lindgren, B. W. *Vector Calculus* (New York: MacMillan Company, 1963, 1964).

[24] Lockwood, E. H. *A Book of Curves* (Cambridge, England: Cambridge University Press, 1963).

[25] Maak, Wilhelm. *An Introduction to Modern Calculus* (New York: Holt, Rinehart and Winston, 1963).

[26] MacKenzie, K. V. "Formulas for the Computation of Sound Speed in Sea Water," *Journal of the Acoustical Society of America* **32**, 100–104 (1960).

[27] Morrey, Charles B. *University Calculus with Analytic Geometry* (Reading, Mass.: Addison-Wesley, 1962).

[28] Morrey, Charles B., and Protter, Murray H. *Modern Mathematical Analysis* (Reading, Mass.: Addison-Wesley, 1964).

[29] Petrovskii, I. G. *Partial Differential Equations* (Philadelphia: W. B. Saunders Co., 1967).

[30] Polya, George. *Mathematics and Plausible Reasoning, Vol. I: Induction and Analogy in Mathematics* (Princeton, N.J.: Princeton University Press, 1954).

[31] Protter, Murray H., and Weinberger, Hans F. *Maximum Principles in Differential Equations* (Englewood Cliffs, N.J.: Prentice-Hall, 1967).

[32] Schwartz, J. "The Formula for Change in Variables in a Multiple Integral," *American Mathematical Monthly* **61**, 81–85 (1954).

[33] Skilling, H. *Fundamentals of Electric Waves*, 2nd ed. (New York: John Wiley & Sons, 1964).

[34] Stiefel, Eduard L. *An Introduction to Numerical Mathematics* [translated by Werner C. Rheinboldt and Cornelie J. Rheinboldt] (New York: Academic Press, 1963).

[35] Walker, Robert J. *Algebraic Curves* (Princeton, N.J.: Princeton University Press, 1950).

[36] Widder, D. V. *Advanced Calculus*, 2nd ed. (Englewood Cliffs, N.J.: Prentice-Hall, 1961).

Answers to
Selected Exercises

Chapter one

Section 1

1.1   *a.* $\langle 2, 1 \rangle$ *c.* $\langle 1, -1 \rangle$ *e.* $\langle 2, 1 \rangle$

1.2   *a.* $|\mathbf{v}| = 3, \ \alpha = 0$ *c.* $|\mathbf{v}| = 2, \ \alpha = \frac{2}{3}\pi$ *e.* $|\mathbf{v}| = \sqrt{2}, \ \alpha = \frac{3}{2}\pi$

1.3   *a.* $\langle -5/\sqrt{2}, \ 5/\sqrt{2} \rangle$ *c.* $\langle -\sqrt{3}, \ -1 \rangle$ *e.* $\langle 3, 4 \rangle$

1.4   *a.* $\langle 0, 0 \rangle$ *c.* 0

1.5   *a.* $\sqrt{13}$ *c.* -10

1.6   *a.* $\frac{1}{2}\pi$ *c.* $\frac{2}{3}\pi$

1.7   *d.* $|ac + bd| \leq \sqrt{a^2 + b^2} \ \sqrt{c^2 + d^2}$
$\Leftrightarrow a^2c^2 + 2acbd + b^2d^2 \leq a^2c^2 + a^2d^2 + b^2c^2 + b^2d^2$
$\Leftrightarrow 2acbd \leq a^2d^2 + b^2c^2 \Leftrightarrow a^2d^2 - 2acbd + b^2c^2 \geq 0$
$\Leftrightarrow (ad - bc)^2 \geq 0$

1.12   *a.* $|\mathbf{v}| = |\mathbf{w}|, \quad |\mathbf{v}| = |\mathbf{u}|, \quad$ or $\quad |\mathbf{u}| = |\mathbf{w}|$
       c. $\mathbf{u} \cdot \mathbf{v} = 0, \quad \mathbf{u} \cdot \mathbf{w} = 0, \quad$ or $\quad \mathbf{v} \cdot \mathbf{w} = 0$

1.13   *b.* $\mathbf{w} + \frac{1}{2}\mathbf{u} = \frac{1}{2}(\mathbf{v} + \mathbf{w})$

1.14   *a.* $\mathbf{v} - \mathbf{u}, \quad \mathbf{v} - \mathbf{w}$ *c.* $\mathbf{u} + \frac{1}{2}(\mathbf{v} - \mathbf{u}) - \frac{1}{2}\mathbf{u} = \frac{1}{2}\mathbf{v}$ *e.* $\frac{1}{2}\mathbf{v}$

1.15   *a.* $\mathbf{u} = \mathbf{v} - \mathbf{w} \Leftrightarrow \mathbf{v} = \mathbf{u} + \mathbf{w} \Leftrightarrow \mathbf{w} = \mathbf{v} - \mathbf{u}$
       c. $\frac{1}{2}\mathbf{v} = \frac{1}{2}(\mathbf{u} + \mathbf{w}) \Leftrightarrow \mathbf{u} = \mathbf{v} - \mathbf{w}$

1.16 $ax + by = c, \ ax_0 + by_0 = c \Rightarrow a(x - x_0) + b(y - y_0) = 0.$
       $\langle a, b \rangle \perp \langle x - x_0, \ y - y_0 \rangle.$

1.19   Speed is 50 miles per hour; angle with direction of train is $\cos^{-1}\frac{4}{5} \sim 37°$.

1.21   130 miles per hour

1.23   The wind is from the Northwest.

1.24   The star always appears to be displaced in the direction of the earth's motion.

1.25   Magnitude $2\sqrt{2}$, in direction of positive y axis.

1.27   *a.*   Equal magnitude and opposite direction.

1.28   *a.*   Equilibrium point is on line joining the two bodies, at distance $d/(1 + \sqrt{\lambda})$ from body of mass m_1.

1.30   *a.*   All lines perpendicular to $\mathbf{v}$.

Section 2

2.1   **a.** $x = 3t + 1, \quad y = 2t + 1, \quad 0 \le t \le 1$

        c. $x = -3t + 2, \quad y = 3, \quad 0 \le t \le 1$

        e. $x = -t, \quad y = t, \quad 0 \le t \le \sqrt{2}$

2.2   **a.** $y = 2x - 5, \quad 2 \le x \le 3$ **c.** $y = x^2, \quad 0 \le x \le 1$

        e. $y = \sqrt{1 - x^2}, \quad 0 \le x \le 1$

2.3 $y = 2x^2 - 1, \quad -1 \le x \le 1$

2.5 $x^2/a^2 - y^2/b^2 = 1$

2.7 $r \ne 0$

2.9   **a.** $\langle 1, 2 \rangle$ **c.** $\langle 2t, 4t^3 \rangle$ **e.** $[2/(1 + t^2)^2] \langle -2t, 1 - t^2 \rangle$

2.10   **a.** $\langle 1, 2 \rangle/\sqrt{5}$ **c.** $\langle 1, 2t^2 \rangle/(1 + 4t^4)^{1/2}$ **e.** $\langle -2t, 1 - t^2 \rangle/(1 + t^2)$

2.11   **a.**   20   **c.** $61/27$ **e.**   8

2.12   **a.** $x^3 - 3xy + y^3 = 0$

2.15   **a.**   Yes   **b.**   all $t \ne 0$ **c.** $y = |x|$

2.16   **a.** $\langle 0, 0 \rangle; \quad -r \langle \cos t, \sin t \rangle = -\langle x - c, y - d \rangle; \quad \langle 0, -g \rangle$

Section 3

3.7   **c.**   Ex. 3.1b, c;   Ex. 3.4a, b;   Ex. 3.5a

3.8   **a.** $f(x, y) = F(\sqrt{x^2 + y^2})$ **c.**   circles $x^2 + y^2 = R^2$

3.9   **a.**   Moved upwards 2 units

        c.   Moved 2 units in the positive y direction

        e.   Reflected in the plane $x = 0$

        g.   Contracted by a factor of 2 in the x direction

        i.   Reflected in the origin

3.10   **a.**   They are the same curves, but correspond to different values of the function.

Chapter two

Section 4

4.1   **a.** $f_x = 4x^3 + 6xy - 5y^3, \quad f_y = 3x^2 - 15xy^2$

        c. $f_x = 3x^2 + 6xy + 3y^2, \quad f_y = 3x^2 + 6xy + 3y^2$

        e. $f_x = \cos(x + y), \quad f_y = \cos(x + y)$

        g. $f_x = x/\sqrt{x^2 + y^2}, \quad f_y = y/\sqrt{x^2 + y^2}$

        i. $f_x = -y/(x^2 + y^2), \quad f_y = x/(x^2 + y^2)$

        k. $f_x = y(y^2 - x^2)/(x^2 + y^2)^2, \quad f_y = x(x^2 - y^2)/(x^2 + y^2)^2$

        m. $f_x = 3 \sec^4 x \tan x \log x + \dfrac{1}{x} \sec^2 x, \quad f_y \equiv 0$

4.2   **a.**   1   **c.** $2e$

4.3   **a.** $3(x^3 + y^3)$ **c.** $3\sqrt{x^6 + y^6}$

4.4   **a.** $f_x = e^x \cos y = g_y, \quad f_y = -e^x \sin y = -g_x$

4.9   **a.** $4(x^2 + y^2)$ **b.** e^{2x} **c.**   0   **d.**   0

4.12   **a.** $f_x(0, 0) = 1, \quad f_y(0, 0) = 0$ **c.** $f_x(0, 0) = 1, \quad f_y(0, 0) = 0$

4.13 *a.* $x^2 + y^2 \leq 1$, $y \leq x$ *c.* $x > 0$, $y > 0$, $x + y < 1$

4.14 *a.* The ellipse $x^2 + 4y^2 = 4$

     c. The ellipses $x^2 + 4y^2 = 4$ and $x^2 + 4y^2 = 16$

     e. The horizontal line segments $y = \pm 2$, $|x| \leq 1$ and the vertical line segments $x = \pm 1$, $|y| \leq 2$

4.15 *a.* Ex. 4.14 a, b

     b. For example, in Ex. 4.14c $(2, 0)$ and in Ex. 4.14e $(1, 0)$

4.17 Ex. 4.14 c, d, f

4.18 *a.* Ex. 4.14 a, d, e

Section 5

5.1 *a.* $z = 3x - 4y + 5$ *c.* $z = 2$ *e.* $25z = 6x + 12y - 20$

5.3 *a.* $z = z_0 - (z_0/x_0)^{1/2}(x - x_0) - (z_0/y_0)^{1/2}(y - y_0)$

5.8 *a.* $f_x(x_0, y_0) = 2x_0 \sin 1/x_0 - \cos 1/x_0$, $f_y(x_0, y_0) = 2y_0$

     b. $f_x(0, y_0) = 0$, $f_y(0, y_0) = 2y_0$

5.10 *a.* $f_x = 2y(y^2 - x^2)/(x^2 + y^2)^2$, $f_y = 2x(x^2 - y^2)/(x^2 + y^2)^2$

Section 6

6.1 *a.* $\langle 3, -7 \rangle$ *c.* $\langle 3, -7 \rangle$

6.2 *a.* 1   *c.* -1

6.3 *a.* $\langle 2, 0 \rangle$; $2, \sqrt{2}, 0, -\sqrt{2}, -2, -\sqrt{2}, 0, \sqrt{2}$

     c. $\langle 1, 1 \rangle$; $1, \sqrt{2}, 1, 0, -1, -\sqrt{2}, -1, 0$

6.4 *a.* 1; $\alpha = \frac{1}{2}\pi$ *c.* 1; $\mathbf{T}_\alpha = \langle \frac{3}{5}, \frac{4}{5} \rangle$

6.5 $\nabla f(x_0, y_0) = 2F'(x_0^2 + y_0^2) \langle x_0, y_0 \rangle$

6.9 $\langle -\sqrt{5}, 2\sqrt{5} \rangle$

6.12 *a.* $(\frac{1}{2}, 1)$; maximum   *c.* $(0, 0)$; minimum

6.14 *a.* $(-2, 1)$ *c.* All points on the line $2x - 3y + 1 = 0$

6.17 $a = \frac{3}{2}$, $b = \frac{10}{3}$; $r_1 = -\frac{1}{6}$, $r_2 = \frac{1}{3}$, $r_3 = -\frac{1}{6}$

Section 7

7.1 *a.* $4 \cos 2t$ *c.* 4   *e.* $50t^3$

7.2 *a.* 0   *c.* 0   *e.* 0

7.3 $x^x(1 + \log x)$

7.5 $\frac{2}{15}$; increasing

7.8 *a.* Degree 7   *c.* Not homogeneous   *e.* Degree 3   *g.* Degree $-\frac{1}{2}$

7.10 *a.* Homogeneous if $k = l$; then, of degree k

     c. Degree $k - l$ *e.* Degree $2k$

7.16 *a.* 21.3   *c.* 0.975

7.17 $21.30 < f(2.1, 3.2) < 21.43$

7.18 *a.* $(\operatorname{Sin} \pi x)/x$ *c.* $(3^x - 2^x)/x$ *e.* $(2e^{-x^2} - e^{-x})/x$

7.19 *a.* π

7.21 *a.* $-4x^3 \int_{\sqrt{x}}^{x^2} \cos(t^2 - x^4)\, dt - \dfrac{1}{2\sqrt{x}} \sin(x - x^4)$

7.25 *a.* $dr/dt = (x\, dx/dt + y\, dy/dt)/r$

Section 8

8.1   *a.* $-\frac{5}{4}$ *c.* $-\frac{7}{4}$ *e.* -1

8.2   *a.* $\dfrac{\sin y + y \sin x}{\cos x - x \cos y}$ *c.* $\dfrac{2(x - y)}{3x^2 - 2y}$

8.4   *a.* $\dfrac{dy}{dx} = -\dfrac{b^2 x}{a^2 y}$; $y = \pm\dfrac{b}{a}\sqrt{a^2 - x^2}$; $\dfrac{dy}{dx} = \mp\dfrac{b}{a}\dfrac{x}{\sqrt{a^2 - x^2}}$

       c. $\dfrac{dy}{dx} = -\dfrac{e^{x+y} + e^x}{e^{x+y} + e^y} = -\dfrac{1 - e^y}{1 - e^x}$; $y = \log\dfrac{1 - e^x}{1 + e^x}$, $\dfrac{dy}{dx} = -\dfrac{2e^x}{1 - e^{2x}}$

8.5   *a.* $-b^4/a^2 y^3$ *c.* $(e^y - 1)(e^y + e^x)/(1 - e^x)^2$

8.8   In Ex. 6.5, level curves are circles about the origin.

8.9   Level curves are: *a.* $x^2 + (y - 1)^2 = 1$ *c.* $x^2 + y^2 = 2$

8.11   *a.* $\nabla f = 2\langle x, -y\rangle$, $\nabla g = \langle y, x\rangle$; $\nabla f \perp \nabla g$

8.12   *a.* $f_x g_x + f_y g_y = 0$

8.14   *a.* $f_y = 0$; cannot solve in the form $y = g(x)$

       c. $f_y = 0$; can solve as $y = g(x)$, but $g(x)$ not differentiable at $x = 0$

8.16   *a.*   Cusp at the origin

       c.   Curve consists of the origin and the circle of radius 1 about the origin

       e.   "Level curve" has no points on it

8.18   *a.*   If $f(t, \varphi) = \varphi - kt - \epsilon \sin\varphi$, then $f_\varphi = 1 - \epsilon\cos\varphi > 0$ since $\epsilon < 1$

       c. $(x, y) = (a, 0) \Leftrightarrow \cos\varphi = 1$; $(x, y) = (-a, 0) \Leftrightarrow \cos\varphi = -1$

       d. $\mathbf{v}(t) = \dfrac{d\varphi}{dt}\langle -a\sin\varphi, \ b\cos\varphi\rangle$

Section 9

9.1   *a.* $\sqrt{5}, -\sqrt{5}$ *c.*   4   *e.*   1

9.2 $(0, -2)$

9.4   2 radians

9.6   *a.* $(3, 4)$ *c.*   Point on the line $3x + 4y = 25$ nearest the origin

9.8   *a.*   Pairs of straight lines parallel to the given one

9.10   *a.*   Pairs of lines, $x + y = \pm c$ *c.*   4; when $x + y = 0$ *e.*   No

9.11   *a.* $f(x, y) = \pi$: the segment L; $f(x, y) = 0$: the y axis except for L; $f(x, y) = c$, $0 < c < \pi$: a pair of circular arcs through the endpoints of L

       b. $(\sqrt{3}, 0)$

9.12   *a.*   1   *c.* $(0, 0)$

Section 10

10.1   *a.*   3, 1   *c.*   3, -1 *e.* -1, -5 *g.* $\sqrt{2}, -\sqrt{2}$

       i. $(\sqrt{13} - 1)/2, (-\sqrt{13} - 1)/2$ *k.* $(5 + 3\sqrt{2})/2, (5 - 3\sqrt{2})/2$

10.3   *a.* $\max(1/\sqrt{2}, 1/\sqrt{2}), (-1/\sqrt{2}, -1/\sqrt{2})$;

          $\min(1/\sqrt{2}, -1/\sqrt{2}), (-1/\sqrt{2}, 1/\sqrt{2})$

       c. $\max(1/\sqrt{2}, -1/\sqrt{2}), (-1/\sqrt{2}, 1/\sqrt{2})$;

          $\min(1/\sqrt{2}, 1/\sqrt{2}), (-1/\sqrt{2}, -1/\sqrt{2})$

10.5 **a.** III **c.** I **e.** V

10.6 $\lambda_1 = \frac{1}{2}[(a + c) + \sqrt{(a - c)^2 + 4b^2}]$, $\lambda_2 = \frac{1}{2}[(a + c) - \sqrt{(a - c)^2 + 4b^2}]$

10.8 **a.** $a^2 - (\lambda_1 + \lambda_2)a + \lambda_1\lambda_2 + b^2 = 0$

10.10 **a.** $\lambda_2 = 0$ and $\lambda_1 + \lambda_2 = a + c \Rightarrow \lambda_1 = a + c$

10.11 **a.** $\lambda_2 = a + c$ **c.** $q(x, y) = -(\alpha x \pm \beta y)^2$, $\alpha = \sqrt{-a}$, $\beta = \sqrt{-b}$

10.14 $A = a \cos^2 \theta + 2b \cos \theta \sin \theta + c \sin^2 \theta$
$B = (c - a) \cos \theta \sin \theta + b(\cos^2 \theta - \sin^2 \theta)$
$C = a \sin^2 \theta - 2b \sin \theta \cos \theta + c \cos^2 \theta$

10.18 **a.** $3X^2 - Y^2$ **c.** $\frac{3}{2}X^2 + \frac{1}{2}Y^2$ **e.** $\frac{1}{2}X^2 - \frac{1}{2}Y^2$

10.19 **a.** Ellipse; $\frac{1}{3}\pi$ **c.** Hyperbola **e.** Parallel straight lines

10.20 **a.** Ellipse if $k > 0$; point if $k = 0$; no locus if $k < 0$
c. Pair of parallel straight lines if $k > 0$; single line if $k = 0$; no locus if $k < 0$
e. Hyperbola if $k > 0$ or $k < 0$; pair of intersecting lines if $k = 0$

10.22 $1, \frac{1}{3}$

10.23 **a.** They are reciprocals.

10.27 $2 - \sqrt{2}$

10.29 **a.** $A = \frac{1}{2}(a - c)$, $B = b$, $C = \frac{1}{2}(a + c)$, $D = \sqrt{A^2 + B^2}$

Section 11

11.1 **a.** $f_{xx} = 30x^4 - 60x^2y^2$, $f_{xy} = -40x^3y + 15y^4$, $f_{yy} = -10x^4 + 60xy^3$

c. $f_{xx} = -\dfrac{y^2}{x^4} \sin \dfrac{y}{x} + \dfrac{2y}{x^3} \cos \dfrac{y}{x}$, $f_{xy} = \dfrac{y}{x^3} \sin \dfrac{y}{x} - \dfrac{1}{x^2} \cos \dfrac{y}{x}$,

$f_{yy} = -\dfrac{1}{x^2} \sin \dfrac{y}{x}$

e. $f_{xx} = -f_{xy} = f_{yy} = 20(x - y)^3$

11.2 **a.** $f_{xx} = -f_{yy} = 0$ **c.** $f_{xx} = -f_{yy} = 2(x^3 - 3xy^2)/(x^2 + y^2)^3$

11.8 **a.** $f_{xx} = 2g'(x^2 + y^2) + 4x^2g''(x^2 + y^2)$, $f_{xy} = 4xyg''(x^2 + y^2)$,
$f_{yy} = 2g'(x^2 + y^2) + 4y^2g''(x^2 + y^2)$

11.11 **a.** $f_{xxx} = e^x \sin y$, $f_{xxy} = e^x \cos y$, $f_{xyy} = -e^x \sin y$, $f_{yyy} = -e^x \cos y$
c. $-e^x \sin y$

11.12 **a.** $f_{xxxx} = y^4e^{xy}$, $f_{xxyy} = (2 + 4xy + x^2y^2)e^{xy}$, $f_{xyyy} = (3x^2 + x^3y)e^{xy}$
c. $(ky^{k-1} + xy^k)e^{xy}$
e. 0, if $k \neq \ell$; $k!$ if $k = \ell$

11.13 **a.** $2(3 \cos^2 \alpha - 8 \cos \alpha \sin \alpha + 2 \sin^2 \alpha)$ **c.** $\alpha = \frac{1}{4}\pi$

11.15 **a.** $11/2$

11.16 $12, 2$

11.17 $x''f_x + y''f_y + x'^2f_{xx} + 2x'y'f_{xy} + y'^2f_{yy}$

11.19 **a.** $r_{xx} = y^2/r^3$, $r_{xy} = -xy/r^3$, $r_{yy} = x^2/y^3$
c. Positive semidefinite

11.20 **c.** Upper hemisphere; all curves are concave downward

11.21 **b.** $R = [f_x^2 + f_y^2]^{3/2} / |f_y^2f_{xx} - 2f_xf_yf_{xy} + f_x^2f_{yy}|$

11.22 $\nabla_\alpha^3 f = f_{xxx} \cos^3 \alpha + 3f_{xxy} \cos^2 \alpha \sin \alpha + 3f_{xyy} \cos \alpha \sin^2 \alpha + f_{yyy} \sin^3 \alpha$

11.26 **a.** $i(i - 1) \cdots (i - k + 1)cx^{i-k}y^j$, if $k \leq i$; 0 if $k > i$
c. $k!\ell!$ if $k = i$, $\ell = j$; 0 otherwise

Section 12

12.1 **a.** $(0, -1)$: saddle point; $(0, 3)$: local minimum
   c. $(0, 3)$, $(-4, -3)$: saddle points; $(0, -3)$: local minimum;
   $(-4, 3)$: local maximum

12.2 **a.** See Ex. 6.15
   c. If $AC - B^2 > 0$, a local maximum if $A < 0$ and a local minimum
   if $A < 0$; if $AC - B^2 < 0$, a saddle point
   e. A straight line

12.3 **a.** $(0, 0)$: local minimum   **c.** $(0, 0)$: neither
   e. All points on the lines $y = x$ and $y = -x$: local minimum

12.6 **a, c.** Local minimum   **e.** Neither

12.7 **a.** $1/(x + 2y) = \frac{1}{3} - \frac{1}{9}(x - 1) - \frac{2}{9}(y - 1) + \frac{1}{27}(x - 1)^2$
   $+ \frac{4}{27}(x - 1)(y - 1) + \frac{4}{27}(y - 1)^2 + R(x, y)$
   c. $\log\sqrt{x^2 + y^2} = (y - 1) + \frac{1}{2}[x^2 - (y - 1)^2] + R(x, y)$

12.8 **a.** $f(x, y) = -11 + 6(x - 1) + 3(y + 2) + 2(x - 1)^2 - (x - 1)(y + 2)$
   c. $f(x, y) = -6 + 3(x - 1) + 5y + 3(x - 1)^2 + 3(x - 1)y$
   $+ y^2 + (x - 1)^3 - 2(x - 1)^2 y + (x - 1)y^2 + 4y^3$
   e. $f(x, y) = 0 + R(x, y)$

12.9 **a.** $e^{x+y} = ee^{x+(y-1)}$
   $= e\{1 + x + (y - 1) + \frac{1}{2}[x + (y - 1)]^2 + \cdots$
   $+ \frac{1}{n!}[x + (y - 1)]^n\} + R(x, y)$

   c. $\sin(x + y) = x + y - \frac{1}{3!}(x + y)^3$

   $+ \frac{1}{5!}(x + y)^5 - \cdots + (-1)^m \frac{1}{(2m + 1)!}(x + y)^{2m+1} + R(x, y),$
   where $2m + 1$ is the largest odd integer which does not exceed n

   e. $e^{xy} = 1 + xy + \frac{1}{2}x^2y^2 + \frac{1}{3!}x^3y^3 + \cdots + \frac{1}{n!}x^n y^n + R(x, y)$

12.10 **a.** $z = 51 - 6x - 8y$; crosses

12.16 Cube

12.18 $10 \times 10 \times 20$ inches

12.20 The absolute maxima and absolute minima are respectively:
   a. $\sqrt{2}$; $-\sqrt{2}$ **c.** $(3 + \sqrt{5})/2$; 0 **e.** 9; 0   **g.** 4; -3
   i. 1; -1 **k.** 2; $4/e^3$

12.21 **b.** $f_{xx} \leq 0$, $f_{yy} \leq 0$, $f_{xx}f_{yy} - f_{xy}^2 \geq 0$

Chapter three

Section 13

13.1 **a.** Uniform stretching by a factor of 3

    c. Reflection in the origin (or rotation through 180°)

    e. Horizontal stretching by factor of 2 and vertical stretching by factor of 3, followed by reflection in the horizontal axis

    g. "Shearing" motion   *i.* Double folding over

    k. Distortion in vertical direction   *m.* Map of plane onto square

13.2   *a.* $R = 2r,\ \varphi = \theta$ *c.* $R = r,\ \varphi = -\theta$

    e. $R = r\sqrt{2},\ \varphi = \theta + \frac{1}{4}\pi$ *g.* $R = r,\ \varphi = \frac{1}{2}\pi - \theta$

13.4   *a.* $x = \frac{1}{2}u,\ y = \frac{1}{2}v$ *c.* $x = u - 1,\ y = u + v - 1$

    e. $x = \frac{1}{2}(2u + v)^{1/3},\ y = -\frac{1}{2}(2u - v)^{1/5}$

13.5   *a.* Bijective   *c.* Surjective

    e. Injective   *g.* None   *i.* None

13.6   *a.* The line $y = x$ *c.* The x and y axes

13.7   *a.* $x = 0$ *c.* $y = (\sqrt{2} - 1)x$ and $y = -(\sqrt{2} + 1)x$

13.8   *a.* $x^2 + y^2 = R^2/2$ *c.* $x^2/4R^2 + y^2/9R^2 = 1$

    e. $(x + 2)^2 + (y - 3)^2 = R^2$ *g.* $x^2 + y^2 = R$

13.9   *a.* The x and y axes

    c. A rectangular hyperbola whose asymptotes are the x and y axes, lying in the first and third quadrants if $d > 0$ and in the second and fourth quadrants if $d < 0$.

    e. They start with the lines $y = \pm x$ and move outward through a family of hyperbolas.

13.10 *a.* $R = e^x,\ \varphi = y$

    c. It moves counterclockwise around the circle $u^2 + v^2 = e^{2c}$

    e. The positive v axis   *g.* The positive u axis

    i. The image ray rotates about the origin in the counterclockwise direction

13.11 *a.* $(1, 1)$ *c.* The line $y = 1$, for example

13.13 *a.* S is upper half-plane: $v \geq 0$

    c. S is fourth quadrant: $u > 0,\ v < 0$; $F^{-1}: x = \log u,\ y = \log(-v)$

    e. S is square $-\pi/2 < u < \pi/2,\ -\pi/2 < v < \pi/2$; $F^{-1}: x = \tan u,\ y = \tan v$

    g. S is right half of parabola: $v = u^2,\ u \geq 0$

13.14 *a.* $x = u - u^2 - v^2 + 2uv,\ y = v - u$

Section 14

14.1   *a.* $\Delta = 1$ *c.* $\Delta = -a^2 \neq 0$ *e.* $\Delta = 0$ *g.* $\Delta = 1$ *i.* $\Delta = 0$

    k. $\Delta = 0$

14.2   *a.* $\Delta = -4;\ x = \frac{1}{2}v,\ y = \frac{1}{2}u;\ \Delta' = -\frac{1}{4}$

    c. $\Delta = -10;\ x = \frac{1}{10}(3u + v),\ y = \frac{1}{10}(u - 3v);\ \Delta' = -\frac{1}{10}$

14.3   *a.* $v = -u$ *c.* $v = -3u$

14.4   *a.* $F^{-1}(e, 2e)$ is the line $x + y = e$

14.6   *a.* $\varphi = -\theta$ *c.* $\varphi = \theta + \pi$ *e.* $\varphi = \theta - \frac{1}{4}\pi$

14.8   *a, c.* $x^2 + y^2 \leq R^2;\ \Delta = 1$ *e.* $x^2 + y^2 \leq R^2/2;\ \Delta = 2$

14.9   *c.* $(0, 0),\ (a, c),\ (b, d)$

14.10 *a.* 2

14.11 *b.* $A = a^2 + c^2,\ B = ab + cd,\ C = b^2 + d^2$ *c.* $AC - B^2 = (ad - bc)^2$

14.14 *a.* $(-1, 0)$

14.17 $\Delta = \pm 1$

14.18 *b.* $\Delta = ad - bc \neq 0$

14.22 *a.* $F(\langle x, y \rangle) = F(x\langle 1, 0 \rangle + y\langle 0, 1 \rangle) = xF(\langle 1, 0 \rangle) + yF(\langle 0, 1 \rangle)$

14.24 *b.* The equation in part *a* satisfied by any characteristic value is quadratic and can have at most two roots.

    d. If $ad - bc < 0$, the equation satisfied by λ has positive discriminant. If a ray in the x, y plane rotates in the counterclockwise direction, and the image ray in the clockwise direction, then at some point they will have the same direction.

14.25 *a.* C is the image of the unit circle under a nonsingular linear transformation.

Section 15

15.1 *a.* $\begin{pmatrix} 1 & 0 \\ 0 & 1 \end{pmatrix}$ *c.* $\begin{pmatrix} a & b \\ c & d \end{pmatrix}$ *e.* $\begin{pmatrix} c & d \\ a & b \end{pmatrix}$ *g.* $\begin{pmatrix} 0 & 2 \\ -2 & 0 \end{pmatrix}$

15.4 The product of the determinants can be zero only if one of the determinants is zero.

15.7 *a.* $u = (x - y)/\sqrt{2}$, $v = (x + y)/\sqrt{2}$ *c.* $u = -y$, $v = -x$

15.8 *a.* $y = x$

15.9 The line $y = (\tan \frac{1}{8}\pi)x$

15.10 *a.* $-\frac{1}{2}\pi$ *c.* Reflection in the line $y = -x$

15.15 *a.* A dilation of $\sqrt{2}$ and rotation through $-\frac{1}{4}\pi$

    c. A dilation of 5 and rotation through $\alpha = \arctan(-\frac{4}{3})$, $0 < \alpha < \pi$

15.17 *a.* Horizontal stretching by a factor of λ_1

    c. Projection of entire plane onto the y axis along lines parallel to the x axis, followed by reflection in the x axis.

15.26 Use the fact that the determinant of a product of matrices is the product of the determinants.

Section 16

16.1 *a.* $\begin{pmatrix} 1 & 0 \\ 0 & 1 \end{pmatrix}$ *c.* $\begin{pmatrix} 3x^2 - 3y^2 & -6xy \\ 6xy & 3x^2 - 3y^2 \end{pmatrix}$ *e.* $\begin{pmatrix} e^{x+y} & e^{x+y} \\ e^{x-y} & -e^{x-y} \end{pmatrix}$

    g. $\begin{pmatrix} ye^{xy} & xe^{xy} \\ 2xe^{x^2} & 0 \end{pmatrix}$ *i.* $\begin{pmatrix} \cos x \sin y & \sin x \cos y \\ -\sin x \cos y & -\cos x \sin y \end{pmatrix}$

    k. $\begin{pmatrix} ye^{-x^2y^2} & xe^{-x^2y^2} \\ y \sin x^2y^2 & x \sin {}^2xy^2 \end{pmatrix}$

16.2 *a.* 1; nowhere *c.* $9(x^2 + y^2)^2$; $(x, y) = (0, 0)$

    e. $-2e^{2x}$; nowhere *g.* $-2x^2e^{x^2+xy}$; $x = 0$

    i. $\sin^2 x \cos^2 y - \cos^2 x \sin^2 y = \sin(x + y) \sin(x - y)$; $x + y = 2\pi n$ or $x - y = 2\pi n$, n an integer

    k. 0; everywhere

16.5   If F_k is given by $u = f_k(x, y)$, $v = g_k(x, y)$, then f_1 and f_2 have the same gradient, and hence differ by a constant. The same is true for g_1 and g_2.

16.6   **a.** dF is the linear transformation with matrix $\begin{pmatrix} a & b \\ c & d \end{pmatrix}$.

16.7   **a.** $\begin{pmatrix} 0 & 0 \\ 0 & 1 \end{pmatrix}$

16.9 F maps $(1, \sqrt{3})$ onto $(-8, 0)$; $dF_{(1, \sqrt{3})}$ is a dilation by factor of 12 composed with rotation through $\frac{2}{3}\pi$.

16.11   The image is a horizontal ray to the right of $(0, 1)$.

16.12   **a.** $\begin{pmatrix} f'(x) & 0 \\ 0 & g'(y) \end{pmatrix}$

   c.   Orientation-preserving if $f'(x_0)$ and $g'(y_0)$ have the same sign; orientation-reversing if opposite signs.

   e. $|f'(x_0)|$ and $|g'(y_0)|$ represent horizontal and vertical stretching at (x_0, y_0).

16.14   **a.** $1/e$ **c.** $e^{-4} \int_1^2 \sin\left(\frac{\pi}{2}t^2\right) dt$

16.16 C regular means that the tangent vector is never zero. F regular means that dF maps nonzero vectors into nonzero vectors, hence vectors tangent to Γ are never zero.

Section 17

17.1   **a.** $z_x(1, 0) = 2$, $z_y(1, 0) = 0$ **c.** $z_x(0, 0) = e$, $z_y(0, 0) = 0$

17.2   **a.** $2cf_u$

17.6   **a.** $J_F = \begin{pmatrix} 2 & 1 \\ 3 & 2 \end{pmatrix}$, $J_G = \begin{pmatrix} 2 & -1 \\ -3 & 2 \end{pmatrix}$, $J_H = \begin{pmatrix} 1 & 0 \\ 0 & 1 \end{pmatrix}$

17.7   **a.**   1   **c.** e^2 **e.** $4e(1 + \pi)$

17.11   **a.** $F(0, 0) = (1, 1)$; $x_u = \frac{1}{2}$, $x_v = \frac{1}{2}$, $y_u = \frac{1}{2}$, $y_v = -\frac{1}{2}$

   c. $F(0, 0) = (0, 0)$; $x_u = 1$, $x_v = 0$, $y_u = 0$, $y_v = 1$

   e. $F(0, 0) = (1, 0)$; $x_u = 1$, $x_v = 0$, $y_u = 0$, $y_v = 1$

   g. $F(0, 0) = (0, 0)$; $x_u = 1$, $x_v = -1$, $y_u = 0$, $y_v = 1$

17.12   **a.** $x = \frac{1}{2}(\log u + \log v)$, $y = \frac{1}{2}(\log u - \log v)$; $u > 0$, $v > 0$

   c. $x = u - v^3$, $y = v$; the whole plane

   e. $x = \frac{1}{2}\log(u^2 + v^2)$, $y = \arctan v/u$; $u > 0$, for example

17.14   **a.**   A differentiable function $f(x)$ has a differentiable inverse in some neighborhood of x_0 if and only if $f'(x_0) \neq 0$.

Section 18

18.1   **a.**   Dependent; $(u - 1)(v - 1) = 1$ **c.**   Independent;
   $\dfrac{\partial(u, v)}{\partial(x, y)} = 2(x + y) \neq 0$.   **e.**   Dependent; $u = \log\dfrac{2 - v}{1 - v}$

g. Dependent; $u = 2v$ **i.** Independent;

$$\frac{\partial(u, v)}{\partial(x, y)} = -2\sin(x + y)\sin(x - y)$$

18.3 **a.** If $f(x, y) \equiv 0$, then Eq. (18.2) holds with $\lambda = 1$, $\mu = 0$.
c. Suppose $f(x, y)$, $g(x, y)$ are linearly dependent. If $f(x, y) \not\equiv 0$, choose $c = 0$, and $f(x, y) \equiv cg(x, y)$; if $g(x, y) \equiv 0$, then $g(x, y) = 0 \cdot f(x, y)$; if neither f nor g is $\equiv 0$, use part b. Conversely, if $f(x, y) \equiv cg(x, y)$ use $\lambda = 1$, $\mu = -c$ in Eq. (18.2).

18.6 **b.** The image lies on the curve $u = F(t)$, $v = G(t)$.

18.7 **a.** $\frac{2}{3}\pi$ **c.** $\cos^{-1}(2/\sqrt{13})$ or $\tan^{-1}\frac{3}{2}$.

18.8 **a.** Diffeomorphism; conformal.   **c.** Diffeomorphism; not conformal
e. Not a diffeomorphism since (x_0, y_0) and $(x_0, y_0 + 2\pi)$ map into same point
g. Not a diffeomorphism since $\partial(u, v)/\partial(x, y) = 0$ at $(\frac{1}{2}, \frac{1}{2})$
i. A conformal diffeomorphism

Section 19

19.1 **a.** 0; solenoidal   **c.** $\varphi'(x) + \psi'(y)$; solenoidal $\Leftrightarrow \varphi(x) = ax$
$\psi(y) = -ay$, a constant
e. $2y$; not solenoidal   **g.** $4xy + 6x^2$; not solenoidal
i. $2\cos x \cos y$; not solenoidal   **k.** $2e^x \cos y$; not solenoidal

19.2 **a.** Conservative; $ax + by$

c. Conservative; $\int_{x_0}^{x} \varphi(t)\, dt + \int_{y_0}^{y} \psi(t)\, dt$ **e.** Conservative; $x^2 y$

g. Not conservative   **i.** Conservative; $-\cos x \cos y$
k. Not conservative

19.3 a, f, l; also part b if $a + b = 0$; part c if of the form $\langle ax, -ay \rangle$; and part d if of the form $\langle ay, ax \rangle$

19.9 **a.** $\langle 1, 0 \rangle$ **c.** $2\langle x, y \rangle$ **e.** $(1/x^2)\langle -y, x \rangle$

Section 20

20.1 **a.** $(a^2 + b^2)u_{XX} + 2(ac + bd)u_{XY} + (c^2 + d^2)u_{YY}$
c. $(X^2 + Y^2)(u_{XX} + u_{YY})$

20.6 **a.** $u_x = u_r \cos\theta - \dfrac{1}{r}u_\theta \sin\theta$, $u_y = u_r \sin\theta + \dfrac{1}{r}u_\theta \cos\theta$,

$$u_x^2 + u_y^2 = u_r^2 + \frac{1}{r^2}u_\theta^2.$$

c. $u_\theta = -rv_r$, $v_\theta = ru_r$

20.9 $u = A\theta + B$, A and B constants, $\theta = \arctan(y/x)$.

20.12 $u_{xy} = u_{XX}f_x f_y + u_{XY}(f_x g_y + f_y g_x) + u_{YY}g_x g_y + u_X f_{xy} + u_Y g_{xy}$
$u_{yy} = u_{XX}f_y^2 + 2u_{XY}f_y g_y + u_{YY}g_y^2 + u_X f_{yy} + u_Y g_{yy}$

20.14 **a.** $A = a\cos^2\alpha + 2b\cos\alpha\sin\alpha + c\sin^2\alpha$
$B = -a\cos\alpha\sin\alpha + b(\cos^2\alpha - \sin^2\alpha) + c\cos\alpha\sin\alpha$
$C = a\sin^2\alpha - 2b\cos\alpha\sin\alpha + c\cos^2\alpha$

20.15 $u(x, y) = G(3x - y) + H(3y - x)$

20.16 *c.* $u(x, y) = g(y) + xh(y)$

20.18 $U_Y = u_y \cos^2 \alpha + (v_y - u_x) \cos \alpha \sin \alpha - v_x \sin^2 \alpha$

$V_X = -u_y \sin^2 \alpha + (v_y - u_x) \cos \alpha \sin \alpha + v_x \cos^2 \alpha$

20.19 *a.* $(u_x - v_y) \cos 2\alpha + (u_y + v_x) \sin 2\alpha$ *c.* $u_y - v_x$

Chapter four

Section 21

21.1 *a.* $-\frac{1}{6}$ *c.* $\frac{1}{3}$

21.3 $\int_C f \, dx = -\int_{\widetilde{C}} f \, dx = \log 2;$ $\int_C f \, dy = -\int_{\widetilde{C}} f \, dy = \frac{\pi}{2}$.

21.4 *a.* $\frac{1}{5}$ *c.* $1 - \frac{1}{2}\pi$ *e.* 0

21.5 *a.* 1 *c.* 1 *e.* -1

21.6 *a.* $\frac{1}{2}\pi$ *c.* $-\frac{3}{2}\pi$

21.7 *a.* 0 *c.* $2\pi ab$ *e.* $-\pi ab$ *g.* 0

21.9 $-2\pi ab$

21.13 *a.* $\sqrt{2}/2$ *c.* $\frac{1}{4}\pi$ *e.* $\frac{1}{3}ab(a^2 + ab + b^2)/(a + b)$

21.14 $ds/d\tau = (ds/dt)|dt/d\tau|$

21.16 *a.* 3π *c.* $\frac{14}{9}$ *e.* 1

21.17 *a.* $\pi r \sqrt{r^2 + h^2}$ *c.* $13\pi/3$ *e.* $2\pi a + \pi \sinh 2a$ *g.* $4\pi a^2$

21.18 *a.* $\frac{3}{2}$ *c.* $\pi\sqrt{2}$

21.19 *a.* $(\bar{x}, \bar{y}) = (\frac{34}{45}, \frac{11}{18})$ *c.* $(\bar{x}, \bar{y}) = \left(\frac{1}{\pi} \sinh \frac{\pi}{2}, \frac{1}{\pi} \sinh \frac{\pi}{2}\right)$

21.20 *a.* $\left(\dfrac{\sin \alpha}{\alpha}, \dfrac{1 - \cos \alpha}{\alpha}\right)$ *c.* $(\pi, \frac{4}{3})$

21.21 *c.* $(2\pi a)(2\pi b)$

21.22 *a.* π *c.* 1

21.23 *a.* $k\left(\dfrac{1}{r_1} - \dfrac{1}{r_2}\right)$ *c.* $k\left(\dfrac{1}{r_1} - \dfrac{1}{\sqrt{r_1^2 \cos^2 \alpha + r_2^2 \sin^2 \alpha}}\right)$

Section 22

22.1 *a.* 0 *c.* 1 *e.* $2^{e+1} - 1$

22.4 *a.* $C_1 : x = 1 + t, \, y = 1, \, 0 \le t \le 2;$ $C_2 : x = 3, \, y = t - 1, \, 2 \le t \le 3$
   c. 11

22.5 *a.* $C_1 : x = 3 - t, \, y = 2, \, 0 \le t \le 2;$ $C_2 : x = 1, \, y = 4 - t, \, 2 \le t \le 3$
   c. -11

22.6 *c.* $2ab$

22.8 *a.* $x^2 y^3 + x^3 y^2$ *c.* $\frac{1}{2}[\log(1 + x^2) - y^2/(1 + x^2)]$

22.9 *a.* 4 *c.* $\log\sqrt{2} - 1$

22.11 *a.* $p_y = 3 \ne q_x = 6; \, f = 3xy^2 - y^3$

22.13 *a.* $v = x + y$ *c.* $v = \cos x \cosh y$ *e.* $v = e^x \sin y$

22.15 *a.* -1 *c.* -14

24.7   a. πt
24.9   b. $2\pi^2 a^2 b$ d. π
24.10 c. $\bar{x} = 4a/3\pi$, $\bar{y} = 0$
24.12 a. $\frac{2}{3}a^2$ c. $(c^5 - a^5 - b^5)/15$, where $c^2 = a^2 + b^2$
       e. $(0, 3\pi a/16)$ g. $\sqrt{10/27}$
24.17 a. $h = x^3 y^2/6$ c. $h = (x^2 + y^2)^{5/2}/15$
24.19 $\frac{1}{2}(e - 1)$
24.22 a. $0 \le y \le 2 - 2x$, $0 \le x \le 1$
       c. $y \le x \le \sqrt{y}$, $0 \le y \le 1$
       e. $e^y \le x \le 1 + (e - 1)y$, $0 \le y \le 1$
24.23 a. $\frac{8}{15}$ c. 2

Section 25

25.1   a. $(c - d)(b - a)$ c. 0
25.3 $(\bar{x}, \bar{y}) = (\frac{1}{2}(a + b), \frac{1}{2}(c + d))$
25.4   a. 12   c. 0
25.5   a. $\displaystyle\int_C \frac{1}{2}x^2 y \, dy = \frac{1}{4}$

       c. $\displaystyle\int_C e^y \cos \pi x \, dx + e^z \sin \pi y \, dy = 2(e - 1)/\pi$
25.6   a. 2π
25.7   a. 2π; the integral over each side is the angle subtended by that side at (X, Y).
25.9   a. $\log \sqrt{x^2 + y^2}$ c. $y \log x + \log y - 1/x$
25.15 a. $\frac{1}{2}ab$ c. 0
25.17 a. $2\pi k$

Section 26

26.1   a. 0   c. πab
26.2   a. $x = 2 \cos t$, $y = 3 \sin t$, $0 \le t \le 2\pi$
       c. $C_1 : x = 3 \cos t$, $y = 3 \sin t$, $0 \le t \le 2\pi$
       $C_2 : x = 1 + \cos t$, $y = -\sin t$, $0 \le t \le 2\pi$
       e. $x = t^2$, $y = -t$, $-1 \le t \le 1$; $x = 1$, $y = t - 2$, $1 \le t \le 3$
26.3   a. 0   c. 5π e. 4
26.4   a. πab c. $a^2(2 - \frac{1}{2}\pi)$ e. $\frac{3}{2}\pi a^2$
26.7   a. $\tilde{F} : 0 \le u \le 1 - v^2$; $A(\tilde{F}) = \displaystyle\iint_F 2x \, dA = \frac{2}{3}$

       c. $\tilde{F} : a^2 \le u^2 + v^2 \le b$, $v \ge 0$; $A(\tilde{F}) = \displaystyle\iint_F x \, dA = \frac{1}{2}\pi(b^2 - a^2)$

Section 27

27.1   a. $\frac{1}{4}$ c. $2/\pi$ e. $a^2/12$ g. $ab/4$
27.2   a. $\frac{2}{3}$

Section 23

23.1 *a.* $h(t) = 1 - t, \quad 0 \le t \le 1$ *c.* $h(t) = t - t^2, \quad 0 \le t \le 1$
e. $h(t) = \sqrt{1 - t^2} - 2t^2 + 1, \quad -\sqrt{3}/2 \le t \le \sqrt{3}/2$
g. $h(t) = 2 \sinh t, \quad 0 \le t \le 2$ *i.* $h(t) = 2(1 - |t|), \quad -1 \le t \le 1$

k. $h(t) = \begin{cases} \frac{10}{3}t, & 0 \le t \le 1 \\ 5 - \frac{5}{3}t, & 1 \le t \le 3 \end{cases}$

m. $h(t) = \begin{cases} 2\sqrt{9 - t^2}, & -3 \le t \le -2 \\ 2\sqrt{9 - t^2} - 2\sqrt{4 - t^2}, & -2 \le t \le 2 \\ 2\sqrt{9 - t^2}, & 2 \le t \le 3 \end{cases}$

o. $h(t) = \begin{cases} 3, & -2 < t \le -1 \\ 3 - 2\sqrt{1 - t^2}, & -1 \le t \le 1 \\ 3, & 1 \le t \le 2 \end{cases}$

23.2 *a.* $\frac{1}{2}$ *c.* $\frac{1}{6}$ *e.* $\frac{1}{3}\pi + \frac{3}{4}\sqrt{3}$ *g.* $2(\cosh 2 - 1)$
i. 2 *k.* 5 *m.* 5π *o.* $12 - \pi$

23.3 *a.* $h(t) = 1 - t, \quad 0 \le t \le 1$ *c.* $h(t) = \sqrt{t} - t, \quad 0 \le t \le 1$
e. $h(t) = \begin{cases} \sqrt{2(1 + t)}, & -1 \le t \le \frac{1}{2} \\ 2\sqrt{1 - t^2}, & \frac{1}{2} \le t \le 1 \end{cases}$

g. $h(t) = \begin{cases} 2 + \log t, & e^{-2} \le t \le 1 \\ 2 - \log t, & 1 \le t \le e^2 \end{cases}$

i. $h(t) = 2 - t, \quad 0 \le t \le 2$ *k.* $h(t) = \begin{cases} \frac{5}{2}(1 + t), & -1 \le t \le 0 \\ \frac{5}{6}(3 - t), & 0 \le t \le 3 \end{cases}$

m. Same answer as Ex. **23.1m** *o.* $h(t) = \begin{cases} 4 - \sqrt{4 - t^2}, & 0 \le t \le 2 \\ 4, & 2 \le t \le 3 \end{cases}$

23.5 *a.* $x + y = t\sqrt{2}$ *c.* $h(t) = \sqrt{2}, \quad -\sqrt{2}/2 \le t \le \sqrt{2}/2$

23.6 *c.* $N_k = 2^{k-1}(2^k - 1), \quad A_k = \frac{1}{2}\left(1 - \frac{1}{2^k}\right)$

23.8 $\frac{2}{3}$
23.9 $\frac{3}{8}\pi a^2$
23.10 *a.* Not bounded
c. Not bounded and does not consist of a domain together with its boundary
e. Does not consist of a domain together with boundary (a domain is connected)
23.14 *a.* $h(t) = \sqrt{2 - 2|t|}, \quad -\sqrt{2}/2 \le t \le \sqrt{2}/2$

Section 24

24.1 *a.* 98 *c.* $\frac{1}{15}(34^{5/2} + 5^{5/2} - 13^{5/2} - 26^{5/2})$
24.2 *a.* $\frac{1}{96}$
24.3 *a.* 0
24.4 *a.* $\varphi_{f,F}(t) = \begin{cases} \frac{3}{2}, & 0 \le t \le 1 \\ 2 - t^2/2, & 1 \le t \le 2 \end{cases}$
24.5 *a.* $\frac{2}{3}$ *c.* $abc/6$
24.6 *b.* $\frac{4}{3}\pi abc$

27.3   *a.* $(e-1)/2$

27.6   Possible choices are:

    a. $(\tfrac{1}{2}, \tfrac{1}{2})$ *c.* $(1, 2/\pi)$ *e.* $(a/4, a/3)$ *g.* $(a/2, b/2)$

27.8 $4 + 1/4k^2$

27.10 *a.* $8x_0(3y_0^2 + 1/k^2)/k^2$

27.11 *a.* $(y_2 - y_1)(x_2^2 - x_1^2)/2$

27.20 *a.* $\displaystyle\int_C v_T \, ds = \int_C v_N \, ds = 0$

    c. $\displaystyle\int_C v_T \, ds = -2\pi r^2, \quad \int_C v_N \, ds = 0$

    e. $\displaystyle\int_C v_T \, ds = 0, \quad \int_C v_N \, ds = 2\pi x_0 r^2$

Section 28

28.1   *a.* $0 \le r \le 1, \ \tfrac{1}{2}\pi \le \theta \le \tfrac{3}{2}\pi$

    c. $0 \le r \le (\sqrt{2}/2)\sec(\theta - \tfrac{1}{4}\pi), \ \ 0 \le \theta \le \tfrac{1}{2}\pi$

    e. $a \le r \le 2a\sin\theta, \ \ \tfrac{1}{6}\pi \le \theta \le \tfrac{5}{6}\pi$

28.2   *a.* $\tfrac{1}{2}\pi$ *c.* $\tfrac{1}{2}$ *e.* $a^2(\tfrac{1}{3}\pi + \tfrac{1}{2}\sqrt{3})$

28.3   *a.* $(-4/3\pi, 0)$

28.4   *a.* $\displaystyle\int_0^{2\pi} \left(\int_0^a r^3 \cos^2\theta \, dr \right) d\theta = \pi a^4/4$

28.5   *a.* $\displaystyle\int_0^{2\pi} \left(\int_0^{\sqrt{c}} (c - r^2) r \, dr \right) d\theta = \pi c^2/2$

    c. $\displaystyle\int_0^{2\pi} \left(\int_0^{\sqrt{\pi/2}} r \cos r^2 \, dr \right) d\theta = \pi$

28.7   *a.* $ab \displaystyle\iint_{\tilde{F}} f(au, bv) \, du \, dv$

28.8   *a.* $\tilde{F} : |u + v| \le 2, \ |u - v| \le 2;$

    $\displaystyle\iint_{\tilde{F}} (u^2 + v^2) \, du \, dv = \int_F 4(x^2 + y^2) \, dx \, dy = \tfrac{32}{3}$

    c. $\tilde{F} : \dfrac{v^2}{9} \le u \le \dfrac{v^2}{9} + 1, \ \ 0 \le v \le 1;$

    $\displaystyle\iint_{\tilde{F}} \sqrt{9u - v^2} \, du \, dv = \iint_F 9\sqrt{x} \, dx \, dy = 6$

28.9 $(e - 1)^2$

28.11 *a.* $u = 2x + y, \ v = x + 3y$

28.12 $\pi(1 - e^{-R^2})/6$

28.13 *c.* $\pi R^4/2\sqrt{AC - B^2}$

28.16 *a.* π *c.* $\tfrac{1}{6}\pi$ *e.* $\pi/\sqrt{AC - B^2}$

Index